Ziheng Sun
Editors

Actionable Science of Global Environment Change

From Big Data to Practical Research

Springer

Editor
Ziheng Sun
Center for Spatial Information Science and Systems
Department of Geography and Geoinformation Science
George Mason University
Fairfax, VA, USA

ISBN 978-3-031-41757-3 ISBN 978-3-031-41758-0 (eBook)
https://doi.org/10.1007/978-3-031-41758-0

This Springer imprint is published by the registered company Springer Nature Switzerland AG
The registered company address is: Gewerbestrasse 11, 6330 Cham, Switzerland

Paper in this product is recyclable.

Contents

Chapter 1
What Is "Actionable" Science for Climate and Environment?

Ziheng Sun

Contents

1 Introduction

The pursuit of science is forever fueled by human curiosity. It begins with individuals like Newton, who pondered why apples descend to the earth, and continues with visionaries like Dirac, who sought to understand the existence of negative energy states. Science, by nature, is designed to address inquiries that delve into the "why" (Inan 2013). Meanwhile, "how" questions, such as the methodology for constructing a dam, tend to find resolution within the realms of engineering disciplines (Pawley 2009). Nevertheless, in recent years, due to the extensive convergence of science and technology fields and the rapid advancement of substantial cross-disciplinary research, the once-clear demarcation between science and engineering has become

Z. Sun (✉)
Center for Spatial Information Science and Systems, Department of Geography and Geoinformation Science, George Mason University, Fairfax, VA, USA
e-mail: zsun@gmu.edu

Z. Sun (ed.), *Actionable Science of Global Environment Change*,
https://doi.org/10.1007/978-3-031-41758-0_1

increasingly blurred (Sara 2018). One illustrative example is the current shift of science into its fourth paradigm – data-driven science – which fuses AI and vast data generated by scientific experiments to produce readily usable AI models and services (Tansley and Tolle 2009). Furthermore, the cycle of science projects is accelerating compared to previous times, and the demands for each research outcome are becoming increasingly detailed and precise resembling the requirements typically associated with engineering projects (Mishra and Mishra 2011). This has imposed a significant responsibility on scientists to stay updated with the latest advancements in both industry and academia. They must also promptly embrace the most recent technological innovations upon their release to ensure that their scientific discoveries remain current and pertinent to today's society, rather than being oriented solely towards a world that may exist one or two centuries from now.

While conventional science aims to address broad questions with minimal spatial and temporal constraints, "actionable" science is is oriented towards answering inquiries such as "What actions should we take in this specific place and time today" (Beier et al. 2017). Actionable science is characterized by a more specific focus on application time and location. For example, we now understand that as polar ice melts and sea levels continue to increase, coastal communities and cities could be engulfed by the ocean (Thompson 2010). Researchers specializing in sea-level studies typically dedicate a significant amount of time to investigating its effects on ecosystems, climate, and the economy, while also projecting future changes based on upward trends. In contrast, actionable science research should prioritize endeavors that can be implemented and accomplished in the present or within the foreseeable future (Sylvester and Brooks 2020). For instance, should homeowners build their homes on a floating foundation? (Johnson et al. 2009). If the trends of sea-level-rising risk are inevitable, what immediate steps should individuals take to mitigate potential property damage? Those residing along coastal areas and waterfront cities will undoubtedly express high concern and seek actionable guidance. The pricing of their property insurance premiums may also be influenced by scientific forecasts. It's worth noting that the evidence indicates sea-level rise is more likely to be a gradual, incremental process rather than a sudden, dramatic change (Kates et al. 2012). What steps should the city government take at present? California, as a prominent advocate for addressing climate change and home to numerous coastal megacities, has outlined its objective of supporting actionable science initiatives and integrating actionable science outcomes directly into their decision-making processes. They have delineated 12 specific action items, one of which involves addressing sea-level predictions and their correlation with California's coastal areas (Arnott et al. 2020).

Throughout history, climate science has consistently revolved around questions like, "What actions can we take today to prevent or mitigate a climate catastrophe in the future?" "How can we ensure a healthier environment for our future generations?" "What lifestyle choices should we avoid in our daily lives to contribute to a more sustainable planet?" Despite years of research, it's clear that there are no universally applicable answers to these complex questions. More recently, public figures like Bill Gates have emphasized that relying on people to lead impoverished lives is not a viable solution to the climate challenge (David 2023). The debate surrounding climate and environmentally friendly personal actions and policies has long been a prominent battleground, consistently raising questions such as, "What precisely constitutes environmentally friendly behavior, both on an individual and policy level?" and "How can we effectively discern and ensure that our present actions are beneficial rather than harmful to the environment?" (Moisander 2007). The central point of contention revolves around the idea that our human history is an exceedingly brief chapter in the Earth's much lengthier history. It's argued that our actions cannot alter the planet's fundamental trajectory. The next Ice Age or the eventual extinction of Earth's species will occur when it is naturally predetermined, and our current actions may merely delay or, in more dire circumstances, accelerate these events through unforeseen reactions that remain beyond our scientific understanding. To respond to all the inquiries, scientific consensus have underscored that the current trend of global warming is largely driven by human activities, particularly the emission of greenhouse gases such as carbon dioxide. This unprecedented and rapid rise in global temperatures is a departure from traditional climate cycles and is a pressing issue that requires immediate attention and proactive efforts to mitigate its potentially catastrophic consequences (Pielou 2008; Büntgen and Hellmann 2013; Trexler and Johns-Putra 2011).

Over the last century, informed by the findings from climate and environmental modeling research, we have formulated a range of policies aimed at steering our behaviors towards greater consideration and friendliness towards both our urban and rural environments (Edwards 2011). A common example in our daily routines is the limited utilization of single-use plastic bags (Wagner 2017). Numerous regions have implemented charges for each disposable plastic bag as a means to support environmental initiatives aimed at mitigating the harm inflicted by plastic waste particles (Kosior and Crescenzi 2020). In the ocean's deepest reaches, researchers have identified the presence of man-made plastic particles, significantly affecting marine life. Incidents involving oil and gas extraction platforms have led to extensive oil spills in the Gulf of Mexico, causing substantial losses to coastal ecosystems, wildlife, and vegetation. These events have resulted in immeasurable environmental damage (Lin and Mendelssohn 2012; Soto et al. 2014). As technology advances, the influence of human activities on the environment becomes increasingly significant and has the potential to fundamentally alter the trajectory of development (Palmer 2012). Global climate change poses significant challenges that affect us all like impacting our food production, and without immediate action, these issues cannot be reversed or averted.

2 Definition of Actionable Science

What is "actionable" science exactly? Like many broadly defined terms, "actionable" science is a frequently employed yet somewhat ambiguous phrase. People constantly ask questions such as "Aren't all scientific disciplines inherently

Fig. 1.1 Position of actionable science

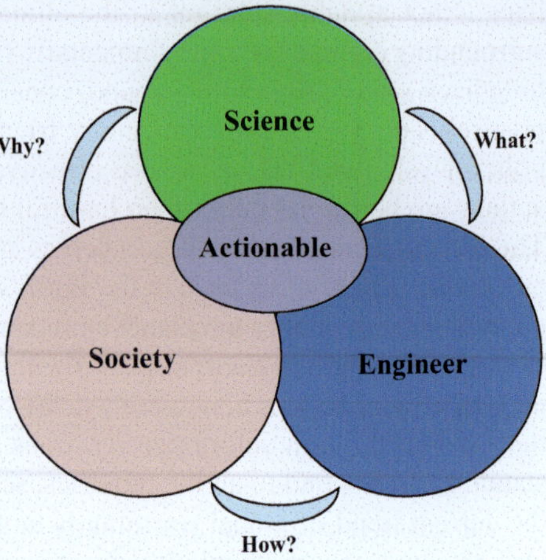

actionable?" After all, every research outcome should, in some way, influence the real world, offering potential guidance for practical actions, either directly or indirectly. Even the most enigmatic and profoundly intricate research, such as the investigation into concepts like wormholes (Morris et al. 1988) and parallel universe (Wolf 1988), aspires to pave the way for the development of spacecraft capable of exceeding the speed of light and inexhaustible sources of energy in the event of their successful realization. Human imagination is never restricted by whether it is actionable or not. But we tend to evaluate all ideas with some form of actionable potential, at least from the perspective of people in the current context. However, despite the ambiguity, misconceptions, and frequent use of the term, it remains essential to refine the concept to the greatest extent possible. This clarity helps elucidate the objectives of all actionable science projects more precisely, ensuring that both scientists and end-users can arrive at a common understanding with fewer points of contention and misunderstandings (Fig. 1.1).

So indeed, scientists can consistently articulate the significance of their research and why the outcomes are poised to become actionable in an ideal scenario of success. However, in practice, numerous research initiatives have not realized the anticipated societal effects as originally envisioned, often due to funding limitations and the cessation of many projects. A majority of these endeavors find their place primarily within the realm of literature, rather than being put into practical operation. Multiple factors can be held responsible for this outcome. Take the controversial carbon dioxide-capturing efforts for power plants as example (Bhown and Freeman 2011; Abu-Zahra et al. 2007). The idea is to capture the carbon dioxide emissions from fossil-fueled power plants and store them underground, aiming to reduce our greenhouse gas emission and mitigate climate change. However, the existing technology is still in its early stage and demanding substantial energy consumption, and the cost of capturing and transporting carbon dioxide is so high that the power plant operators are biting their fingernails at it. Another concern pertains to the storage sites and methods. Assume we successfully transport CO_2 underground, the safety and the impact on the environments around the storage sites are still worrisome. It is critical to permanently prevent them from leaking and released into the atmosphere someday, which will cause even more damage (e.g., one approach to store them is keeping the captured CO_2 in supercritical fluid under pressure of over 72.9 standard atmosphere) (Leung et al. 2014; Nikolaj 2021). The operation of these storage systems and sites further intensifies energy consumption, potentially necessitating increased fossil fuel usage to bridge the energy demand gap. Therefore, this question slowly transitioned from scientific endeavor to an engineering project, focused on finding solutions to lower the costs of capturing and securely storing CO_2 underground, allaying concerns of unintended major leaks and additional CO_2 emissions throughout the process. Until these concerns are adequately addressed, carbon capture projects are likely to remain "nonactionable" for the majority of power plants.

Actionable science requires a meticulous examination of ideas within the confines of practical constraints and a holistic assessment of their feasibility for stakeholders. It goes beyond providing binary responses, such as a simple "yes" or "no," as in the question, "Is it beneficial to capture CO_2?" Instead, it engages in a systematic exploration of hypothetical scenarios, addressing a cascade of "what-if" inquiries like "In the event of CO_2 capture, what are the potential outcomes? Is there a risk of future leakage? Might the expenses associated with capturing and storing CO_2 result in additional CO_2 emissions, potentially undermining the environmental efforts in progress?" (Stanger et al. 2015; Miller et al. 2019; Roussanaly et al. 2021). More importantly, actionable science addresses critical "how" queries, such as "How can we reduce the cost of underground CO_2 storage to make it economically viable?" While it remains a subset of the broader scientific domain, actionable science takes into account all these inquiries and fosters close collaboration and engagement with engineers and society, primarily involving stakeholders and intended end users. With considerations encompassing the "why," "what," and "how" aspects, actionable science projects aim to furnish concrete solutions to the primary obstacles encountered by both engineers and users. They also assist operators in seamlessly practical implementation strategies for incorporating research concepts into their operational workflows. In simpler terms, unlike traditional scientific research that primarily delves into unraveling the "why" questions, actionable science projects allocate a substantial portion of their efforts to the deductive tasks that logically follow when people aim to translate these ideas into practice. Typically, this is not a focal point for scientists or something they are inclined to undertake beforehand. However, the advent of the Internet and knowledge-sharing technologies has made scientific knowledge easily accessible to the public. With research cycles progressing at an accelerated pace, the enigmatic questions demanding a lifetime of dedicated study are diminishing, and people are increasingly interested in exploring creative ways to employ this knowledge in addressing pressing climate challenges (Tan 2021; Collins 2018). The role of scientists has greatly changed since to a more mixed role of not only thinking about "yes or no" questions, but also the need to prototype and do interdisciplinary work involving engineering questions like "what if." If you examine the tools employed in scientific research, such as

Python, R, Linux servers, and cloud computing, you'll find that they closely align with those used by engineers in their daily efforts to tackle real-world issues. The traditional demarcations between science and engineering are gradually eroding, fostering continuous collaboration between scientists and engineers. In the course of a research project, roles frequently shift, reflecting the dynamic and cooperative nature of this partnership. This collaboration is one of the primary reasons why it's crucial to differentiate actionable science research from conventional science, as it reflects a prevalent and natural form of cooperation in contemporary research. The identification of factors that render science projects actionable can significantly enhance the success rate of efficiently translating scientific concepts into practical applications, allowing society to reap the benefits of scientific advancements more promptly and effectively.

Climate science at most times is a collaborative effort, with thousands of scientists working collectively on answering different facets of the overarching problems. Each individual project might only focus on a small segment problem in the question chains. "Actionable" is not only for the end users, but also pertains to the scientists working on the downstream tasks who encounter obstacles while awaiting solutions. In other words, actionable science can also mean those research that can remove road blocks for other scientists. We can envision a multitude of such actionable research endeavors that, when combined, render the overall effort actionable for society as a whole. This can empower society to comprehend climate changes comprehensively and make informed daily decisions based on research-driven guidance. It's akin to the manufacturing of a car, where thousands of suppliers diligently work on individual components, eventually assembling them into a functional vehicle that people rely on daily (e.g., Fig. 1.2). The absence of any component can jeopardize the completeness and utility of the final product. Therefore, each part is indispensable, and the work behind it is practical and unquestionably beneficial for end-users, as it forms an integral part of the daily products they depend on.

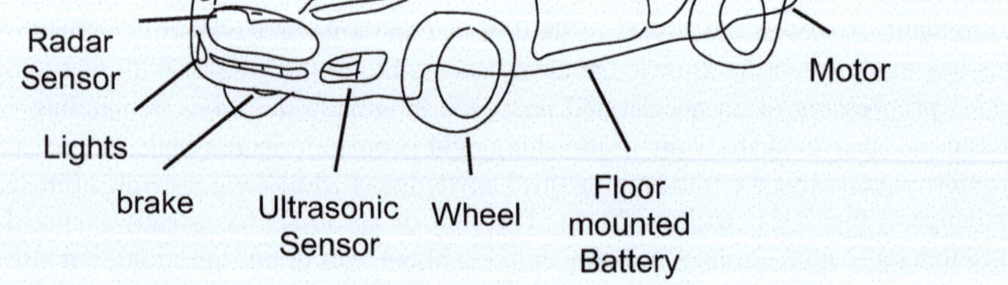

Fig. 1.2 Brief breakdown of the parts of an electric car

Based on these observations, it should be simple to identify the actionable science projects now. Please consider posing the following three questions:

- Is the project's objective aligned with real-world challenges and part of a broader community effort aimed at future operational applications? The project's design should revolve around generating knowledge and insights that contribute to informed decision-making processes regarding environmental concerns. This form of science centers on producing information and data that can be transformed into actionable advice and practical solutions, subsequently embraced by stakeholders in relevant fields. An actionable science endeavor should not run counter to the overarching consensus goals shared by both the scientific community and society at large.
- Will the project's outcomes find practical application and serve as essential tools for society in addressing significant climate-related challenges? For example, in the field of weather forecasting, scientists use sophisticated computer models to project future climate scenarios, aiding stakeholders in decision-making regarding adaptation to changing conditions. They use mathematical equations to simulate climate processes, such as the movement of air and water in the atmosphere and oceans, the interaction between the land surface and the atmosphere, and the exchange of energy between the Earth and space. These equations are then programmed into computer models that can simulate the behavior of the climate system over time. These models have demonstrated their effectiveness and have gained widespread acceptance within society, serving as regularly utilized resources in daily life.
- Does the project consider tackling the potential "what-if" questions from engineers? For example, developing climate models is a complex and multidisciplinary process that involves a combination of observations, theory, and computational modeling (Clune and Rood 2011). Engineers ask "What if the model got the wrong results? What if the result accuracy is very bad?" Actionable science modeling projects should consider proposing practical plans for these what-if questions. Once the models are built, scientists test them against historical climate data to see how well they reproduce past climate conditions. This process helps to identify any errors or biases in the model and improve its accuracy (Oreskes 2018). Once the models pass validation, scientists can employ them to project future climate scenarios under various greenhouse gas emission scenarios. This can help answer the questions from the stakeholders about how to operationally use the model. As new data and observations become available, scientists refine their models to improve their accuracy and account for any new processes or feedback that may be important for understanding climate dynamics. This process can control the uncertainty and address the errors and calibrate the model to keep them on the track during operation.

If the answer is "yes" to all three questions, the research falls within the high basket of actionable science. If the response is "no" to any of them, the actionability ranking of the research will be degraded, potentially render it as nonactionable for the time being. Actionable climate science should yield knowledge and insights that can be promptly applied to mitigate or adapt to climate change. It should be able to generate information and data that can be translated into practical solutions and recommendations readily embraced by stakeholders in the field. Actionable science for climate can include research on climate impacts, vulnerability and risk assessments, mitigation and adaptation strategies, and the development of new technologies and practices to reduce greenhouse gas emissions and build resilience to climate change. The overarching goal is to provide decision-makers with the tools and information they need to make well informed choices on how to build a more sustainable future for all.

3 How to Measure the Actionableness of Climate and Environment Projects?

Since we are identifying actionable science projects from all science projects, it would be easier to quantitatively measure the actionability of a research project by considering a few factors to get an understandable index value so that we can cross-compare them using a unified standard. This section introduces a mathematical model we tried to build for evaluating actionability of science projects. According to previous studies (Meinke et al. 2006; Kirchhoff et al. 2013), the actionability of science projects is impacted by several major factors: the positive and negative impacts on the society, feasibility and cost of implementation, level of clarity for public understanding and acceptance, impacts for society, knowledge transferring to operators, and engagement and approval from the stakeholders. Here is a simple formula that can be created as a starting point:

$$A(s,i) = \sum_{n=6}^{x} f_x(s,i) + e$$

where s represents a science project, i represents the target individuals that the science project will impact and could be anyone who comes across the research project or, $f_x(s, i)$ is the function to calculate the impact on aspect x, and we have six ($n = 6$) factors to consider right now; e means any error or unaccounted uncertainties that might cause offsets. The factors are calculated using the following equations. Each formula will be explained in the following sections.

3.1 *Relevance Between Research and Action*

Among the six factors, the initial consideration is the relevance between the research and real-world actions. Relevance is often misconstrued as its impact within the scientific community and quantified using metrics such as researcher publication citations, funding sources from industry and government, patent generation resulting from the research, and collaborations between research teams and industrial entities. However, these metrics primarily look at the relevance from the researchers' standpoint. This book will predominantly explore how the end users perceive what constitutes relevant research. For individuals, the relevance of the research outcomes is typically defined by whether they are aligned with their personal/corporate goals, work, and daily routines. We have drawn upon prior research on knowledge exchange between universities and industries, as well as the contemporary knowledge transfer framework (Chai et al. 2003; Agrawal 2001; Goh 2002; Ward et al. 2009) to formulate the subsequent equation for assessing the comprehensive alignment between scientific research and practical, real-world applications:

$$f_1(s,i) = C(i) \times \sum_{3}^{x=1} R_x(s) \times I(s) \times \sum_{3}^{x=1} P_x(i)$$

where

- $f_1(s, i)$ is the overall relevance score of research s to individual i.
- $C(i)$ represents connectivity score of the research for individual i (value range 0 ~ 1), and measures the level of accessibility between researchers and the individual. This can normally be seen that regardless of the impacts of the research on the society as a whole, its end products are sometimes not accessible or reachable by the individuals. $C(i) = 1$ means the individual can access the research-derived products; $C(i) = 0$ means it is completely inaccessible.
- $R_1(s)$ is the relevance score for industry, and its value ranges 0 ~ 1. The minimum value means there is no way to implement the research idea in industry context in the foreseeable future, and the maximum value 1 means the research is very practical and can be very easily converted into products in industries that can be replicated millions of times and sold to any consumers.
- $R_2(s)$ is the relevance score for society, and ranges 0 to 1 as well. 0 means the research is not relevant to the human society in any way; 1 means the research is strongly related and highly applicable in the human society.
- $R_3(s)$ is the relevance score for academia, and also ranges from 0 to 1. 0 means the research is not related to any other research and interests other researchers, while 1 means the research is a high-profile topic and related to many other researchers' work and receives broad intention for application from other scientists.
- $I(s)$ is the impact score of the research s. It is a coefficient from 0 to 1 to measure the impacts of the research. It is estimated based on the number of patents, number of spin-off companies, the value of the new products, services, or policies, the estimated economic values by reducing poverty, improving health outcomes, or mitigating climate change. 0 means the research has not resulted in any of these impactful results; 1 means the research has touched all the points and has achieved real social and economic values.
- $P_1(i)$ is the relevance score for individual i to industry. For individual i (could be a person or an entity with representatives), this score measures how relevant an

industry sector is. For example, if the individual focuses on transportation, the highly relevant industry sector could be automobile manufacturing, and other industries like real estate will be low relevant.

- $P_2(i)$ is the relevance score for individual i to society. It measures the level of the individual i to the social issues we are trying to fix, for example, environment, health, climate inequality, or safety. 0 means highly relevant, and 1 is irrelevant. This can calculate the score based on how the individual's expertise contributes to addressing the needs and concerns of society. For example, if the individual focuses on developing new medical treatments, then their research could be highly relevant to addressing health issues in society.
- $P_3(i)$ is the relevance score for individual i to academia. Identify the research field or discipline that is most relevant to individual i's expertise and assess the relevance of the research field to the individual i's work.

This formula is designed to measure the relevance from two sides: researchers and individuals. Actionable science is never a one-sided thing and must consider both slides' background, motivation, goal, and expertise (Eyal 2019). The transmissional relevance between research and action is a complex process that involves multiple factors and perspectives from both researchers and impact targets. Both sides must be considered to ensure that the research is not only scientifically rigorous, but also relevant and actionable to the real-world problems and needs. Impact targets should be open to engaging with researchers and providing input on the research questions and needs. They can also provide valuable feedback on the relevance and practicality of the research findings and recommendations. Nonetheless, in certain instances, not every target is accurately directed, leading to a surplus of inconsequential dialogue directed at individuals who are not the most pertinent. This formula will help people estimate their interests in a research, and, on the other hand, help researchers to find the most relevant groups. A good research-action conversion requires a high relevance score.

3.2 *Feasibility for Implementation*

Another important factor for actionable research is the feasibility of implementing the results into real-world products or applications. Quantitatively measuring feasibility requires evaluating multiple factors, such as resource availability, scalability, and political and economic factors. Here we give a simple formula to calculate the factor quantitatively:

$$f_2(s,i) = \frac{A(s) + S(s) + P_4(s) + E(s)}{4}$$

where

- *A* (*s*) means the availability of resources. The resources for a research are critical components for its success to be converted to usable products, generally including funding, personnel, equipment, materials, infrastructure, and time restriction. Among them, funding is the prerequisite for research. It is important to assess whether the project has sufficient funding to cover all aspects of the research process, including personnel, equipment, and materials. Other factors like qualified and trained personnel, necessary equipment and materials, infrastructure such as laboratory facilities, or access to study populations are essential for research. Additionally, the availability of time is also another factor to consider when assessing the feasibility of a research project. Time might be the only restriction that cannot be easily addressed, and the assessors need to check whether the project has sufficient time to complete the research within the desired timeline.
- *S*(*s*) means the scalability of the research findings and results. The important factors include reproducibility, generalizability, adaptability, and sustainability. We can rate each factor on a scale of 0–1, with 0 indicating low scalability and 1 indicating high scalability. The final score can sum up the following four and calculate the average to be the final score.

  1. Reproducibility refers to the ability to reproduce research findings in different settings, populations, or contexts (National Academies of Sciences, Engineering, and Medicine 2019). Reproducibility allows other researchers to validate the original research findings and build upon them. If a study's findings cannot be reproduced, it may indicate that the original research was flawed or that the findings were not accurate. Without reproducibility, the implementation of research findings becomes difficult as it may be challenging to apply the findings in real-world settings. One possible approach is to calculate the degree of agreement between the original study and the replicated study, and conduct a meta-analysis of the original study and any replications. The meta-analysis can be used to determine the degree of consistency between the original study and replications, and identify any factors that may be influencing the reproducibility of the study.
  2. Generalizability refers to the ability to generalize research findings and results to a larger population or context. It determines whether the research findings are applicable beyond the specific population or context studied. If research findings are generalizable, it means that they can be applied to other populations or contexts with similar characteristics. This makes it easier to implement research findings across a wider range of settings. To measure the generalizability, we first need to determine the number of different contexts (such as different populations, settings, or situations) in which the research has been tested. Then, we divide this number by the total number of contexts tested. For example, if a research study has been tested in 5 different contexts out of a total of 10 contexts tested, the generalizability score would be 0.5.
  3. Adaptability refers to the ability to adapt research findings and results to different situations or contexts. It allows researchers to modify or adapt their findings to meet the needs of different populations or contexts. Adaptable

research findings mean that they can be adjusted to suit different situations without losing their effectiveness. This makes it easier to implement research findings across a wide range of contexts. It can be measured by the proportion of potential adaptations that are actually feasible and implemented. For example, suppose a group of researchers has conducted a study on the impacts of climate change on the agricultural sector in a specific region, using a particular set of methods and assumptions. To assess the adaptability of their research findings, the researchers might consider changes in technology and practices, changes in future climate patterns, and differences in local conditions like soil types, water availability, or historical agricultural practices (Altieri et al. 2015).

4. Sustainability refers to the ability to sustain the implementation of research findings and results over time. If research findings are not sustainable, they may not be implemented properly or their impact may diminish over time. Key factors include whether the research findings and interventions have the potential for long-term impact, and whether they can continue to be effective given changing conditions such as lack of funding and unstable resources. The involvement of local communities and stakeholders, policy and regulatory support, and economic sustainability are also factors that influence sustainability. For example, let us take a look at the Coastal Storms Awareness Program (CSAP) that is funded by the NOAA sea grant, focusing on helping coastal communities prepare for and respond to the impacts of coastal storms and sea level rise (Rezaie et al. 2020). The program involves ten social science research activities, including data collection and analysis, stakeholder engagement, and development of decision-support tools. The project developed a coastal storm and sea level rise scenario planning tool to enable communities to evaluate potential future scenarios and identify adaptive management strategies. The evaluators can assess the costs and benefits of the proposed coastal management strategies and estimate whether they are financially feasible and sustainable. Also, we can engage with local communities and stakeholders to gather their feedback and assess their support for the projects. They can also conduct social impact assessments to identify any potential negative impacts and develop strategies to mitigate them, and conduct environmental impact assessments to evaluate the potential environmental impacts, while developing plans to mitigate any negative impacts, such as by minimizing the use of nonrenewable resources, conserving habitats, and promoting biodiversity. Additionally, the team can evaluate the potential for the strategies to enhance the resilience of the local ecosystem to future climate change impacts. More evaluation like assessing the existing institutional arrangements and governance mechanisms, and identifying any gaps or challenges that may hinder the implementation and maintenance of the proposed strategies, such as by strengthening local institutions, building capacity, and promoting collaboration and coordination among different stakeholders, will more precisely get understanding about the sustainability of the CSAP projects (https://seagrant.noaa.gov/Portals/0/Documents/funding_fellowship/sandy_supplemental/CSAP%20research%20results.pdf).

- $P_4(s)$ means the political factor focusing on whether there is a restriction on the policy that prevents from implementing the research ideas into real-world applications. One typical example is the energy industry sector. Suppose some research projects develop renewable energy technology, but the local government policy does not support the implementation of the technology into real operation as the government has close ties with the fossil fuel industry, and they believe that renewable energy technologies are not economically feasible and could potentially harm the local economy (Mey et al. 2016). As a result, the society in the area is unable to implement the green technology into real-world applications. The lack of political support and policy changes makes it difficult to secure funding and support for the adoption of the technology. Without the necessary resources and support, it is unlikely that the society will be able to implement the research findings and recommendations in a meaningful way. To measure the political factor, we can assess the level of government support and policy alignment toward renewable energy technologies by analyzing government policies, finding the level of historical funding and support for renewable energy projects, and evaluating the political climate and stakeholders' interests. The political feasibility score can range from 0 (no support for renewable energy technologies) to 1 (strong political support and alignment with renewable energy technologies). In the case of many oil and gas major economy cities, the political feasibility score would be low due to the lack of government support and policy alignment toward renewable energy technologies.
- $E(s)$ means the economic factor. Implementing research outputs into industry products requires a lot of economic support from public and private funding sources. For obvious reasons, the success of converting research into actionable products relies on the macro- and microeconomic status. For example, suppose a research project invented a promising new technology for reducing greenhouse gas emissions from transportation such as carbon capture, utilization, and storage (CCUS). However, the implementation of this technology requires significant investment in infrastructure and manufacturing facilities, as well as changes to existing regulations and policies. The macroeconomy can play a significant role in determining the feasibility. If there is a global economic recession and governments are cutting back on investments in infrastructure and technology, it may be difficult to secure the necessary funding for implementing the technology. On the other hand, if the economy is booming and there is a strong focus on sustainability and climate action, it may be easier to garner support for the necessary investments and policy changes. Not only macroeconomics, microeconomic factors can also impact feasibility. If the new technology is more expensive than existing options, consumers may be reluctant to purchase it, and companies may be reluctant to invest in producing it. Additionally, if the technology requires specialized skills or knowledge, there may be a shortage of trained workers, which could limit its widespread adoption. To measure $E(s)$, it is easy to calculate the financial costs of implementing CCUS that can vary depending on various factors such as the size of the project, the type of technology used, and the location. Generally, the costs can be broken down into three categories: capital costs,

operational costs, and maintenance costs. Capital costs refer to the initial investment required to build the CCUS infrastructure, such as the construction of the capture facility, pipelines, and storage sites. Capital costs can range from several million dollars to over a billion dollars, depending on the scale and complexity of the project. Operational costs include ongoing expenses related to the operation of the CCUS infrastructure, such as the cost of energy and chemicals required to capture and transport the carbon dioxide. These costs can also vary depending on the type of technology, the energy efficiency of the system, and the specific operational requirements of the project. Maintenance costs refer to the ongoing maintenance and repair of the CCUS infrastructure, including equipment replacement and upgrades, inspection and monitoring, and general upkeep. These costs can be significant, particularly for larger and more complex projects. In a downhill economy, this factor will be very low for this project.

- The final overall score of feasibility will be the average value by summing all the above factors and dividing by 4.

3.3 *Public Understanding*

Another often overlooked factor is public understanding, which refers to how well the general public can comprehend and make sense of the research project's goals, methods, and findings. It affects the level of support, engagement, and trust the public has for the project and its outcomes. Effective communication and transparency are key components of achieving a high level of public understanding. Here is a simple formula to evaluate the level quantitatively:

$$f_3(s,i) = \frac{C(s) + T(s)}{2}$$

where

- $C(s)$ means the level of clarity in communication, or how well the research project is communicated to the public in a clear and concise manner. This involves using clear language, avoiding jargon, and presenting information in a way that is accessible and engaging for the intended audience. Effective communication is essential for promoting public understanding and support for the project, and for ensuring that the research findings are used to drive positive change in society. Each factor could be rated on a scale of 0–1, with 0 indicating poor clarity and 1 indicating excellent clarity. The scores for each factor could then be averaged to obtain an overall score for clarity. However, it is important to note that measuring clarity in communication is subjective and context-dependent, and may require additional qualitative assessments to gain a deeper understanding of the public's perception of the research project's communication.

- $T(s)$ refers to how transparent the research project is with the public regarding its goals, methods, and findings. First, the openness of data, which is the degree to which research data is made publicly available and accessible, can be measured by the percentage of data made open to the public, the level of detail provided, and the accessibility of the data. Then, the openness of methodology is the degree to which the methodology of the research is open to public scrutiny. This can be measured by the level of detail provided in the research methodology section, the extent to which research methodology is discussed with the public, and the level of clarity in the methodology. Another factor is the conflict of interest disclosure that refers to the degree to which conflicts of interest are disclosed by the research team. We can measure the number and extent of disclosures made in the research article, the transparency of the disclosures, and the severity of the conflicts of interest. The public engagement can be measured by the level of public participation in the research process, the number of public consultations held, and the extent to which public feedback is incorporated into the research. Also, reproducibility can be measured by the number of attempts made to reproduce the findings, the level of success in reproducing the findings, and the transparency. These metrics can be used together to estimate the transparency of a research.

3.4 Impact on Society

This factor considers the direct and indirect impacts of the research on human society.

$$f_4(s,i) = \frac{I_1(s) + I_2(s) + I_3(s) + I_4(s) + I_5(s)}{5}$$

where

- $I_1(s)$ means the economic benefits or costs resulting from the implementation of the research project (measured in monetary terms). Many researchers will claim the saved costs or estimated increased economic values, which, however, is usually just a rough estimate and very unreliable when it is really used in production. Here is a better solution. First, we need to identify the economic impacts of the research project, such as changes in production costs, revenue, employment, and overall economic growth. The next step is to quantify the economic impacts in monetary terms. This can be done through cost–benefit analysis, input–output analysis, or other economic impact assessment methods. Economic impacts may occur over different time horizons, so it is important to consider the duration of the impacts and the timing of costs and benefits. Future economic impacts should be discounted to reflect their present value as future costs and benefits are less valuable than present ones. Finally, the net economic impact can be evaluated by comparing the total benefits to the total costs. If the benefits exceed the costs, then the project is economically viable, and vice versa. A sensitivity analysis can also be performed to assess how the economic impacts would change under different assumptions or scenarios.

- $I_2(s)$ means the environmental benefits or costs resulting from the research (measured using established environmental impact assessment tools). This needs to conduct a comprehensive assessment to identify the potential environmental impacts resulting from the implementation that should cover all aspects from its design and construction to its operation and decommissioning. Then use established tools for measuring the environmental impact of a project, such as life cycle assessment (LCA) and environmental risk assessment (ERA) (Muazu et al. 2021). These tools use scientific methods to assess the potential impacts and provide a quantitative estimate of the impacts. Next is to assign a value to quantify the environmental impacts in monetary terms using established valuation methods. Once the environmental impacts have been quantified, the costs and benefits can be compared. This comparison can help decision-makers evaluate the environmental sustainability of the project and make informed decisions. Finally, it is important to monitor and evaluate the actual environmental impacts resulting from the implementation of the research project. This will help ensure that the environmental impacts are in line with the initial assessment and that any necessary corrective actions are taken to mitigate any negative impacts. Take the investigation project of ocean acidification and its impact on marine ecosystems as example. Ocean acidification is caused by the increase in atmospheric carbon dioxide, which is absorbed by seawater and reacts with it to form carbonic acid. This process lowers the pH of seawater and can have significant impacts on marine life, particularly shell-forming organisms. The research can lead to the development of new technologies and strategies to mitigate the impacts of ocean acidification on marine life. This can result in a reduction of environmental damage, leading to economic benefits such as increased seafood production and tourism. A recent study estimated that the economic damages from ocean acidification to the global shellfish industry alone could reach $230 billion by 2100 if carbon emissions are not reduced (US EPA 2017) and the economic impacts of ocean acidification on coral reefs, including losses in tourism and fisheries, could reach $1 trillion by 2100 (Hennige et al. 2014). These all can be used to estimate the potential mitigated costs of the research.
- $I_3(s)$ means the positive or negative impact on the social fabric of the affected community, including changes in social cohesion, inequality, and access to resources (measured using surveys, interviews, and other social impact assessment tools). For example, let us consider a research project that aims to develop a new renewable energy technology in a rural community. The project would involve constructing wind turbines in the area to generate electricity and reduce the community's reliance on nonrenewable energy sources. A baseline survey should be conducted to establish the existing social conditions in the community. This may involve gathering information on income levels, employment rates,

access to education and healthcare, and other relevant social factors. This may involve conducting focus groups, stakeholder consultations, and other forms of community engagement to gather feedback and input on the proposed project. Impact indicators should be developed to help measure the changes in social cohesion, inequality, and access to resources resulting from the project. These may include measures such as changes in income levels, employment rates, educational attainment, and access to healthcare. Once the project has been completed, a post-project assessment should be conducted to evaluate the actual impacts of the project on the community. This may involve conducting follow-up surveys and interviews to measure the changes in the identified impact indicators. The baseline and post-project data should be compared to determine the actual impact of the project on the social fabric of the community. This will help to identify the positive and negative social impacts of the project and inform future project planning and implementation.

- $I_4(s)$ means the impact on public health, including changes in morbidity, mortality, and quality of life (measured using established public health assessment tools). This needs to define the health outcomes that will be impacted by the research project, including changes in morbidity, mortality, and quality of life. Identify the population who live in a certain geographic area or those who are exposed to certain environmental factors. Before the research project is implemented, collect baseline data on the health outcomes of interest. Once the research project is implemented, collect data on the health outcomes of interest from the same population, like collecting data from medical records, surveys, or other sources. Compare the data collected before and after the implementation of the research project to determine whether there were any changes in the health outcomes of interest. Statistical analysis can determine whether any changes were statistically significant. Then based on the data analysis, interpret the results and draw conclusions about the impact of the research project on public health. These steps can help precisely measure the health impacts of the research. The correlation result can be used as the final score for this factor.
- $I_5(s)$ means the impact on cultural heritage, traditions, and practices of the affected community (measured using established cultural heritage assessment tools). There is cultural heritage impact assessment (CHIA) (Rogers 2011), which is a systematic and comprehensive method for assessing the impacts of development projects on cultural heritage resources. The assessment typically involves identifying and evaluating the significance of cultural heritage resources, assessing the potential impacts of the project on those resources, and identifying ways to avoid, minimize, or mitigate any adverse impacts. The method usually involves consultation with the affected community to provide valuable insights into the potential impacts of a project on cultural heritage. This can involve engaging with community members through public meetings, focus groups, or one-on-one interviews to gather information on their cultural practices, traditions, and values. This is usually an important factor in research in remote regions. For instance, one example is the study of melting glaciers in the Andes Mountains. The Andean region is home to many indigenous communities that

have a deep cultural connection to the glaciers and the water they provide. As the glaciers melt due to climate change, this can impact the cultural heritage of these communities, including their traditional practices and beliefs. The CHIA method could be used to assess the cultural heritage values and significance of the glaciers and the water they provide to the indigenous communities. We must engage with the local communities and gather information on their traditional practices and beliefs related to the glaciers and the water. The HIA would also assess the potential impacts of the melting glaciers on these cultural heritage values and significance. With established cultural heritage assessment tools such as the HIA, researchers and project incubators can better understand the impact and take steps to mitigate any negative impacts.

3.5 *Practicality by Operators*

Sounds similar but this factor is not the same as $f_2(s, i)$ – feasibility of implementation. It focuses on the practicability of the research. Practicability focuses on the ability of the operators or end users to implement a project, while feasibility focuses on the overall ability of the project to be successfully completed. It considers the issue from the perspective of operators. There are also a few factors to measure it, and the formula is

$$f_5(s,i) = \frac{E_{\text{impl}}(s) \times C_{\text{infra}}(s) \times A_{\text{res}}(s)}{C_{\text{hard}}(s) \times C_{\text{money}}(s)}$$

where

- $E_{\text{impl}}(s)$ refers to how easy it is for operators to integrate the project into their existing systems or processes. This needs to find any changes or modifications that will need to be made to the existing systems or processes. Develop a plan for implementing these changes, including a timeline, resource requirements, and potential roadblocks, assessing the time, cost, and effort required, and user feedback from operators and stakeholders to identify the areas for improvement and make adjustments as necessary. For instance, given the project "the development of a low-carbon transportation system for a city" (e.g., Price et al. 2013), first find the key stakeholders and their roles in the transportation system (e.g., city government, transportation companies, commuters), then conduct interviews and surveys with stakeholders to assess their current transportation systems and infrastructure, as well as their willingness and ability to adopt a low-carbon transportation system. Next, analyze the existing transportation infrastructure and identify the potential areas for improvement or modification to accommodate a low-carbon transportation system, estimate the financial and operational costs associated with the implementation of the low-carbon transportation system, including any necessary upgrades or modifications to infrastructure, and evaluate the technical feasibility of the low-carbon transportation system, taking into account the availability of technology and resources. The practicality score (ranging from 0 to 1) can be calculated using a formula that takes into account factors such as stakeholder buy-in, infrastructure compatibility, financial feasibility, and technical feasibility. The score can then be used to identify the areas where improvements or modifications can be made to increase the practicality of the low-carbon transportation system and prioritize implementation efforts.

- $C_{\mathrm{infra}}(s)$ refers to how well the project aligns with the existing infrastructure and technology of the operators. This needs to take an inventory of the current equipment, software, and systems that the operators use to carry out their operations. Analyze how the project can integrate with the current infrastructure and technology of the operators, considering hardware compatibility, software compatibility, and system interoperability. If the project is not fully compatible with the existing infrastructure and technology of the operators, modifications or upgrades may be required. This includes the potential disruption of operations during the implementation of the project, as well as the potential benefits and drawbacks of integrating the project into the existing operations. The overall practicality score can be calculated by weighing the above factors and assigning scores based on their importance and impact. The practicality score can range from 0 to 1, with higher scores indicating better alignment with the existing infrastructure and technology of the operators.
- $A_{\mathrm{res}}(s)$ refers to the availability and accessibility of the resources required to implement the project. This includes finding the specific materials, equipment, technology, and human resources needed to implement the project, assessing whether the required resources are available in the location where the project will be implemented. This can be done through surveys, interviews, or other data collection methods. Assess whether the required resources can be easily accessed by the project implementers. Factors such as distance, transportation, and infrastructure should be considered.
- $C_{\mathrm{hard}}(s)$ refers to the level of complexity of the project, including the technology and processes involved. This can be done through a detailed review of the project documentation, including technical reports, schematics, and workflow diagrams. The metrics used to measure complexity include the number of steps involved in the process, number of interdependent components and subsystems, level of technical expertise required to implement and maintain the system, degree of customization required to integrate the project with existing systems and infrastructure, level of automation and the amount of manual intervention required, time and effort required for system integration and testing, and level of redundancy and fail-safe mechanisms required. Here is an example: suppose we want to develop a new climate model to simulate the interactions between the ocean, atmosphere, and ice sheets in order to predict future sea level rise. Factors to estimate complexity include the number and types of variables in the new model, such as temperature, salinity, ocean currents, and ice thickness, the level of detail and resolution of the model, such as the size of the grid cells, the computational requirements of the model, such as the processing power and memory needed to

run the simulations, and the level of uncertainty in the model's predictions (e.g., the level of confidence in those outcomes). To measure the complexity changes, we need to analyze the level of detail and resolution of the model, such as the size of the grid cells used to simulate the ocean and atmosphere. More importantly, we need to measure the computational requirements of the model, such as the processing power and memory needed to run the simulations. Evaluating the level of uncertainty in the model's predictions, such as the range of possible outcomes and the level of confidence in those outcomes, through statistical analysis and sensitivity testing.

- $C_{\text{money}}(s)$ refers to the financial cost of implementing the project. We have mentioned cost several times in other factors as well, but here this cost mainly emphasizes the resources that are spent during the project implementation stage. This needs to calculate all of the costs associated with implementing the project, including the initial investment, ongoing maintenance and operation costs, and any potential environmental or social costs. The benefits of the project should also be calculated, such as increased efficiency, reduced emissions, or improved environmental outcomes.

3.6 Engagement with the Stakeholders and End Users

This factor estimates the level of engagement of target users and stakeholders in the research project. Many climate and environment research projects conduct surveys and interviews with stakeholders and end users that can provide valuable insights into their level of engagement with the project. Questions can be designed to understand the level of understanding, interest, and support for the project. Meanwhile, in recent years, social media has been widely used in engaging potential users as well (Kujur and Singh 2017). There is also research that monitors social media activity related to the project and can provide insight into the level of engagement and sentiment among stakeholders and end users. This can include tracking mentions, hashtags, and sentiment analysis of posts related to the project. Other common methods for engagement include organizing focus groups and workshops with stakeholders and end users that can provide an opportunity for open discussion and collaboration. This can help identify the areas for improvement and increase engagement with the project. Measuring the participation rates of stakeholders and end users in project-related activities, such as meetings, workshops, and surveys, can provide insight into their level of engagement.

An easy formula to calculate the engagement score is

$$f_7(s,i) = \frac{N_p(s)}{N_{\text{total}}(s)}$$

where $N_p(s)$ is the number of participants in project-related activities, while $N_{total}(s)$ is the total number of stakeholders and end users. The score ranges from 0 (no outside participant at all) to 1 (the project covers basically all the intended users and stakeholders).

4 The World Is Real Time, Should Science Be Real Time Too?

Climate change is an issue of rapidly evolving significance with profound global repercussions, and the availability of current information is imperative for decision-making across all levels (Bannayan and Hoogenboom 2008; Lemos et al. 2012). Incorporating real-time data can enrich the foundations of policies and strategies geared towards both mitigating and adapting to climate change. These real-time data can be procured through a variety of channels, including satellite imagery, remote sensing technologies, and on-site monitoring networks. They enable the monitoring of fluctuations in temperature, precipitation, sea levels, and other environmental parameters, which serve as pivotal indicators of climate change. Additionally, real-time information deployment can expedite responses to climate-related crises and emergencies, such as hurricanes, floods, and wildfires. Early warning systems, relying on real-time data, contribute to reducing the impact of these events and safeguarding lives.

Climate scientists can harness near-real-time data from satellites, buoys, ground stations, and other sources to continuously monitor the latest changes in the climate system. They can further improve modeling capabilities to better simulate and predict climate conditions in real time. Meanwhile, scientists must forge closer collaboration with decision-makers and stakeholders, gaining insights into their needs and delivering timely, easily accessible information. Scientists also need to adopt more agile research frameworks that allow for rapid response to emerging climate issues and changes. For instance, with the rapid development in machine learning (Sun et al. 2022), climate scientists are prompted to incorporate new artificial intelligence (AI) techniques to analyze large amounts of data and identify the patterns and trends in real time (Sun and Cristea 2023).

Let us take the example of a climate science project aimed at predicting and mitigating the impact of heatwaves in urban areas. Traditionally, such projects involve collecting and analyzing data over a period of time, developing models and simulations, and then providing recommendations to stakeholders such as city officials and emergency responders. However, this approach may not be sufficient to address the real-time needs of the community during a heatwave event. Several steps can be taken:

- Establish real-time monitoring: Install sensors and monitoring systems in urban areas to collect real-time data on temperature, humidity, and other relevant parameters. This data can be used to predict the onset and severity of heatwaves, and inform emergency response plans.

- Develop predictive models: Use machine learning and other advanced techniques to develop predictive models that can forecast heatwave events with high accuracy (Cho et al. 2020). These models should be updated in real time based on the latest data and made accessible to stakeholders through user-friendly interfaces.
- Deploy targeted interventions: Use the predictive models to identify the areas most at risk during a heatwave and deploy targeted interventions such as cooling centers, water stations, and outreach campaigns to inform vulnerable populations (VanderMolen et al. 2022).
- Engage with stakeholders: Work closely with city officials, emergency responders, and community groups to ensure that the project meets their real-time needs and addresses their concerns. Establish regular communication channels to enable quick dissemination of information and feedback.

It involves leveraging cutting-edge technology, engaging with stakeholders, and developing innovative solutions that can be quickly deployed during a crisis. This approach ensures that the project is aligned with real-world needs and can have a meaningful impact on the community.

5 Controlling Sunk Costs

As actionable science will inevitably be interwoven into the economic lives of the society, it becomes imperative to approach each project from an economic perspective, evaluating factors like return on investment. This will aid researchers in gaining a clearer understanding of their results and the actionability, especially when faced with resource constraints. In this section, our focus is on studying the sunk costs, which mean the costs spent and cannot be recovered regardless of future actions (Perignat and Fleming 2022). Being aware of the potential sunk costs empowers scientists to allocate their resources judiciously, minimizing wastage while maximizing their impact. This strategic resource allocation strategy helps scientists steer clear of becoming overly committed to a specific course of action, ensuring greater flexibility and adaptability in the face of swiftly evolving developments in environmental sciences. By conducting a meticulous evaluation of risks and potential benefits associated with various approaches, research projects can manage their exposure to risk, diminish the likelihood of costly errors and setbacks, and increase the probability of attaining their objectives and delivering tangible outcomes.

However, implementing these principles is more challenging than articulating them, primarily because many research projects fall into the high-risk, high-reward category where sunk costs can loom large. Exercising control over sunk costs in climate science projects presents a formidable task. It includes pre-project budget and resource planning, regular monitoring and adjusting, maintaining project flexibility, keeping external collaboration and partnership, and conducting in-time cost–return analysis. A research project normally lasts from 1 year to 5 years, and scientists need to make many decisions on whether some tasks are worth doing and allocating resources. Take a research project aiming to study the impacts of urbanization on the local ecosystems as example. It is originally planned to study the changes in vegetation, wildlife, and water quality in and around urban areas, and develop science-based recommendations for sustainable urban development. Before launching the project, scientists need to outline the objectives contributing to the goal. Objectives are itemized smaller tasks that constitute the final goal. This project's objectives include identifying changes in vegetation cover in urban and suburban areas, evaluating the effects on wildlife abundance and diversity in different habitat types (e.g., forests, wetlands, grasslands), assessing the water quality to compare the levels of pollutants and nutrients, analyzing the socioeconomic drivers of urbanization, like population growth, land use change, and industry and market development. As there are so many ways available to achieve these objectives, scientists have to conduct a small-scale pilot study to compare methods and gather initial benchmark data. This will help identify potential challenges and adjust their approach before scaling up the project. Also, we should reasonably evaluate the actionability of each objective and precisely estimate the costs, mainly in how many person-hours each task will take. Making reasonable assignments by splitting the tasks will help team members stay focused, like avoiding investing time in regions that are not relevant to the project region and study periods. Another useful approach is to leverage the existing data whenever possible, like gathering vegetation cover information from the satellite and aerial imagery that are either publicly available at NASA and USGS or purchased from private sources like PlanetLabs and BlackSky. The water quality data could reuse the USGS National Water Information System (https://waterdata.usgs.gov/nwis) data that are professionally collected and calibrated following strict data quality protocols. In addition, scientists should avoid doing research behind locked doors, and reach out and collaborate with local organizations, especially nonprofits and local environment groups to share resources and engage with local community members.

On the other hand, for the research stakeholders and end users, managing sunk costs can reduce the associated financial risks (Kardes et al. 2013). Carefully tracking expenses and ensuring that resources are allocated efficiently can help decision-makers and funders mitigate the risk of cost overruns and budget shortfalls. Many research projects are funded by government agencies like the National Science Foundation using taxpayers' money. Keeping an eye on sunk costs could ensure research projects are aligned with the needs and goals of the broader community and serve the best interests of the society. Sharing sunk cost analysis reports with them can let all parties be aware of how resources are being allocated to help build trust and optimize outcomes.

For example, let us continue to assume that the urbanization impact study above identified that urban development has caused a significant reduction in tree canopy cover, which is negatively impacting local air quality and the health of residents. One potential solution would be to plant more trees in the area. The scientists should spend time on finding and providing local decision-makers with the following information: (1) the specific areas in the community where tree planting would have the greatest benefit on improving air quality and public health; (2) the most effective tree species for improving air quality and mitigating the negative impacts of urbanization; and (3) the estimated cost of tree planting and maintenance in the identified areas. With this project proposal and the expected results, local decision-makers could make an informed decision on whether to invest in the research project, or initiate a tree-planting program and allocate the necessary resources. They could weigh the costs and benefits of this research and resulting intervention and compare it to other potential solutions identified in the study. For example, if the cost of the research project is $100,000, implementing a tree planting program is estimated to be $200,000 and the estimated benefits (in terms of improved air quality, public health, and ecosystem services) are valued at $1,000,000, the decision-makers could conclude that the program is a cost-effective solution. If they finally decide to not move forward with the tree planting program, the sunk cost will be $100,000. Furthermore, even if the resources are unavailable at the moment, the stakeholders could use the research findings to work with local decision-makers in the future to develop an implementation plan for the tree planting program, such as identifying funding sources, engaging with the community to increase public support, and establishing maintenance protocols to ensure the longevity and success of the program. Eventually, the goal is to achieve the $1 million benefits for all the community residents in environmental benefits and justice.

6 Why Do We Write This Book?

We firmly believe that all the ongoing research endeavors make valuable contributions, and it is essential to clarify that this book is not intended to diminish the significance of any research that may be currently nonactionable. Instead, the goal of this book is to propose a framework that all research teams can readily implement to enhance the actionability of their work. Give advice to decision-makers, put all information in front transparently and offer them clear insights in the consequences of taking action or maintaining the status quo. This book will help bridge the gap between scientific research and real-world decision-making. Scientific research is often too complex for non-experts, including decision-makers to comprehend fully. They may not always be aware of the latest research and likely struggle to translate it into actionable steps. This book will serve as a tool to help scientists present their research in a clear, accessible way and focus on developing concrete recommendations for actions, to ensure that scientific insights are well-prepared to be effectively communicated to decision-makers and readily applied in real-world contexts.

Action is always a compromised result, even when based on the most robust and well-established scientific findings. It's important to acknowledge that even scientifically proven facts can be subject to distortion when communicated to the public, as documented in previous research (McGarity et al. 2010). Many people may not fully grasp the gravity or urgency of environmental problems confronting the world today. The intention behind crafting this book is to present scientific research on these critical issues, shed light on the potential repercussions of inaction, and contribute to heightened awareness and inspire proactive measures. There are numerous individuals and organizations eager to address environmental issues, but they may be unsure of where to commence or the specific steps to undertake. This book seeks to provide them with tangible guidance, effective strategies, and practical tools to initiate action, thereby enabling them to effect meaningful change. Addressing global environmental problems necessitates collaboration across disciplines, sectors, and borders. Meanwhile, bringing together insights from a range of scientific disciplines and showcasing examples of successful collaboration can promote knowledge sharing and encourage more effective collaboration in the future.

Additionally, the science research in universities is often disconnected from the real-world problems, and researchers publish papers that are the final products of many projects, yet the conversion of their findings into actionable solutions is typically not well-funded or financially incentivized. In other words, the universities and research institutions prioritize the production of research papers and academic recognition over practical and actionable solutions to real-world problems. The review committees for promotion and raises usually add more weight to the science-specific metrics that could diverge from the actionable results. Many research domains are small communities and highly specialized. The insular nature can make it difficult for scientists to effectively communicate with and engage stakeholders outside of their field. Many scientists are compelled to prioritize paper publication over the pursuit of actionableness. This can result in the gradual separation between the scientific community and the stakeholders who could benefit from their work. Moreover, the separation will result in the lack of funding and financial incentives for turning research into actionable items and continue to discourage scientists from pursuing practical applications for their work. This book is also sending a signal to academia to draw the attention of everyone to the importance of actionable science.

Funding agencies like the NSF need more actionable science community building. The NSF's broad impact emphasizes internal impacts like mentoring students and next-generation researchers. However, external impacts like research to action conversion are usually not emphasized or mentioned. This book tries to guide scientists to translate the research results into more FAIRable, tangible materials (Sun et al. 2020) and allows new scientists like an undergraduate to easily reuse or replicate with a small amount of effort. For people outside university or the research institutes hosting the projects, it can be considerably challenging to comprehend the significance of research results, both their positive and negative aspects, as well as the necessity for action. The good first step is always to make science credible and transparent to the public. This book will touch on those points and give suggestions and examples on how to cultivate a science community with a focus on making science results actionable.

7 What Can the Readers Expect from the Book?

This book can offer readers a clear and comprehensive overview of the key principles and methods of actionable science, including how to design and conduct research to be easily translated into real-world solutions. People will gain a better understanding of the challenges and barriers in the translation of scientific research into actionable solutions for global environmental issues. You will have insights into the principles and methods of actionable science, including case studies and examples of successful applications in various environmental contexts. Readers can learn about the strategies for effective communication and engagement with stakeholders and decision-makers, ensuring that scientific research aligns seamlessly with practical needs and priorities. The book will include practical guidance on how to incorporate actionability into research projects and proposals, including techniques for measuring impact and assessing feasibility. The book will feature chapters that outline a roadmap for navigating the complex and interdisciplinary nature of global environmental issues, including understanding of the role of policy, economics, and social factors in shaping sustainable solutions.

8 Summary

This chapter serves as an exploration of the concept of actionable science and its relevance in addressing global climate and environmental issues. It discusses the traditional approach of academic research, where researchers focus on answering "why" questions without necessarily considering the practical applications of their findings. The chapter clarifies that this approach is inadequate in addressing complex global problems such as climate change and environmental degradation, where actions are needed to mitigate or adapt to the effects of these issues. Then we redefine the concept of actionable science, which refers to scientific research that is designed to produce actionable outcomes or recommendations for addressing real-world problems. We emphasize that actionable science should involve collaboration between scientists and stakeholders, including policymakers, industry leaders, and local communities. This collaboration helps ensure that research findings are relevant, accessible, and applicable to the challenges faced by the society.

The chapter also highlights the challenges of implementing actionable science, including the need for interdisciplinary collaboration and effective communication between scientists and stakeholders. We provide a framework for quantitatively measuring the actionability of a science project and examples to explain how to make science projects actionable, such as the carbon-capturing technologies, and the evaluation of urbanization stress on local ecosystems. This chapter lays the foundation for the rest chapters by clarifying the need for collaboration and communication between scientists and stakeholders to ensure that research findings are relevant and actionable, and provides examples of successful actionable science projects.

References

Abu-Zahra, Mohammad R.M., Léon H.J. Schneiders, John P.M. Niederer, Paul H.M. Feron, and Geert F. Versteeg. 2007. CO2 capture from power plants: Part I. A parametric study of the technical performance based on monoethanolamine. *International Journal of Greenhouse Gas Control* 1 (1): 37–46.

Agrawal, Ajay K. 2001. University-to-industry knowledge transfer: Literature review and unanswered questions. *International Journal of Management Reviews* 3 (4): 285–302.

Altieri, Miguel A., Clara I. Nicholls, Alejandro Henao, and Marcos A. Lana. 2015. Agroecology and the design of climate change-resilient farming systems. *Agronomy for Sustainable Development* 35 (3): 869–890.

Arnott, James C., Christine J. Kirchhoff, Ryan M. Meyer, Alison M. Meadow, and Angela T. Bednarek. 2020. Sponsoring actionable science: What public science funders can do to advance sustainability and the social contract for science. *Current Opinion in Environmental Sustainability* 42: 38–44.

Bannayan, Mohammad, and Gerrit Hoogenboom. 2008. Weather analogue: A tool for real-time prediction of daily weather data realizations based on a modified k-nearest neighbor approach. *Environmental Modelling & Software* 23 (6): 703–713.

Beier, Paul, Lara J. Hansen, Lynn Helbrecht, and David Behar. 2017. A how-to guide for coproduction of actionable science. *Conservation Letters* 10 (3): 288–296.

Bhown, Abhoyjit S., and Brice C. Freeman. 2011. Analysis and status of post-combustion carbon dioxide capture technologies. *Environmental Science & Technology* 45 (20): 8624–8632.

Büntgen, Ulf, and Lena Hellmann. 2013. The Little Ice Age in scientific perspective: Cold spells and caveats. *Journal of Interdisciplinary History* 44 (3): 353–368.

Chai, Kah-Hin, Mike Gregory, and Yongjiang Shi. 2003. Bridging islands of knowledge: a framework of knowledge sharing mechanisms. *International Journal of Technology Management 25 (8): 703–727.*

Cho, Dongjin, Cheolhee Yoo, Jungho Im, and Dong-Hyun Ch. 2020. Comparative assessment of various machine learning-based bias correction methods for numerical weather prediction model forecasts of extreme air temperatures in urban areas. *Earth and Space Science* 7 (4): e2019EA000740.

Clune, Thomas, and Richard Rood. 2011. Software testing and verification in climate model development. *IEEE Software* 28 (6): 49–55.

Collins, Hilary. 2018. *Creative research: The theory and practice of research for the creative industries*. Bloomsbury Publishing.

David, Knowles. 2023. Bill Gates criticizes idea of leading 'impoverished lifestyle' to address climate change. https://news.yahoo.com/bill-gates-climate-change-criticizes-impoverished-lifestyle-idea-200125212.html. Accessed on 30 May 2023.

Edwards, Paul N. 2011. History of climate modeling. *Wiley Interdisciplinary Reviews: Climate Change* 2 (1): 128–139.

Eyal, Gil. 2019. *The crisis of expertise*. John Wiley & Sons.

Goh, Swee C. 2002. Managing effective knowledge transfer: An integrative framework and some practice implications. *Journal of Knowledge Management* 6 (1): 23–30.

Hennige, Sebastian, J. Murray Roberts, and Phillip Williamson. 2014. An updated synthesis of the impacts of ocean acidification on marine biodiversity. *CBD Technical Series* 75.

Inan, Ilhan. 2013. *The philosophy of curiosity*. Routledge.

Johnson, T.R., Joe Letter, and Judith Kemerait Livingston. 2009. Floating foundations: Kairos, community, and a composition program in post-Katrina new Orleans. *College English* 72 (1): 29–47.

Kardes, Ilke, S. Ayse Ozturk, Tamer Cavusgil, and Erin Cavusgil. 2013. Managing global megaprojects: Complexity and risk management. *International Business Review* 22 (6): 905–917.

Kates, Robert W., William R. Travis, and Thomas J. Wilbanks. 2012. Transformational adaptation when incremental adaptations to climate change are insufficient. *Proceedings of the National Academy of Sciences* 109 (19): 7156–7161.

Kirchhoff, Christine J., Maria Carmen Lemos, and Suraje Dessai. 2013. Actionable knowledge for environmental decision making: Broadening the usability of climate science. *Annual Review of Environment and Resources* 38: 393–414.

Kosior, Edward, and Irene Crescenzi. 2020. Solutions to the plastic waste problem on land and in the oceans. In *Plastic waste and recycling*, 415–446. Academic Press.

Kujur, Fedric, and Saumya Singh. 2017. Engaging customers through online participation in social networking sites. *Asia Pacific Management Review* 22 (1): 16–24.

Lemos, Maria Carmen, Christine J. Kirchhoff, and Vijay Ramprasad. 2012. Narrowing the climate information usability gap. *Nature Climate Change* 2 (11): 789–794.

Leung, Dennis Y.C., Giorgio Caramanna, and M. Mercedes Maroto-Valer. 2014. An overview of current status of carbon dioxide capture and storage technologies. *Renewable and Sustainable Energy Reviews* 39: 426–443.

Lin, Qianxin, and Irving A. Mendelssohn. 2012. Impacts and recovery of the Deepwater horizon oil spill on vegetation structure and function of coastal salt marshes in the northern Gulf of Mexico. *Environmental Science & Technology* 46 (7): 3737–3743.

McGarity, Thomas O., and Wendy E. Wagner. 2010. *Bending science: How special interests corrupt public health research.* Harvard University Press.

Meinke, Holger, Rohan Nelson, Phil Kokic, Roger Stone, Ramasamy Selvaraju, and Walter Baethgen. 2006. Actionable climate knowledge: From analysis to synthesis. *Climate Research* 33 (1): 101–110.

Mey, Franziska, Mark Diesendorf, and Iain MacGill. 2016. Can local government play a greater role for community renewable energy? A case study from Australia. *Energy Research & Social Science* 21: 33–43.

Miller, Melanie, William E. Robinson, Ana Rita Oliveira, Nina Heidary, Nikolay Kornienko, Julien Warnan, Inês AC Pereira, and Erwin Reisner. 2019. Interfacing formate dehydrogenase with metal oxides for the reversible electrocatalysis and solar-driven reduction of carbon dioxide. Angewandte Chemie International Edition 58 (14): 4601–4605.

Mishra, Deepti, and Alok Mishra. 2011. Complex software project development: Agile methods adoption. *Journal of Software Maintenance and Evolution: Research and Practice* 23 (8): 549–564.

Moisander, Johanna. 2007. Motivational complexity of green consumerism. *International Journal of Consumer Studies* 31 (4): 404–409.

Morris, Michael S., Kip S. Thorne, and Ulvi Yurtsever. 1988. Wormholes, time machines, and the weak energy condition. *Physical Review Letters* 61 (13): 1446.

Muazu, Rukayya Ibrahim, Rachael Rothman, and Lorraine Maltby. 2021. Integrating life cycle assessment and environmental risk assessment: A critical review. *Journal of Cleaner Production* 293: 126120.

National Academies of Sciences, Engineering, and Medicine. 2019. *Reproducibility and replicability in science*. National Academies Press.

Nikolaj, Skydsgaard, and David Evans.2021. World's largest plant capturing carbon from air starts in Iceland. https://www.reuters.com/business/environment/worlds-largest-plant-capturing-carbon-air-starts-iceland-2021-09-08/. Accessed on 30 May 2023.

Oreskes, Naomi. 2018. *The scientific consensus on climate change: How do we know we're not wrong?* Springer International Publishing.

Palmer, T.N. 2012. Towards the probabilistic earth-system simulator: A vision for the future of climate and weather prediction. *Quarterly Journal of the Royal Meteorological Society* 138 (665): 841–861.

Pawley, Alice L. 2009. Universalized narratives: Patterns in how faculty members define "engineering". *Journal of Engineering Education* 98 (4): 309–319.

Perignat, Elaine, and Fraser F. Fleming. 2022. Sunk-cost bias and knowing when to terminate a research project. *Angewandte Chemie* 134 (36): e202208429.

Pielou, Evelyn C. 2008. *After the ice age: The return of life to glaciated North America*. University of Chicago Press.

Price, Lynn, Nan Zhou, David Fridley, Stephanie Ohshita, Lu Hongyou, Nina Zheng, and Cecilia Fino-Chen. 2013. Development of a low-carbon indicator system for China. *Habitat International* 37: 4–21.

Rezaie, Ali Mohammad, Jarrod Loerzel, and Celso M. Ferreira. 2020. Valuing natural habitats for enhancing coastal resilience: Wetlands reduce property damage from storm surge and sea level rise. *PLoS One* 15 (1): e0226275.

Rogers, Ayesha Pamela. 2011. Cultural heritage impact assessment: Making the most of the methodology. In *International conference on heritage conservation*, 12–13.

Roussanaly, Simon, Niels Berghout, Tim Fout, Monica Garcia, Stefania Gardarsdottir, Shareq Mohd Nazir, Andrea Ramirez, and Edward S. Rubin. 2021. Towards improved cost evaluation of Carbon Capture and Storage from industry. *International Journal of Greenhouse Gas Control* 106: 103263.

Sara, Laux. 2018. Blurring the boundaries between science and engineering. https://brighterworld.mcmaster.ca/articles/blurring-the-boundaries-between-science-and-engineering/. Accessed 30 May 2023.

Soto, Luis A., Alfonso V. Botello, Sergio Licea-Durán, Marcial L. Lizárraga-Partida, and Alejandro Yáñez-Arancibia. 2014. The environmental legacy of the Ixtoc-I oil spill in Campeche Sound, southwestern Gulf of Mexico. *Frontiers in Marine Science* 1: 57.

Stanger, Rohan, Terry Wall, Reinhold Spörl, Manoj Paneru, Simon Grathwohl, Max Weidmann, Günter Scheffknecht, et al. 2015. Oxyfuel combustion for CO2 capture in power plants. *International Journal of Greenhouse Gas Control* 40: 55–125.

Sun, Ziheng, and Nicoleta Cristea. 2023. Introduction of artificial intelligence in earth sciences. In *Artificial intelligence in earth science*, 1–15. Elsevier.

Sun, Ziheng, Liping Di, Annie Burgess, Jason A. Tullis, and Andrew B. Magill. 2020. Geoweaver: Advanced cyberinfrastructure for managing hybrid geoscientific AI workflows. *ISPRS International Journal of Geo-Information* 9 (2): 119.

Sun, Ziheng, Laura Sandoval, S. Robert Crystal-Ornelas, Mostafa Mousavi, Jinbo Wang, Cindy Lin, Nicoleta Cristea, et al. 2022. A review of earth artificial intelligence. *Computers & Geosciences*: 105034.

Sylvester, Zephyr T., and Cassandra M. Brooks. 2020. Protecting Antarctica through co-production of actionable science: Lessons from the CCAMLR marine protected area process. *Marine Policy* 111: 103720.

Tan, Oon-Seng. 2021. *Problem-based learning innovation: Using problems to power learning in the 21st century*. Gale Cengage Learning.

Tansley, Stewart, and Kristin Michele Tolle. 2009. In *The fourth paradigm: Data-intensive scientific discovery*, ed. Anthony J.G. Hey, vol. 1. Redmond, WA: Microsoft Research.

Thompson, Lonnie G. 2010. Climate change: The evidence and our options. *The Behavior Analyst* 33: 153–170.

Trexler, Adam, and Adeline Johns-Putra. 2011. Climate change in literature and literary criticism. *Wiley Interdisciplinary Reviews: Climate Change* 2 (2): 185–200.

US EPA (US Environ. Prot. Agency). 2017. Multi-model framework for quantitative sectoral impacts analysis: A technical report for the fourth national climate assessment.

VanderMolen, Kristin, Nicholas Kimutis, and Benjamin J. Hatchett. 2022. Recommendations for increasing the reach and effectiveness of heat risk education and warning messaging. *International Journal of Disaster Risk Reduction* 82: 103288.

Wagner, Travis P. 2017. Reducing single-use plastic shopping bags in the USA. *Waste Management* 70: 3–12.

Ward, Vicky, Allan House, and Susan Hamer. 2009. Developing a framework for transferring knowledge into action: a thematic analysis of the literature. *Journal of Health Services Research & Policy 14 (3): 156–164.*

Wolf, Fred Alan. 1988. *Parallel universes*. Simon and Schuster.

Chapter 2
Data Foundation for Actionable Science

Ziheng Sun

Contents

1 Introduction

Good quality data is essential in climate and environment research as the foundation for making informed decisions and taking accurate action (Wiens et al. 2009). Our climate and environmental systems are complex and highly interconnected, and changes in one area can have cascading effects on others (Lawrence et al. 2020).

Z. Sun (✉)
Center for Spatial Information Science and Systems, Department of Geography and Geoinformation Science, George Mason University, Fairfax, VA, USA
e-mail: zsun@gmu.edu

Z. Sun (ed.), *Actionable Science of Global Environment Change*,
https://doi.org/10.1007/978-3-031-41758-0_2

Accurate and comprehensive data is required for understanding these systems and predicting their behavior under different scenarios. Without reliable data, it is difficult to make climate and environmental policies, and forecast the potential consequences of those decisions. Climate and environmental research often involves large amounts of data from multiple sources, including remote sensing, in situ measurements, and modeling. The integration and analysis of these data require advanced computational tools and techniques, which can only be applied effectively if the data is high quality and properly documented. Poor quality or incomplete data can lead to inaccurate or biased results, which can have serious consequences if utilized in real-world decision-making and policy development (Ruiz-Benito et al. 2020). For example, consider the case of sea level rise, a concerning issue for coastal communities, and accurate predictions are essential for developing effective adaptation strategies. However, sea level is a complex variable that is influenced by many different factors, including thermal expansion, melting ice sheets and glaciers, and changes in ocean currents (Golledge 2020). To accurately model sea level rise, researchers must integrate data from multiple sources, including satellite measurements of ocean height, in situ measurements of ocean temperature and salinity, and models of ice sheet dynamics (Meyssignac et al. 2019; Cook et al. 2023). If any of this data is inaccurate or incomplete, it can lead to incorrect predictions of sea level rise and potentially disastrous consequences for coastal communities and public distrust of the coastal sciences.

Without accurate and reliable data, it is difficult to identify the root causes of environmental problems and develop effective solutions (Sun et al. 2021). Climate data can be sophisticated, large in scale, and come from various sources, such as satellites, weather stations, or citizen science projects. Managing, cleaning, and integrating the data is a daunting task, requiring significant expertise and resources. There is often a lack of standardized protocols and tools for collecting and processing data, and can lead to inconsistencies, making it difficult to compare and combine datasets from different sources (Zimmerman 2008). Issues often arise around data access and sharing, particularly when data is collected by government agencies or private companies. Access to data may be restricted due to privacy concerns, national security issues, or commercial interests, making it difficult for researchers to obtain the data they need to conduct their studies. Moreover, funding for climate and environmental research may be limited, leading to a lack of investment in necessary data infrastructure with incomplete or outdated datasets, which may not accurately reflect current conditions and trends.

Collecting and analyzing high-quality data requires specialized equipment, trained personnel, and infrastructure support. For example, a single oceanographic research vessel (Brett et al. 2020) can cost tens of millions of dollars, and operating costs can run into the millions per year. The cost of satellite missions can also be significant, with some missions costing several hundred million dollars. These costs can be a significant barrier to entry for researchers and institutions, particularly those in developing countries or with limited funding.

The chapter introduces the importance of collecting, storing, and processing data in an accessible, transparent, and replicable way. It overviews best practices in environmental data management and interdisciplinary collaboration and open data sharing. For example, assume there is a research project aiming to assess the impacts of deforestation on biodiversity in a tropical rainforest. In order to effectively conduct this research, the scientists need access to high-quality data on the forest ecosystem, including information on the tree species, soil composition, water cycles, and animal populations (Giam 2017). This data must be collected and stored in a way that allows for easy access and analysis. To ensure the reliability of the data, the researchers must use standardized methods for collecting and analyzing the data, and they must carefully document their procedures to ensure transparency and replicability. They must also consider ethical considerations, such as obtaining informed consent from any individuals or communities affected by the research. Once the data is collected, it must be stored in a secure and accessible database that allows for easy sharing and collaboration among researchers. This promotes interdisciplinary collaboration and enables scientists from different disciplines to work together to address complex environmental challenges. In summary, the chapter emphasizes the importance of good quality data in actionable science for climate and environment, and provides guidance on best practices for data collection, storage, and processing. This is critical for ensuring that our efforts to address climate change and environmental degradation are based on sound scientific data and can effectively inform decision-making and policy development.

2 Data Categories and Availability for Actionable Science

Based on the sources and the collection methods, scientific datasets can be divided into the following categories: satellite data, in situ data, model simulation data, citizen science data, and social media data. This section will overview each category and analyze their current availability for actionable science.

2.1 Satellite Data

The sensors onboard satellites can provide a lot of information on land use, vegetation, ocean temperature, atmospheric composition, and many Earth's surface, ocean, and atmosphere processes. Some widely used satellite datasets include Moderate Resolution Imaging Spectroradiometer (MODIS) (Justice et al. 2002), Landsat series (Tucker et al. 2004), Sentinel series (Spoto et al. 2012), Suomi National Polar-orbiting Partnership (Suomi NPP) (Weng et al. 2012), ICESat (Schutz et al. 2005), Global Precipitation Measurement (GPM), SWOT (Biancamaria et al. 2016), etc. Table 2.1 lists some popular available satellite datasets that can be publicly retrieved and used for climate and environment research.

Many reasons draw scientists to usually first turn to satellite image datasets to look for those that can fit their research purposes. Satellites can cover vast areas of

Table 2.1 Available satellites for actionable science

Satellite name	Operator	Specs	Application
GRACE and GRACE-FO	NASA and DLR	Revisit: 30 days	Earth gravity field, hydrology, and ice
Landsat	USGS	Revisit: 16 days Resolution: 30 m	Land use and land cover, forest, and environmental studies
Sentinel-2	ESA	Revisit: 5 days Resolution: 10–20 m	Vegetation, crop, and land use change detection
MODIS (Aqua and Terra)	NASA	Revisit: 1–2 days Resolution: 250–1000 m	Atmospheric composition, land cover, sea surface temperature, and vegetation indices
GOES	NOAA	Resolution: 0.5 km Revisit: 5–15 min	Weather forecasting, storm, hurricane, and atmospheric studies
Suomi NPP	NASA	Resolution: 750 m Revisit: 16 days	Weather, climate, sea surface temperature, vegetation indices, and aerosol concentrations

the Earth's surface and provide data on a global scale (Wulder et al. 2008). It can monitor changes in the environment over time, identify patterns and trends, and make predictions about future conditions. While circling Earth by themselves (some satellites stay unmoving relative to Earth if they are in Earth's synchronous orbit) (Boain 2004), satellites can provide continuous monitoring of the environment over the years, which is especially important for identifying areas of high risk for natural disasters like floods, hurricanes, and wildfires. Another unbeatable advantage is that satellite data can cover remote areas difficult to access on foot or by vehicle and very important for monitoring changes in biodiversity, forest cover, and other critical ecosystem services. Also, for projects with a limited budget for data collection, publicly available satellite datasets could be a lifesaver. While satellite data can be expensive to acquire and process, it is often more cost-effective than ground-based monitoring. The data providers like NASA, NOAA, and USGS often preprocessed their datasets into different levels, and if scientists know about them and when to use them and how to connect the data to their data analytics, that would avoid a lot of duplicated work and is a financially wise choice (Ebert-Uphoff et al. 2017). For example, the Landsat satellites operated by NASA and the US Geological Survey have been providing images of the Earth's surface since 1972 (Wulder et al. 2016), and used for mapping land use, monitoring deforestation, and tracking changes in the cryosphere. The derived information has been used by governments, NGOs, and individuals to make informed decisions about land use, natural resource management, and climate change mitigation and adaptation strategies. Another example is the European Space Agency's Sentinel-1 satellite, which provides radar imagery to monitor changes in land cover, detect oil spills, and track changes in sea ice, and disaster response efforts and help governments and NGOs respond to environmental emergencies.

The availability of satellite data varies depending on the mission and data type. Many government space agencies provide free satellite datasets for scientific and environmental research purposes. For example, NASA, the US Geological Survey (USGS), and the European Space Agency (ESA) all have missions that provide freely accessible data to the public. Generally, the low-to-medium resolution data are less sensitive and more easily available than the high-to-very-high resolution data. The Landsat program, a joint mission between NASA and USGS provides free and open access to over 40 years of satellite imagery data for land-use monitoring, natural resource management, and environmental monitoring and generated billions of dollars of benefits to the public and greatly advanced the Earth science developments. Other NASA missions that provide free data include the Moderate Resolution Imaging Spectroradiometer (MODIS), the Atmospheric Infrared Sounder (AIRS), and the Ozone Monitoring Instrument (OMI). The ESA also provides a range of freely available satellite data through their Sentinel missions, which are part of the Copernicus program (Thépaut et al. 2018). The Sentinel-1, Sentinel-2, and Sentinel-3 missions provide data on land cover, vegetation health, sea level, and sea ice. The Japan Aerospace Exploration Agency (JAXA) provides access to their Advanced Land Observing Satellite (ALOS) and the Global Precipitation Measurement (GPM) mission (Shimada et al. 2009). Compared to government-funded missions offering free satellite data, there are commercial satellite operators whose data will require a fee such as DigitalGlobe, Planet Labs, and DarkSky. These companies provide very-high-resolution (<1 m) satellite imagery for commercial and government purposes, such as disaster responses. In some cases, satellite data is completely restricted and only accessible to authorized users or purchasers due to national security concerns. For example, data from spy satellites operated by military and intelligence agencies will not (or never) be publicly available. China has a large and growing satellite program, including civilian satellites such as the Gaofen (Li et al. 2017) and ZY Earth observation satellites partially available for purchase, while others, such as the military reconnaissance satellites, are completely restricted and openly sharing is a very serious offense. Russia has a similar program like the Resurs-P Earth observation satellites. The India Remote Sensing program has launched several EO satellites like Resourcesat-2 (Rao et al. 2006), Cartosat-3, and Oceansat-3, and manages them via the Indian Space Research Organization. Several download portals are provided like the National Remote Sensing Centre (NRSC) and the Indian Earth Observation Data Gateway (IEODG). The Canadian Space Agency (CSA) Earth Observation program includes the RADARSAT Constellation satellites (Thompson 2015) and provides spatial data infrastructure (SDI). For scientific purposes, most datasets are good to use. However, please note that access to RADARSAT-2 and RCM data may be subject to licensing requirements and restrictions, and users may need to obtain permission or pay for commercial use.

2.2 In Situ Data

It refers to the data collected on the ground by in situ devices, such as weather station data, stream gauges, soil moisture probes, water quality sensors, carbon dioxide sensors, seismometers, spectrometers, GPS receivers, fluorometers, acoustic Doppler current profilers, air quality sensors, and thermometers. These sensors are critical to measure temperature, humidity, air pressure, wind speed and direction, precipitation, water content in soil, etc. Table 2.2 presents a nonexhaustive list of the common sensor types and the variables they are designed to monitor.

You can see that all these data are highly valuable by providing detailed information about specific locations and thus are often used for validation and calibration of satellite data or model simulation as "ground truth." A single sensor has too many uncertainties and could be inaccurate, and the data is unreliable without

Table 2.2 In situ sensor types and their target monitoring variables

Sensor type	Variables
Weather station	Temperature, humidity, air pressure, wind speed and direction, and precipitation
Soil moisture probe	Water content in soil
Stream gauge	Flow of water in rivers and streams
Water quality sensor	Water pH, dissolved oxygen, and turbidity
Carbon dioxide sensor	Atmospheric CO_2 concentrations
Seismometers	Ground motion caused by earthquakes and other seismic events
Spectrometers	Spectrum of light reflected or emitted by the Earth's surface
GPS receivers	Location, velocity, and direction of motion of objects
Fluorometers	Fluorescence of chlorophyll in plants
Acoustic Doppler current profiler	Water velocity and direction in the ocean
Air quality sensor	Concentration of pollutants such as ozone, nitrogen oxides, and particulate matter in the atmosphere
Thermometers	Temperature
LiDAR	The height and structure of vegetation, terrain, and buildings
Flux towers	Exchanges of carbon, water, and energy between the Earth's surface and the atmosphere
Buoys	Floating devices to monitor weather conditions, water temperature, salinity, and ocean currents
Ice core drills	Extract ice cores from glaciers and ice sheets
Tree cores	Extract cores from trees to study tree growth
Ocean gliders	Autonomous underwater vehicles to collect temperature, salinity, and currents
Animal tagging devices	Track the movements and behavior of animals in their natural habitats
Ground-penetrating radar	Subsurface geology and soil structure
Photometers	Amount of light in a specific location
Soil respiration chambers	Amount of carbon dioxide released by soil

professional calibration. To make solid and reliable observations, one common practice is for scientists to usually create ground observation networks with similar sensors and calibrate and clean the data into standard products. Ground observation networks can provide spatially extensive and long-term coverage of environmental parameters over large areas, and often use standardized protocols and equipment to ensure consistency and reliability of the data collected. Examples of in situ datasets include the Global Historical Climatology Network (GHCN) (Menne et al. 2012) and the National Water Quality Assessment (NAWQA) program (Gilliom et al. 1995). Table 2.3 contains the well-known ground observation networks we have set up so far.

These are just some dedicated well-known networks. Many other ground observation networks are also set up for many other purposes, such as Landsat has its own ground stations for satellite calibration. NASA operates the Near-Earth Network (NEN) (Schaire et al. 2016), a global network of ground stations that provides support for the agency's Earth-observing missions to transmit data from satellites to Earth. It is worthwhile to search for any available networks and reach out to the operator to check for availability.

2.3 Model Simulation Data

Besides the real observations collected remotely and in the field, there is a great amount of data generated by climate and environmental models, such as global circulation models (GCMs) (Dosio et al. 2015), regional climate models (RCMs) (Rummukainen 2010), and ecosystem models. These models can simulate the behavior of complex systems and provide projections of future climate and environmental conditions. Examples of model output datasets include the Coupled Model Intercomparison Project (CMIP) (Eyring et al. 2016) and the North American Regional Climate Change Assessment Program (NARCCAP) (Mearns 2009). We list some commonly seen models in Table 2.4. Model-generated data can help make science more actionable by providing projections of future climate and environmental conditions based on different initial conditions of greenhouse gas emissions and other human activities. These projections can inform decision-making in various sectors, such as agriculture, water management, and infrastructure planning, by providing insights into potential future risks and opportunities. Compared to satellite and in situ data, model-generated data can provide detailed information on processes that cannot be directly measured and also give consistent, continuous, and long-term records of environmental conditions. Besides, model output does not have the hard restrictions on looking into the future and can make projections of future environmental conditions, allowing for better planning and decision-making. Another benefit is that models can fill in the gaps in spatial and temporal coverage where there are no satellite or in situ observations, and is a great resource to test hypotheses and validate observations. Satellite data provides global coverage and can monitor changes over time, but may have limited resolution and accuracy. In

Table 2.3 Ground observation network list

Name	Operator	Sensors	Variables	Availability	Coverage
Global Historical Climatology Network (GHCN)	NOAA	Weather stations	Temperature, precipitation, pressure, and humidity	GHCN-Monthly and GHCN-Daily are on the NCEI website	100,000 stations in 180 countries 1–175 years
National Water Quality Assessment (NAWQA)	USGS	Gauges	Nutrients, pesticides, sediment, water temperature, pH, and more	USGS National Water Information System (NWIS) website	Specific watersheds and aquifers in the United States 1970 to present
FLUXNET	Various	Eddy covariance towers	Carbon dioxide, water vapor, and energy fluxes between the land surface and atmosphere	https://fluxnet.org/	Global coverage 1990 to present
OceanSITES	Various	Moored buoys, profiling floats, gliders, etc.	Temperature, salinity, currents, nutrients, carbon dioxide, and more	OceanSITES website	Global coverage 1990s to present
NEON	NSF	Various	Climate, geology, hydrology, and vegetation	NEON data portal	United States 2013 to present
ICOS	European research infrastructure consortium ERIC)	Various sensors	Carbon dioxide, methane, and nitrous oxide	ICOS carbon portal	Europe 2015 – present
Ocean observatories initiative (OOI)	NSF	Moored, autonomous, and cabled sensors	Temperature, salinity, pressure, dissolved oxygen, pH, chlorophyll, turbidity, currents, and seismic activity	https://ooinet.oceanobservatories.org/	2014 – present Pacific and Atlantic oceans
International Soil Moisture Network (ISMN)	Various	Soil moisture sensors	Soil moisture at various depths	https://ismn.geo.tuwien.ac.at/	Global 2007 – present

Name	Operator	Sensors	Variables	Availability	Coverage
Snow Telemetry (SNOTEL)	USDA	Automatic monitoring station with many sensors	Snow water content, snow depth, and barometric pressure, soil moisture, solar radiation, etc.	https://www.nrcs.usda.gov/wps/portal/wcc/home/	Western United States, Alaska, and Puerto Rico 1978 – present
Atmospheric Radiation Measurement (ARM)	U.S. DOE	Various sensors	Atmospheric radiation and cloud properties	https://www.arm.gov/	The United States and internationally1989 – present
Global Atmosphere Watch	World Meteorological Organization (WMO)	Various sensors	Greenhouse gas and air pollutants	https://public.wmo.int/en/programmes/global-atmosphere-watch-programme	>65 countries 1960s – present
AirNow	EPA	Various	Particulate matter, ozone, nitrogen dioxide, sulfur dioxide, and carbon monoxide	airnow.gov	The United States Real time
CHARM (Canadian High Arctic Research Station Monitoring)	Polar Knowledge Canada (POLAR)	Weather stations	Air quality, water quality, permafrost temperature, and biodiversity	https://arcticportal.org/maps/download	Canada's high arctic Real time

Table 2.4 A list of some example climate and environment models

Name	Provider	Description	Coverage
Coupled Model Intercomparison Project (CMIP)	WCRP	A framework for coordinated climate model experiments	Globe
North American Regional Climate Change Assessment Program (NARCCAP)	NARCCAP	A collaboration of climate modeling groups to produce high-resolution regional climate model simulations over North America	North America
Community Earth System Model (CESM)	NCAR	A global climate model that simulates the Earth's system by integrating atmosphere, ocean, land, and sea ice components	Globe
Global Land Data Assimilation System (GLDAS)	NASA	A land surface modeling system that assimilates observational data to produce a suite of land surface variables	Globe
Community Earth System Model Large Ensemble	NCAR	A large ensemble of simulations from the CESM climate model that provides a range of possible future climate scenarios	Globe

situ data provides high-resolution and accurate measurements of environmental parameters, but may have limited spatial coverage. Model data can fix both issues and produce both high spatial and temporal resolution datasets with the best continuity and completeness.

Despite the advantages, the existing models are still improving and cannot say they are already good enough (in many cases they are not). One common issue is that model results are calculated based on a simplified simulation of the real-world complex systems and may not fully capture all relevant processes or interactions. The results are sensitive to input parameters and theory assumptions, and contain biases or errors at local and regional scale, which can impact their effectiveness in decision-making. Also, the model outputs are sometimes confusing and hard to visualize and interpret, and not as easily accessible or understandable as satellite or in situ observations for nonexpert users. The uncertainty in numerical models mainly comes from three places: the model inputs, parameterization, and model structure. Also, the spatial and temporal resolution of numerical models is usually coarse because running complex numerical models can require significant computational resources, which may limit the ability to generate results at high resolution or with large ensembles of simulations. These uncertainties and drawbacks can affect the use of model output in real-world decision-making and planning. In particular, decision-makers may be hesitant to rely on model output when there are significant uncertainties in the results. Additionally, the limited spatial and temporal resolution of models may make it difficult to apply model results at the local scale, where decisions are often made.

The availability of model outputs is usually good as long as the model operators have a place to store the data in a publicly accessible server for people to download. It depends on the specific model and the producer. Some models have well-established processes for making their output available to the public, while others may have limited accessibility or require special permissions to access. Many numerical models used in climate and environmental science have established data distribution systems, such as the Earth System Grid Federation (ESGF) (Cinquini et al. 2014), which provides access to a wide range of model output data from the CMIP and other modeling efforts. However, there can still be challenges in accessing and using the data, particularly for nonexperts or those with limited computational resources. Additionally, the sheer volume of data generated by some models can make it difficult to store and distribute the data efficiently. The data publishers usually use a moving-window approach, meaning the data can be stored for a limited amount of time, such as 2 weeks or a month, before being automatically deleted to free up storage space. This approach ensures that the most recent and relevant data are available to users while also managing the storage and maintenance of the data repository. However, it is important to note that this approach may not be suitable for all applications, and the specific needs of each user group should be carefully considered before implementing any data management strategy.

2.4 *Citizen Science Data*

This includes data collected by members of the public, often through crowdsourcing and citizen science projects (Table 2.5). Citizen science data is valuable to make science actionable as it enables a large number of people to contribute to scientific research and monitoring. These projects not only increase the volume of data but also provide a way to engage and educate the public on environmental issues. Citizen science projects can collect data on a wide range of variables such as temperature, precipitation, air quality, water quality, and dust storms. Famous projects like the eBird project (Sullivan et al. 2014) encourage citizen scientists to collect data on bird sightings, and the iNaturalist project (Nugent 2018) invites people to collect data on species to study biodiversity. This data can be used to address some of the grand challenges, such as climate change, biodiversity loss, and habitat degradation. For example, USA-NPN (Betancourt et al. 2005) data can help to identify patterns and changes in phenology, the timing of plant and animal life cycle events, or help identify the areas where pollution levels are high, where invasive species are spreading. They can foster a sense of ownership and stewardship of the environment for the public, which is critical for ensuring long-term sustainability efforts and public engagement.

Similar to the other data sources, some disadvantages are commonly attached to citizen science datasets. Although most citizen science projects have well planned for all the details and keep protocols up for participants to follow to ensure the data quality, there are still many things that would go wrong in the fields. It is common

Table 2.5 A list of sample citizen science projects and their data topics

Name	Description	Data topic
eBird	A project by the Cornell Lab of Ornithology and the National Audubon Society that invites birdwatchers to contribute their observations to a global database of bird sightings	Bird species
Globe Observer	A project by NASA that invites citizen scientists to contribute data on cloud cover, land cover, and mosquito habitats	Cloud cover, land cover, and mosquito habitats
iNaturalist	A project by the California Academy of Sciences that invites people to document and share observations of plants, animals, and other living things	Species identification
AirVisual	A project by IQAir that invites people to contribute data on air quality by sharing information on air pollution in their area	Air quality
Community Collaborative Rain, Hail, and Snow Network (CoCoRaHS)	A project that invites citizen scientists to measure and report precipitation data in their area	Precipitation
Old Weather	A project by the National Oceanic and Atmospheric Administration (NOAA) that invites people to transcribe historical weather data from ship logs dating back to the nineteenth century	Historical weather data
Citizen Weather Observer Program (CWOP)	A project that invites citizen scientists to contribute weather data from their personal weather stations	Weather data
Floating forests	A project by the University of California, Santa Barbara, that invites people to identify kelp forests from satellite imagery	Kelp forest identification
PhytoMap	A project by NASA's Goddard Space Flight Center that asks citizen scientists to help identify phytoplankton in ocean-color images	Phytoplankton distribution
Project BudBurst	A project by the Chicago Botanic Garden that asks citizen scientists to observe and record the timing of leafing, flowering, and fruiting of plants in their area	Vegetation growth
Snow Tweets	A project by the National Weather Service that asks citizens to report snowfall amounts in their area via Twitter	Snowfall
Dust Bowl Diary	A project by the University of Nebraska-Lincoln that asks citizens to transcribe diaries kept by people who lived through the Dust Bowl of the 1930s	Dust Bowl incidents
Hurricane Watch Net	A project by amateur radio operators that provides communication support during hurricanes	Meteorological data

(continued)

Table 2.5 (continued)

Name	Description	Data topic
PenguinWatch	A project by the University of Oxford that asks citizens to help identify penguins in camera trap images from the Antarctic	Penguin populations
IceWatch USA	A project by the University of Maine that asks citizens to observe and report on ice conditions in their area	Ice data
Earthwatch	A nonprofit organization that provides opportunities for citizens to participate in environmental research projects around the world	Climate change, wildlife conservation, and marine ecology
USA National Phenology Network (NPN)	A project engages citizen scientists in collecting data on phenology, or the timing of life cycle events such as flowering, leaf emergence, and migration	Phenology data including flowering, fruiting, leafing, migration, and hibernation

that citizen science data may not meet the same rigorous standards as data collected by professional scientists. Using citizen science could introduce uncertainty and errors in the analysis, make the models divert from the correct path, and can be difficult to draw valid conclusions. Another major issue is that the participants for one project might not be completely random (sampling bias) and distribution is biased toward certain areas or certain types of observations (Kosmala et al. 2016). For example, many citizen science projects have more data in urban areas, and lack of data in rural or remote areas. Compared to the ground observation network, citizen science projects mostly do not cover all regions or all variables of interest. Also, as citizen scientists have not gone through strict academic training, the participants do not have the same level of expertise as professional scientists and might lead to misunderstanding or wrong labels.

Despite these drawbacks, citizen science data can still be a valuable resource for actionable science in climate and environment. It is important to carefully consider the strengths and limitations of the data, and to use appropriate methods to analyze and interpret the results before considering fully adopting the data in serious actionable science projects. Almost all the citizen science datasets are freely available on their project website. The datasets are made available to the public with the goal of promoting scientific research, education, and environmental conservation. Some projects even make their data available in real time through web portals or mobile apps. However, some projects may have restrictions on data access, particularly if the data involves sensitive information or endangered species.

2.5 *Social Media Data*

Social media platforms are such powerful tools that have played a significant role in our society and touch all the aspects of our lives and the environment we are living in (Lewandowsky et al. 2019). It can be a significant tool to make science more

Table 2.6 A list of popular social media datasets and their application

Name	Launch	Active user	Media format	Open data policy
Facebook	2004	2.85 billion (2021)	Text, images, and videos	Restricted
Twitter (X)	2006	330 million (2019)	Text, images, and videos	Developer account for academia
Instagram	2010	1 billion (2021)	Images and videos	Developer API
YouTube	2005	2 billion (2021)	Videos	Developer API for research
Reddit	2005	52 million (2021)	Text, images, and videos	Developer API
TikTok	2016	1.1 billion (2021)	Short-form videos	Developer API for research
LinkedIn	2002	740 million (2021)	Text, images, and videos	Developer API for research

actionable in tackling climate and environmental challenges. It can monitor and track environmental events in real time, such as wildfires, floods, and storms, and help provide early warning and response to natural disasters, and mitigate their impact on communities. In addition, it can engage the public in discussions about climate and environmental issues, raise awareness and educate people about the impacts of climate change, and encourage them to take action. If a filter and algorithm is applied, social media could be another source to collect data on environmental conditions, such as air quality or hurricane wind speed. Similar to citizen science projects, scientists can use social media to find volunteers and allow them to engage in scientific research. The data is definitely the most important asset of the social media companies, from which a lot of useful information can be concluded on public opinions and behaviors related to climate and environmental issues. Table 2.6 contains the major social media platforms that are most popular at present and their open data policy for academic research.

Similar to citizen science datasets, social media data has issues with data quality, biases in sampling, and also some new problems like data privacy concerns, accessibility to the full-size datasets, and other legal and ethical considerations. For example, people may use different hashtags or keywords to describe the same phenomenon, or may post misleading information intentionally or unintentionally. During the 2018 California wildfires, social media users shared images and videos that were not related to the fires (Du et al. 2019), which led to confusion and hampered response efforts. Furthermore, social media may not capture environmental issues in areas with limited Internet access or low social media use. It tends to be more prevalent among younger and more tech-savvy individuals. The user demographics can vary across different regions and countries, which could lead to an incomplete picture of environmental or climate issues in those areas. The availability of social media datasets varies across platforms, but most of them have developer API opened for academic research purposes. Also, there are legal and ethical considerations associated with using social media data, such as copyright laws and data ownership, and the researchers need to consider all these factors before they decide to involve the data to do actionable science. The social media companies' attitude toward the data openness also plays an important role, for example, Twitter recently changed its API policies, which has limited the availability of free historical tweet data for researchers (Bruns 2019).

3 Data Discovery and Retrieval

It is important to have a clear understanding of the scientific question or problem being addressed and the relevant data sources. To identify the data needs, we first need to scope the project's variables, the spatial and temporal extents, to make the data discovery process easy. For example, if the scientific question is related to understanding the impact of climate change on a particular species of plant, the relevant variables might include temperature, precipitation, and CO_2 concentrations. The spatial extent might be defined by the range of the plant species, while the temporal extent might be defined by the historical record of climate data. Once the data needs are defined, scientists need to search for the data to fill in them. There are various data portals, databases, and repositories that can provide access to climate and environmental data. Some example data portals are the National Oceanic and Atmospheric Administration (NOAA), National Centers for Environmental Information (NCEI), NASA Earth Observing System Data and Information System (EOSDIS) (Ramapriyan et al. 2010), and the European Centre for Medium-Range Weather Forecasts (ECMWF) Climate Data Store (Palmer et al. 1990).

After relevant data sources are identified, scientists need to assess their quality to ensure they meet the project's needs. Data quality assessment involves examining various factors like accuracy, completeness, consistency, and reliability. After the data quality assessment is complete, the next step is to access and retrieve the data. The process of data retrieval may vary depending on the data source, and data may be available in various formats like text, image, or binary. The data may be downloaded directly from the portal or may require registration and authentication. Once the data is retrieved, it needs to be stored and managed properly. This involves organizing the data, ensuring proper metadata and documentation, and implementing appropriate data backup and security measures.

4 Data Preprocessing and Cleaning

Data preprocessing and cleaning are essential in preparing data for actionable science to ensure that the data is accurate, complete, and in the right format for analysis. Otherwise, the data may contain errors, missing values, or inconsistencies that could affect the results of the analysis and ultimately the actions taken. Data preprocessing involves a variety of techniques to transform the data into a format suitable for analysis. This may include techniques such as data normalization, scaling, and feature extraction. Data cleaning involves identifying and correcting errors or inconsistencies in the data, such as missing values, outliers, and duplicates. In the context of climate and environmental data, preprocessing and cleaning may involve correcting for measurement errors, identifying and filling in missing values, and removing outliers or erroneous data points. For example, in satellite data, preprocessing might involve correcting for atmospheric interference or adjusting for changes in instrument calibration over time. In citizen science data, cleaning might involve identifying and removing data points that are clearly incorrect or identifying patterns of data that suggest errors or biases. Here are some common steps:

- Check for missing values, incomplete data or gaps in the time-series data.
- Ensure that the data is of high quality and free from errors, outliers, or any inconsistencies that may affect the analysis.
- When using different sources of data, they may have different scales and units. Normalization and scaling techniques can be used to convert the data to a standard scale, facilitating comparisons between variables.
- The data should be formatted consistently to make it easier to read and understand. For example, consistent date formats and time zones should be used to make sure the data is easily comparable and to avoid errors.
- Transformations like log, power, and square root can be used to reduce the influence of outliers and make the data more normally distributed, which can make statistical analysis more reliable.
- Sometimes datasets may be too large to handle or may contain unnecessary data. In such cases, data reduction techniques like principal component analysis and singular value decomposition or clustering can be used to reduce the dimensionality of the data and remove noise.
- Document all the preprocessing and cleaning steps taken, including any decisions made, code used, and any changes made to the original data. The documentation is essential for reproducibility and transparency.
- Ensure that all the data layers have the same spatial reference system (SRS) and projection. Mismatched SRS can cause alignment and accuracy issues.
- The spatial resolution of the data should be appropriate for the analysis being performed. For example, if analyzing changes in land cover, a coarse resolution may not be suitable.
- The data format should be compatible with the software being used for analysis. Common geospatial data formats include shapefiles, GeoTIFFs, and netCDF.
- Missing or null values should be identified and handled appropriately. They can be replaced with interpolated values or removed depending on the context of the analysis.
- Outliers should be identified and evaluated for their impact on the analysis. They can be removed or adjusted if necessary.
- Different data layers should be integrated appropriately, accounting for differences in resolution, scale, and data type.
- Quality control checks should be performed to ensure data accuracy and consistency, like comparing data with ground truth observations or other reference data sources.

- Metadata should be reviewed and updated as necessary to ensure that data is properly documented and can be easily understood and reused by others.
- Version control should be used to track changes to the data over time, especially if the data is being updated or modified regularly.

The expected output of data preprocessing and cleaning is a clean and consistent dataset that can be directly used for analysis and modeling. The cleaned dataset should be free of missing values, outliers, duplicates, and errors, and have consistent formatting and units. Once the data has been preprocessed and cleaned, it can be directly used for analysis and modeling. This saves time and resources that would otherwise be spent manually addressing data quality issues during analysis. Additionally, a clean and consistent dataset enables scientists to compare and combine data from multiple sources more easily, which can lead to more comprehensive and insightful analysis.

5 Data Integration and Management in Environmental Sciences

Data integration involves combining data from multiple sources to produce a comprehensive dataset. Data management usually refers to the storage, retrieval, and manipulation of large volumes of data. Data integration and management is usually comprised of several steps: data preparation, data transformation, and data fusion. Data preparation involves cleaning and preprocessing data to ensure that it is accurate, complete, and consistent. Data transformation is responsible for converting data into a format that is compatible with the analysis tools and methods that will be used. Data fusion means combining data from different sources to produce a comprehensive dataset that can be used for analysis.

Data management is critical for ensuring that data is accessible and usable over the long term. Common practices involve the use of data repositories and archives that can store large volumes of data and provide access to it over time. It also uses standards and protocols for data formatting, storage, and sharing to ensure that data is interoperable and can be used by others. Metadata is widely used to provide information about the data, such as the location, time, and method of collection, and helps to ensure that the data is accurately interpreted and used. Metadata also facilitates data discovery and sharing as it allows other researchers to understand the context and quality of the data.

6 Continuous Operation and Maintenance of Data Stream

Continuous operation and maintenance of data streams are essential for ensuring the quality and reliability of data used in actionable science for climate and the environment (Becker et al. 2015). Data streams are typically collected over long periods of time and from multiple sources, and it is important to ensure that the data is continuously monitored, validated, and maintained to ensure its accuracy and consistency.

To achieve continuous operation and maintenance, several steps can be taken:

- Data streams need to be monitored on a continuous basis to ensure that they are operating correctly and that the data being collected is valid and reliable. This can be done using various monitoring tools, such as dashboards, alerts, and automated checks.
- Data quality control involves reviewing the data for errors, inconsistencies, and outliers, and taking corrective action if necessary. This can involve data cleaning, data validation, and data verification to ensure that the data is accurate and reliable.
- Regular maintenance and upkeep of data streams are critical to ensure that the data is continuously available and up to date. This includes ensuring that the data collection system is functioning correctly, replacing any faulty equipment, and ensuring that the data is backed up and securely stored.
- Continuous improvement includes identifying and addressing any gaps or issues in the data collection process and making improvements to ensure that the data is of the highest quality possible. This can involve improving data collection methods, upgrading equipment, and implementing new data analysis techniques.

For example, the Global Forest Watch provides near-real-time information on forest loss and gain around the world (Harris et al. 2016) using satellite imagery and data from local sources. The information is updated monthly and can be used to inform policies and interventions to prevent deforestation and promote reforestation. Another project like Smartfin involves equipping surfboards with sensors to collect data on ocean conditions such as temperature, salinity, and pH. The data is uploaded to a cloud-based platform in near real time and can be used by researchers and policymakers to better understand the impacts of climate change on the ocean. Also, many cities around the world have installed air quality sensors to continuously monitor the levels of pollutants such as particulate matter, ozone, and nitrogen dioxide. Policymakers can use the data to identify areas of high pollution and implement interventions to improve air quality.

However, the challenges for maintaining continuous data streams include

- Regularly updating data sources requires regular calibration, validation, and verification of the data, as well as monitoring and detection of any anomalies or errors that may occur.
- Hard to manage the large volume of data generated by continuous data streams. The data needs to be stored, processed, and analyzed in a timely and efficient manner, and this requires a robust data management infrastructure.
- The technical infrastructure required for continuous data streams can be complex and expensive to maintain. The equipment and sensors used to collect the data need to be regularly maintained and calibrated to ensure accurate and reliable data. Additionally, data transmission and storage systems need to be reliable and secure to prevent data loss or corruption.

- It requires a significant investment in funding and resources. This includes the cost of equipment, infrastructure, and personnel. Funding may be limited, and there may be a shortage of skilled personnel available to manage and operate the data streams.
- Continuous data streams are often collected from multiple sources, and integrating this data into a single usable format can be challenging. Data may be collected at different resolutions, time intervals, and spatial scales, and may need to be normalized or transformed before integration.
- Making data accessible and sharing it with others can also be a challenge. Data owners need to ensure that appropriate data sharing agreements are in place, and that the data is properly documented and annotated with metadata to facilitate its use by others.

7 Challenges in Data Community to Support Actionable Science

From the data provider perspective, challenges in supporting actionable science in climate and environment include:

- Data providers need to make sure that the data is reliable, accurate, and accessible to researchers and stakeholders via investing in data infrastructure and maintenance, data documentation, and data standardization efforts.
- Consider the different data formats, standards, and metadata used by various data sources and ensure that their data can be integrated with other datasets.
- Check that the to-be-released datasets are compliant with privacy regulations and that the data they provide is secure.
- Facilitate data sharing and collaboration among researchers and stakeholders. This can involve developing policies and incentives to encourage data sharing and collaboration, as well as developing platforms and tools that enable data sharing and collaboration.
- Secure long-term funding to support data infrastructure, maintenance, and management. This involves developing funding models that ensure sustainable data provision and management, as well as engaging stakeholders to secure support for ongoing data management efforts.
- Keep pace with advances in technology and new data sources. This can involve investing in new data collection methods and technologies, as well as updating data storage and management systems to accommodate new data types and formats.

8 Conclusion

In this chapter, we discussed the importance of data in actionable science for climate and environment and covered various aspects, including data sources, data discovery, retrieval, preprocessing, cleaning, integration, and continuous operation and maintenance. We also discussed the challenges faced by the climate data community in supporting actionable science. We highlighted the crucial role of data in actionable science, enabling researchers and decision-makers to make informed decisions about climate and environmental challenges. Different data sources available like citizen science data and social media data are introduced. The details of various data portals, databases, and repositories that can provide access to climate and environmental data are also provided.

The importance of metadata in data integration and management, providing essential information about the data and ensures accurate interpretation and use, is also mentioned. The challenges in continuous operation and maintenance of data streams and the importance of addressing them for the success of actionable science are discussed. In addition, the challenges faced by the data community in supporting actionable science are outlined too, including the need for interdisciplinary collaboration, data quality and consistency, data sharing, and data security and privacy. In the future, it is foreseeable mandatory path for entire climate science community to collectively address these challenges to ensure that actionable science can continue to provide innovative and practical solutions to climate and environmental challenges.

9 Lessons Learnt

- Data collection and preprocessing could be tedious and challenging for actionable science that relies on accurate and reliable data to achieve successful implementation of climate and environmental actions.
- Citizen science data can provide valuable insights but may come with limitations and challenges that need to be addressed before use in actionable science.
- Metadata is essential for data management and sharing. Properly documented metadata helps ensure data accuracy, reproducibility, and facilitates data discovery and sharing (Sun et al. 2013).
- Continuous operation and maintenance of data streams are important to ensure long-term availability of data and prevent data degradation.
- Collaboration between data providers and end users is critical for actionable science. End users should also be involved in the process to ensure that the data meet their specific requirements.
- Open data policies and data sharing platforms facilitate the access and sharing of data.

- Data integration and management enable the combination of data from different sources, providing a comprehensive and holistic view of climate and environmental issues.
- The challenges in data management and sharing require interdisciplinary collaboration and coordination between scientists, policymakers, and data providers to work closely on.

10 Open Questions and Brainstorming Solutions for Future

As we move toward a more data-driven approach to tackling climate and environmental challenges, there are still several open questions and areas for improvement in the data foundation for actionable science. Here are some potential solutions and ideas in the future:

- While there are many open data portals available, not all datasets are easily accessible in a user-friendly manner. We need to work toward improving open data access and promoting standardization in data formats and utilization.
- To make the most out of the data available, we need to focus on integrating and fusing multi-source data. This can be challenging due to the differences in data formats and metadata, and efforts are much needed to improve data interoperability.
- The field of data science is constantly evolving, and emerging technologies such as machine learning and artificial intelligence can be utilized to quickly extract insights from large datasets. It is important for data providers to stay up to date with emerging technologies and incorporate them into actionable science projects.
- Citizen science can be a valuable source of data, especially in areas where official data collection is difficult or expensive. To engage more citizen scientists, we need to improve data literacy and promote open data access and friendly collection toolkits.
- As more data becomes available, it is mandatory to strengthen data management and sharing practices. This includes developing common metadata standards and promoting open data sharing policies.

References

Becker, David, Trish Dunn King, and Bill McMullen. 2015. Big data, big data quality problem. In *2015 IEEE international conference on big data (big data)*, 2644–2653. IEEE.

Betancourt, Julio L., Mark D. Schwartz, David D. Breshears, Daniel R. Cayan, Michael D. Dettinger, David W. Inouye, Eric Post, and Bradley C. Reed. 2005. Implementing a US national phenology network. Eos Trans. *AGU 86(51), 539–539,* https://doi.org.in/10.1029/2005EO510005

Biancamaria, Sylvain, Dennis P. Lettenmaier, and Tamlin M. Pavelsky. 2016. The SWOT mission and its capabilities for land hydrology. *Remote Sensing and Water Resources*: 117–147.

Boain, Ronald J. 2004. AB-Cs of sun-synchronous orbit mission design. 14 AAS/AIAA Space Flight Mechanics Conference, Maui, Hawaii, February 8–12, https://citeseerx.ist.psu.edu/document?repid=rep1&type=pdf&doi=43ffd02cbb61b0381f6b4ae3e18d0c9a536d446a

Brett, Annie, Jim Leape, Mark Abbott, Hide Sakaguchi, Ling Cao, Kevin Chand, Yimnang Golbuu, Tara J. Martin, Juan Mayorga, and Mari S. Myksvoll. 2020. Ocean data need a sea change to help navigate the warming world. *Nature* 582 (7811): 181–183.

Bruns, Axel. 2019. After the 'APIcalypse': Social media platforms and their fight against critical scholarly research. *Information, Communication & Society* 22 (11): 1544–1566.

Cinquini, Luca, Daniel Crichton, Chris Mattmann, John Harney, Galen Shipman, Feiyi Wang, Rachana Ananthakrishnan, et al. 2014. The earth system grid federation: An open infrastructure for access to distributed geospatial data. *Future Generation Computer Systems* 36: 400–417.

Cook, Sue, Keith W. Nicholls, Irena Vaňková, Sarah S. Thompson, and Benjamin K. Galton-Fenzi. 2023. Data initiatives for ocean-driven melt of Antarctic ice shelves. *Annals of Glaciology*: 1–6.

Dosio, Alessandro, Hans-Jürgen Panitz, Martina Schubert-Frisius, and Daniel Lüthi. 2015. Dynamical downscaling of CMIP5 global circulation models over CORDEX-Africa with COSMO-CLM: Evaluation over the present climate and analysis of the added value. *Climate Dynamics* 44: 2637–2661.

Du, Hanxiang, Long Nguyen, Zhou Yang, Hashim Abu-Gellban, Xingyu Zhou, Wanli Xing, Guofeng Cao, and Fang Jin. 2019. Twitter vs news: Concern analysis of the 2018 California wildfire event. In *In 2019 IEEE 43rd Annual Computer Software and Applications Conference (COMPSAC)*, vol. 2, 207–212. IEEE.

Ebert-Uphoff, Imme, David R. Thompson, Ibrahim Demir, Yulia R. Gel, Anuj Karpatne, Mariana Guereque, Vipin Kumar, Enrique Cabral-Cano, and Padhraic Smyth. 2017. A vision for the development of benchmarks to bridge geoscience and data science. In *17th international workshop on climate informatics*.

Eyring, Veronika, Sandrine Bony, Gerald A. Meehl, Catherine A. Senior, Bjorn Stevens, Ronald J. Stouffer, and Karl E. Taylor. 2016. Overview of the Coupled Model Intercomparison Project Phase 6 (CMIP6) experimental design and organization. *Geoscientific Model Development* 9 (5): 1937–1958.

Giam, Xingli. 2017. Global biodiversity loss from tropical deforestation. *Proceedings of the National Academy of Sciences* 114 (23): 5775–5777.

Gilliom, Robert J., William M. Alley, and Martin E. Gurtz. 1995. *Design of the National Water-Quality Assessment Program: Occurrence and distribution of water-quality conditions*. Vol. 1112. US Geological Survey.

Golledge, Nicholas R. 2020. Long-term projections of sea-level rise from ice sheets. *Wiley Interdisciplinary Reviews: Climate Change* 11 (2): e634.

Harris, Nancy, Rachael Petersen, Crystal Davis, and Octavia Payne. 2016. "INSIDER: Global Forest Watch and the Forest Resources Assessment, Explained in 5 Graphics." World Resources Institute. Accessed on Sep 8, 2023. https://www.wri.org/technical-perspectives/insider-global-forestwatch-and-forest-resources-assessment-explained-5-graphics

Justice, C.O., J.R.G. Townshend, E.F. Vermote, E. Masuoka, R.E. Wolfe, Nazmi Saleous, D.P. Roy, and J.T. Morisette. 2002. An overview of MODIS land data processing and product status. *Remote Sensing of Environment* 83 (1–2): 3–15.

Kosmala, Margaret, Andrea Wiggins, Alexandra Swanson, and Brooke Simmons. 2016. Assessing data quality in citizen science. *Frontiers in Ecology and the Environment* 14 (10): 551–560.

Lawrence, Judy, Paula Blackett, and Nicholas A. Cradock-Henry. 2020. Cascading climate change impacts and implications. *Climate Risk Management* 29: 100234.

Lewandowsky, Stephan, John Cook, Nicolas Fay, and Gilles E. Gignac. 2019. Science by social media: Attitudes towards climate change are mediated by perceived social consensus. *Memory & Cognition* 47: 1445–1456.

Li, Feng, Lei Xin, Yi Guo, Dongsheng Gao, Xianghao Kong, and Xiuping Jia. 2017. Super-resolution for GaoFen-4 remote sensing images. *IEEE Geoscience and Remote Sensing Letters* 15 (1): 28–32.

Mearns, Linda O. 2009. North American Regional Climate Change Assessment Program (NARCCAP). In *Presentation at workshop on water infrastructure sustainability and adaptation to climate change, USEPA*. Slide presentation available at: http://www.epa.gov/nrmrl/wswrd/wqm/wrap/pdf/workshop/A1_Mearns.pdf.

Menne, Matthew J., Imke Durre, Russell S. Vose, Byron E. Gleason, and Tamara G. Houston. 2012. An overview of the global historical climatology network-daily database. *Journal of Atmospheric and Oceanic Technology* 29 (7): 897–910.

Meyssignac, Benoit, Tim Boyer, Zhongxiang Zhao, Maria Z. Hakuba, Felix W. Landerer, Detlef Stammer, Armin Köhl, et al. 2019. Measuring global ocean heat content to estimate the earth energy imbalance. *Frontiers in MarineScience* 6: 432.

Nugent, Jill. 2018. iNaturalist. *Science Scope* 41 (7): 12–13.

Palmer, T.N., C. Brankovic, F. Molteni, S. Tibaldi, L. Ferranti, A. Hollingsworth, U. Cubasch, and E. Klinker. 1990. The European Centre for Medium-range Weather Forecasts (ECMWF) program on extended-range prediction. *Bulletin of the American Meteorological Society* 71 (9): 1317–1330.

Ramapriyan, Hampapuram K., Jeanne Behnke, Edwin Sofinowski, Dawn Lowe, and Mary Ann Esfandiari. 2010. Evolution of the earth observing system (EOS) data and information system (EOSDIS). *Standard-Based Data and Information Systems for Earth Observation*: 63–92.

Rao, M., J.P. Venkata, Ram Rattan Gupta, and K. Thyagarajan. 2006. RESOURCESAT-2: A mission for earth resources management. In *GEOSS and next-generation sensors and missions*, vol. 6407, 119–126. SPIE.

Ruiz-Benito, Paloma, Giorgio Vacchiano, Emily R. Lines, Christopher P.O. Reyer, Sophia Ratcliffe, Xavier Morin, Florian Hartig, et al. 2020. Available and missing data to model impact of climate change on European forests. *Ecological Modelling* 416: 108870.

Rummukainen, Markku. 2010. State-of-the-art with regional climate models. *Wiley Interdisciplinary Reviews: Climate Change* 1 (1): 82–96.

Schaire, Scott, Bill Horne, James Schier, Harry Shaw, George Bussey, Peter Celeste, Yen F. Wong, et al. 2016. NASA near earth network (NEN) and space network (SN) CubeSat communications. In *14th international conference on space operations*, 2598.

Schutz, Bob E., H. Jay Zwally, Christopher A. Shuman, D. Hancock, and John P. DiMarzio. 2005. Overview of the ICESat mission. *Geophysical Research Letters* 32 (21).

Shimada, Masanobu, Takeo Tadono, and Ake Rosenqvist. 2009. Advanced Land Observing Satellite (ALOS) and monitoring global environmental change. *Proceedings of the IEEE* 98 (5): 780–799.

Spoto, Francois, Omar Sy, Paolo Laberinti, Philippe Martimort, Valerie Fernandez, Olivier Colin, Bianca Hoersch, and Aime Meygret. 2012. Overview of sentinel-2. In *2012 IEEE international geoscience and remote sensing symposium*, 1707–1710. IEEE.

Sullivan, Brian L., Jocelyn L. Aycrigg, Jessie H. Barry, Rick E. Bonney, Nicholas Bruns, Caren B. Cooper, Theo Damoulas, et al. 2014. The eBird enterprise: An integrated approach to development and application of citizen science. *Biological Conservation* 169: 31–40.

Sun, Ziheng, Peng Yue, Hu Lei, Jianya Gong, Liangpei Zhang, and Lu Xianchang. 2013. GeoPWProv: Interleaving map and faceted metadata for provenance visualization and navigation. *IEEE Transactions on Geoscience and Remote Sensing* 51 (11): 5131–5136.

Sun, Ziheng, Liping Di, Hui Fang, Liying Guo, Xicheng Tan, Lili Jiang, and Zhongxin Chen. 2021. Agro-geoinformatics data sources and sourcing. *Agro-Geoinformatics: Theory and Practice*: 41–66.

Thépaut, Jean-Noël, Dick Dee, Richard Engelen, and Bernard Pinty. 2018. The Copernicus programme and its climate change service. In *IGARSS 2018–2018 IEEE international geoscience and remote sensing symposium*, 1591–1593. IEEE.

Thompson, Alan A. 2015. Overview of the RADARSAT constellation mission. *Canadian Journal of Remote Sensing* 41 (5): 401–407.

Tucker, Compton J., Denelle M. Grant, and Jon D. Dykstra. 2004. NASA's global orthorectified Landsat data set. *Photogrammetric Engineering & Remote Sensing* 70 (3): 313–322.

Weng, F., X. Zou, X. Wang, S. Yang, and M.D. Goldberg. 2012. Introduction to Suomi national polar-orbiting partnership advanced technology microwave sounder for numerical weather prediction and tropical cyclone applications. *Journal of Geophysical Research: Atmospheres* 117 (D19).

Wiens, John A., Diana Stralberg, Dennis Jongsomjit, Christine A. Howell, and Mark A. Snyder. 2009. Niches, models, and climate change: assessing the assumptions and uncertainties. *Proceedings of the National Academy of Sciences* 106 (Supplement_2): 19729–19736.

Wulder, Michael A., Joanne C. White, Samuel N. Goward, Jeffrey G. Masek, James R. Irons, Martin Herold, Warren B. Cohen, Thomas R. Loveland, and Curtis E. Woodcock. 2008. Landsat continuity: Issues and opportunities for land cover monitoring. *Remote Sensing of Environment* 112 (3): 955–969.

Wulder, Michael A., Joanne C. White, Thomas R. Loveland, Curtis E. Woodcock, Alan S. Belward, Warren B. Cohen, Eugene A. Fosnight, Jerad Shaw, Jeffrey G. Masek, and David P. Roy. 2016. The global Landsat archive: Status, consolidation, and direction. *Remote Sensing of Environment* 185: 271–283.

Zimmerman, Ann S. 2008. New knowledge from old data: The role of standards in the sharing and reuse of ecological data. *Science, Technology, & Human Values* 33 (5): 631–652.

Chapter 3
Technology Landscape for Making Climate and Environmental Science "Actionable"

Ziheng Sun

Contents

Z. Sun (✉)
Center for Spatial Information Science and Systems, Department of Geography and Geoinformation Science, George Mason University, Fairfax County, VA, USA
e-mail: zsun@gmu.edu

Z. Sun (ed.), *Actionable Science of Global Environment Change*,
https://doi.org/10.1007/978-3-031-41758-0_3

1 Introduction

Technology serves as a powerful ally in actionable climate science, empowering researchers to unravel the complexities of our changing planet. From satellite imagery revealing the retreating glaciers (Paul 2015) to data-driven models predicting the impacts of extreme weather events on coastal communities (Haggag et al. 2022), technology enables us to take informed steps toward a sustainable and resilient future. Technology has enabled us to gather high-resolution climate data through advanced satellite systems and deploy sophisticated weather monitoring instruments like Doppler radars (Gao et al. 1999), which help us understand and forecast severe storms with unprecedented accuracy. Innovative data visualization tools and computational models allow us to simulate and analyze complex climate processes (Ali et al. 2016), aiding in the development of effective mitigation and adaptation strategies for a changing world. This chapter will dive into these heroes behind the scene and reveal the amazing toolkit behind the amazing discoveries in climate and environment science. Meanwhile, as the role of technology is vital to support daily research, it is an essential component to complete the successful transition from research to action.

The technology landscape in actionable climate and environmental science is vast and diverse. It encompasses a wide range of tools and systems that enable researchers and decision-makers to gather, analyze, and interpret data for informed decision-making. Usually we divide the technologies into various categories according to the stage of the data information cycles: data collection and processing, modeling and simulation tools, communication and collaboration techniques, and other emerging or breakthrough technologies that have the potential for making science actionable.

One key technology is remote sensing, which uses satellites and aircraft to collect valuable information about the Earth's surface and atmosphere (Campbell et al. 2011). This data helps monitor changes in land cover, vegetation health, and atmospheric conditions, aiding in the assessment of climate impacts and the identification of vulnerable regions. However, remote sensing platforms (except LiDAR) usually rely on clear skies and favorable weather conditions for accurate data collection. For example, during periods of heavy cloud cover, satellite imagery may be obscured in spatial-temporal coverage, limiting our ability to monitor the changes in land cover or track storm systems. Remote sensing approach also has limitations in capturing fine-scale details, such as the precise conditions within small localized areas, which can be important for assessing the impacts of climate change on specific ecosystems or communities. So besides remote sensing, scientists and engineers built a lot of ground-based observation networks, such as weather stations and water quality monitoring systems (Schaefer et al. 2001), to provide critical data on local environmental conditions. These networks help us understand the dynamics of weather patterns, track changes in air and water quality, and inform sustainable resource management practices.

In addition to direct observation, mathematical modeling is another important tool for scientists. Numerical modeling, powered by high-performance computing, simulates complex climate systems and provides projections of future conditions (Lynch 2008). It aims to create a virtual world where we can simulate and understand complex environmental processes by building a detailed, digital replica of Earth, where we can test different scenarios and predict how the climate might change in the future. By inputting data and equations into the model, we can see how factors like temperature, rainfall, and wind patterns interact, helping us make informed decisions about climate-related issues such as flood risk, crop yields, or the spread of diseases. The public can consider mathematical models as the "crystal ball" of scientists. These models help us understand the drivers of climate change, predict future climate scenarios, and assess the potential impacts on ecosystems and communities. In the future, numerical models will become increasingly powerful tools in climate science by incorporating higher resolutions, comprehensive Earth system components, and real-time data assimilation. These models will enable us to better understand and address climate and environmental challenges, supporting sustainable development and resilience efforts.

With remote sensing, ground observation networks, and numerical models, scientists have captured and generated massive amounts of data. Naturally, data management systems are crucial in actionable science as they provide the infrastructure and tools necessary to collect, organize, and store large volumes of climate and environmental data. By ensuring data integrity, accessibility, and interoperability, these systems enable researchers and decision-makers to effectively utilize and analyze data, leading to more informed and impactful actions addressing climate and environmental challenges. They facilitate the integration of multiple data sources, allowing researchers to uncover patterns, identify trends, and derive actionable insights. For example, NASA provides a comprehensive platform for accessing and managing diverse environmental datasets with volumes reaching petabytes (Ma et al. 2015). They can leverage cloud computing and distributed storage solutions to handle the large volume and velocity of data generated, which allows for scalable and flexible infrastructure that can accommodate the growing demands of big data. Additionally, data management systems can incorporate advanced data processing frameworks, such as Apache Hadoop (2011) or Apache Spark (Salloum et al. 2016), which provide powerful tools for distributed data processing and analytics. These frameworks enable parallel computation and distributed storage, allowing for efficient processing of large datasets. Data management systems can utilize machine learning and artificial intelligence algorithms to automate data processing, analysis, and decision-making (Sun et al. 2022, 2023). These technologies can assist in identifying patterns, trends, and correlations within the data, providing valuable insights for actionable science. The future development will focus on improving data integration, standardization, and accessibility, enabling scientists and decision-makers to effectively analyze and utilize data for addressing climate-related challenges. By harnessing these latest technologies and best practices, data management systems can effectively handle big data and enable advanced analytics, empowering scientists and decision-makers in their endeavors.

Citizen science platforms provide great assistance for actionable science by engaging the public in data collection and analysis, allowing for large-scale data gathering across various geographical areas (Buytaert et al. 2014). This crowd-sourced approach not only enhances data coverage but also fosters public participation and awareness, leading to more inclusive and impactful actions to address climate and environmental challenges (e.g., observer.globe.gov in Fig. 3.1). These platforms provide user-friendly interfaces and mobile applications that allow individuals to contribute data through various means, such as uploading photos, recording observations, or participating in data collection campaigns. Advanced technologies like GPS, image recognition algorithms, and data validation systems are utilized to ensure the accuracy and reliability of the collected data. This combination of user-friendly interfaces and technological tools allows citizen scientists to actively contribute to scientific research and enables researchers to harness the collective power of the public for data collection and analysis. This participatory approach not only expands data collection efforts but also enhances public awareness and understanding of climate and environmental issues. With the increasing availability of smartphones and Internet access, citizen science can reach a wider

Fig. 3.1 The citizen science mobile app of the globe program

audience and involve more participants in real-time data collection. The future development of citizen science projects lies in enhancing data quality, expanding project scalability, and enabling more meaningful and actionable insights for addressing climate and environmental challenges.

Social media platforms, such as Twitter, Facebook, and Instagram, allow researchers to tap into a vast pool of user-generated content, including text, images, and videos, which can provide valuable information about people's attitudes, behaviors, and concerns regarding climate change (Pearce et al. 2019). Scientists can leverage social media platforms for actionable science by actively monitoring and engaging with the public discourse on climate change. They can analyze social media data to identify emerging trends, public sentiments, and misinformation, and then use this information to develop targeted interventions and communication strategies. For example, scientists can track hashtags related to climate change on Twitter to gauge public interest and concerns and respond with evidence-based information or engage in conversations to address misconceptions. They can also collaborate with social media influencers or organizations to amplify accurate information and promote positive actions for climate and environmental sustainability. Additionally, scientists can use social media platforms to crowdsource data collection and engage citizen scientists in monitoring environmental phenomena, such as tracking wildlife sightings or documenting weather events, which can contribute to large-scale datasets for research and conservation efforts. By applying natural language processing, sentiment analysis, and network analysis techniques, researchers can extract meaningful patterns and trends from social media data, helping to inform decision-making processes, shape public engagement campaigns, and drive policy changes. However, it is important to acknowledge the limitations of social media data, such as the potential for biases, lack of representativeness, and privacy concerns, which should be carefully considered when interpreting and utilizing these data sources in actionable science for climate and environment.

Artificial intelligence (AI) and machine learning (ML) technologies have rapidly advanced in recent years, revolutionizing actionable climate science (Kaack et al. 2022; Sun et al. 2022). Currently, AI and ML are being used to analyze large datasets from remote sensing, ground observations, and numerical models, allowing scientists to extract valuable insights, identify patterns, and make accurate predictions. For example, ML algorithms can analyze satellite imagery to monitor deforestation rates or detect changes in sea ice extent (Ali et al. 2023). AI and ML models are used to improve climate and weather forecasts, enabling more precise predictions of extreme events like hurricanes and heatwaves. By analyzing historical data and incorporating real-time observations, these models can provide early warnings, helping communities and authorities prepare and mitigate the impacts of such events. For instance, AI-based flood prediction models can analyze rainfall data, river levels, and terrain characteristics to forecast flood risks, assisting in evacuation planning and resource allocation (Munawar et al. 2021). In the future, AI and ML technologies will continue to evolve and thrive in climate science to enhance our understanding of complex climate processes, facilitate scenario analysis for policy-making, and support decision-making in adaptation and mitigation strategies. For example, AI algorithms can optimize energy systems and improve energy efficiency, contributing to the transition to renewable energy sources (Ahmad et al. 2021). The importance of AI and ML technologies is to handle vast amounts of data, identify patterns, and make predictions that are beyond human capacity. They enable us to extract valuable insights from complex climate datasets, supporting evidence-based decision-making and empowering stakeholders to take timely and effective actions. By harnessing the power of AI and ML, we can better understand the impacts of climate change, develop sustainable solutions, and work toward a more resilient and livable future.

2 Data Collection and Processing Technologies

2.1 Sensors and Internet of Things (IoT) Devices

As mentioned in Chap. 2, sensors and IoT devices are widely used to collect real-time, high-resolution data on various environmental parameters such as temperature, humidity, air quality, and precipitation (Sungheetha et al. 2020). These devices can be deployed in remote and inaccessible areas, providing valuable insights into localized climate conditions and facilitating targeted interventions. The future of sensor and IoT technology in climate science holds promise for even greater data collection capabilities, enabling more accurate and localized decision-making to address climate and environmental challenges.

(a) Chemistry Sensors, Optical Sensors, and Semiconductor Sensors

Take the air quality sensors as examples. The underlying mechanism of the air quality sensors is either physics-based or chemistry-based. Air quality sensors employ various technologies to detect specific pollutants such as carbon dioxide (CO_2), particulate matter (PM), and volatile organic compounds (VOCs) (Riediker et al. 2003). One type of chemistry-based sensors are electrochemical sensors that use a chemical reaction between the target pollutant and an electrode to generate an electrical signal. By monitoring this electrical signal, the sensor can accurately detect and quantify the presence of pollutants in the air. The electrode of the sensor is coated with a material called the sensing layer to interact with the specific pollutant of interest while minimizing interference from other gases. This selectivity to particular pollutants ensures accurate and reliable measurements. When the pollutant molecules bind to the sensing layer, they cause a change in the electrochemical properties of the electrode. This change in electrochemical properties results in the generation of an electrical signal that we are looking for, typically a small current or voltage. The magnitude of this signal is usually proportional to the concentration of the pollutant in the air. The sensors are usually calibrated by makers using known pollutant concentrations to establish a relationship between the electrical signal and the actual pollutant concentration. Electrochemical sensors are a cost-effective and portable solution for monitoring air quality. They are widely used in various applications, such as indoor air quality monitoring, outdoor pollution monitoring, and personal exposure assessment.

Optical air quality sensors, one type of physics-based sensors, use light scattering or absorption techniques to detect and quantify particulate matter in the air (Hagan et al. 2020). If using light scattering, the sensor emits a beam of light into the air, and when particles are present, they scatter the light in different directions. Light can be scattered by particles due to a phenomenon called Rayleigh scattering. When light encounters particles in the air, such as dust, smoke, or tiny droplets, it interacts with them and changes its direction. This occurs because the particles are much smaller in size compared to the wavelength of the light. More specifically, when the light wave encounters a particle, it causes the electrons in the particle to oscillate, creating tiny electric fields. These electric fields then reradiate the light in different directions, resulting in scattering. The scattering of light depends on the size of the particles and the wavelength of the light. Smaller particles, such as those found in the atmosphere, tend to scatter shorter wavelengths of light, such as blue and violet, more effectively. This is why the sky appears blue during the day as the shorter blue wavelengths are scattered more by the atmospheric particles. The scattered light is then detected by the sensor, which measures its intensity or analyzes its pattern. About how to detect the scattered light, one common approach is to use a photodetector, such as a photodiode or a photomultiplier tube, to convert light energy into an electrical signal. When the scattered light reaches the sensor's receiving parts, it interacts with the photodetector, causing the release of electrons or the generation of a photocurrent. This electrical signal can then be recorded. Another approach is to employ optical filters that are sensitive to specific wavelengths of light. These filters can be used to selectively capture the scattered light within a certain range of wavelengths that corresponds to the targeted particulate matter. Optical sensors may also utilize additional components such as lenses or mirrors to focus or direct the scattered light onto the detection element, enhancing the sensitivity and accuracy of the measurements. By analyzing the characteristics of the scattered light, such as its intensity or angle of scattering, the sensor can estimate the concentration of particles in the air.

If using light absorption, the sensor emits a beam of light with a specific wavelength into the air. As the light passes through the air, the particles absorb some of the light. Specifically, when a sensor emits a beam of light with a specific wavelength into the air, it is essentially sending out a focused stream of light particles called photons. These photons travel through the air until they encounter particles suspended in the air, such as dust, smoke, or pollutants. As the photons interact with the particles, some of the particles absorb a portion of the light energy. This absorption occurs because the particles have certain physical properties that allow them to capture and retain some of the energy from the passing photons. The amount of light absorbed depends on the properties and composition of the particles, including their size, shape, and chemical composition. By measuring the amount of light absorbed by the particles, the sensor can infer the presence and concentration of those particles in the air. The sensor analyzes the decrease in light intensity after it has passed through the air sample, comparing it to the initial emitted intensity. It measures the amount of light that is absorbed and determines the concentration of particles. The optical sensors are designed to be sensitive to a specific size range of particles, such as PM2.5 or PM10 (Delp et al. 2020). Optical sensors are very common on remote sensing platforms like satellites and airborne observing platforms.

Semiconductor-based sensors rely on the change in electrical conductivity of a material, such as tin oxide or tungsten oxide, that undergoes a change in its electrical properties when exposed to specific gases, allowing them to detect the presence of VOCs and other gases (Eranna et al. 2004). Some advanced air quality sensors may combine multiple sensing technologies to provide a more comprehensive analysis of the air composition. These sensors are typically small and compact, making them suitable for deployment in both indoor and outdoor settings. In these sensors, the semiconductor material is typically heated to a specific temperature. When the material is at an elevated temperature, it acts as a resistive element, meaning it has a certain level of electrical resistance. However, when targeted gases, such as volatile organic compounds (VOCs), come into contact with the surface of the semiconductor material, they cause chemical reactions on the material's surface. These chemical reactions result in a change in the electrical conductivity of the semiconductor material. More specifically, the gas molecules adsorb onto the surface of the material, altering the number of charge carriers and, consequently, the material's electrical resistance. This change in electrical resistance can be measured and correlated to the presence and concentration of the target gases. By monitoring the electrical conductivity changes in the semiconductor material, the sensor can detect the presence of specific gases, including VOCs, and provide an indication of their concentration. This allows for the identification and monitoring of air quality, gas leaks, and other environmental factors. They are essential in monitoring air pollution levels, enabling researchers, organizations, and individuals to assess air quality and identify pollution sources.

(b) Internet of Things (IoT)

It is another buzzword of the era and could mean anything that is connected by the Internet. In the context of actionable climate and environmental science, IoT (Madakam et al. 2015) refers to a network of interconnected devices and sensors that gather and share data about the environment. First, it requires all the sensors to be connected to the live Internet in real time. To connect your sensors to the Internet and transmit data, here is a simplified explanation. Take the air quality sensors as example. Connect the air quality sensors to a microcontroller or development board using wires or connectors. The sensors may have specific pins or interfaces for data transmission. The microcontroller acts as the brain of the system and allows the sensors to communicate with other components. Air quality sensors often output analog signals that need to be converted to digital format for further processing. Use an analog-to-digital converter (ADC) (Walden 1999) to convert the analog sensor readings into digital data that can be understood by the microcontroller. Write code or use a programming platform compatible with the microcontroller to program it. The code instructs the microcontroller to read the digital sensor data, apply calibration factors if necessary, and prepare the data for transmission. Connect a wireless communication module to the microcontroller. This module allows the air quality sensor system to connect to the Internet wirelessly. Examples of wireless communication modules include Wi-Fi, Bluetooth, or cellular modules. Configure the wireless communication module to connect to an available Wi-Fi network or insert a SIM card for cellular connectivity. This allows the air quality sensor system to access the Internet and transmit data. Once the system is connected, it can transmit data to a designated server or cloud platform. The microcontroller code should include instructions to format the sensor data into packets and send them over the Internet using specific protocols, such as HTTP or MQTT. The server or cloud platform receives the data packets and stores them in a database, which enables the collected air quality data to be accessed and analyzed later. To maintain a reliable connection, ensure that the air quality sensor system remains within the range of the wireless communication module or within a cellular coverage area. The operators need to monitor the power source, whether it 'is a battery or a continuous power supply, to ensure uninterrupted operation.

Imagine a network of sensors deployed in various locations, continuously monitoring temperature, humidity, air quality, and other environmental parameters. These sensors are connected to the Internet, allowing the data they collect to be instantly transmitted and accessible to scientists and decision-makers. With this interconnected system, scientists can gather vast amounts of data from multiple sources simultaneously. For example, they can monitor air quality in different regions, track changes in temperature and precipitation patterns, and measure water levels in rivers and lakes. This real-time data provides valuable insights into the state of the environment and helps identify climate-related challenges and their impacts. By analyzing the collected data, scientists can detect trends, patterns, and anomalies that inform decision-making processes. For instance, they can identify areas with poor air quality and devise strategies to reduce pollution, or they can track changes in sea levels and develop adaptation plans for coastal communities. The IoT enables scientists to monitor environmental conditions more comprehensively and make data-driven decisions to address climate-related issues effectively. Moreover, the IoT facilitates the integration of diverse data sources, including satellite imagery, ground-based sensors, and citizen science initiatives. This integration allows for a holistic understanding of the climate system and enables the development of predictive models and early warning systems. For example, by combining weather data, soil moisture readings, and crop health information, scientists can provide farmers with actionable insights on irrigation practices, optimizing agricultural productivity while conserving water resources.

While the Internet of Things (IoT) offers numerous benefits in actionable climate and environment science, there are also some drawbacks and challenges to consider. One challenge is the need for robust connectivity and infrastructure. Reliable Internet connections are essential for seamless data transmission and real-time monitoring. However, in remote or underdeveloped areas, access to stable Internet connectivity may be limited, making it challenging to establish and maintain IoT networks effectively. Another significant drawback is the potential for data privacy and security breaches. With the vast amount of data collected by IoT devices, there is a risk of unauthorized access or misuse of sensitive information, which could compromise privacy and hinder trust in the system. Furthermore, the scalability and interoperability of IoT devices and platforms present challenges. As the number of IoT devices increases, managing and integrating them into a cohesive system becomes more complex. Ensuring compatibility and seamless communication between devices from different manufacturers can be challenging, hindering the overall efficiency and effectiveness of the IoT network. Power management is another consideration. IoT devices require a continuous power supply to operate and transmit data. In remote locations or areas with limited access to electricity, ensuring a sustainable power source for the devices can be a challenge. Power constraints may impact the frequency of data collection and the reliability of the IoT system. Lastly, the cost of implementing and maintaining IoT infrastructure can be a barrier, particularly for smaller organizations or communities with limited resources. The expenses associated with purchasing and deploying IoT devices, as well as the ongoing maintenance and data management costs, can be significant. Addressing these challenges requires collaboration among stakeholders, including researchers, policymakers, and technology providers. It involves developing robust data protection measures, expanding Internet infrastructure, promoting interoperability standards, exploring alternative power sources, and finding cost-effective solutions to make the benefits of IoT accessible to all.

2.2 Remote Sensing Technologies

Remote sensing technologies have revolutionized our ability to observe and understand the Earth's systems on a global scale. They provide a comprehensive and continuous view of our planet, enabling us to monitor changes over time, identify trends, and evaluate the impact of human activities on the environment. The data obtained through remote sensing contributes to actionable science by providing the foundation for evidence-based decision-making and policy development to address climate and environmental challenges. Today they play a crucial role by providing valuable information about Earth's surface and atmosphere from a distance. These technologies involve the use of specialized sensors and instruments mounted on satellites, aircraft, or drones to capture data without direct physical contact.

Satellite remote sensing involves the use of satellites orbiting the Earth to collect data from space. These satellites are equipped with various sensors that can capture images, measure electromagnetic radiation, and detect specific signals emitted by the Earth's surface and atmosphere (Yang et al. 2013). For example, optical sensors capture visible and infrared light reflected or emitted by the Earth, while radar sensors use radio waves to penetrate clouds and obtain information about surface features.

Aircraft-based remote sensing utilizes similar sensors mounted on aircraft to capture data at lower altitudes. This allows for higher-resolution imagery and more targeted observations compared to satellite-based sensing. Aircraft remote sensing is often used for detailed mapping, monitoring specific regions of interest, and conducting specialized research missions (Chadwick et al. 2020).

Drone-based remote sensing is a rapidly evolving field that involves the use of unmanned aerial vehicles (UAVs) equipped with sensors to collect data at a smaller scale and closer proximity. Drones can be deployed in various environmental monitoring applications, such as mapping land cover, tracking wildlife (Afghah et al. 2019), or assessing the health of vegetation.

2.3 Big Data Analytics

With the increasing availability of data, traditional data processing methods often fall short in efficiently handling and extracting meaningful information from these large datasets (Hariri et al. 2019). The technologies in big data analytics focus on handling and manipulating large volumes of data efficiently (Acharya & Ahmed 2016). These technologies work by breaking down complex data processing tasks into smaller, manageable tasks that can be executed in parallel. They leverage distributed computing frameworks and storage systems to handle data across multiple machines or nodes. By distributing the workload and processing tasks across a cluster of interconnected computers, big data analytics technologies can significantly speed up the data processing and analysis (Tantalaki et al. 2020). Let u's consider the monitoring of real-time weather data from multiple sources. The technologies can utilize stream processing frameworks like Apache Kafka (Garg 2013) or Apache Flink (Carbone et al. 2015) to ingest and process high-velocity data streams in real time. These frameworks enable parallel processing and distributed computing, allowing for efficient data processing. For instance, as weather sensors continuously send data, the streaming framework can apply algorithms for data filtering, aggregation, and anomaly detection in real time, enabling timely responses to changing weather conditions. They enable the integration of diverse data sources, including structured and unstructured data, to gain a comprehensive understanding of climate and environmental phenomena. Climate and environmental data come in diverse formats, such as structured data from weather stations, satellite imagery, and textual data from scientific papers (Chi et al. 2016). Big data analytics technologies employ techniques like data integration, schema mapping, and data transformation to handle this variety. For example, data integration platforms like Apache Nifi (Kim et al. 2019) or Talend (Barton 2013) can facilitate the extraction, transformation, and loading (ETL) process (Vassiliadis et al. 2002) by integrating different data sources, converting data into a unified format, and storing it in a centralized repository for analysis. Large-scale climate and environmental datasets pose challenges due to their volume. To handle this, big data analytics technologies employ distributed storage systems like Apache Hadoop or Apache Cassandra. These systems distribute and replicate data across multiple nodes, enabling parallel processing and fault tolerance. In addition, technologies like Apache Spark can perform distributed computing on these datasets, allowing for efficient data processing and analysis. For instance, researchers can leverage these technologies to perform complex computations on large volumes of climate model outputs or satellite data, extracting meaningful insights and patterns.

Distributed file systems, such as Hadoop Distributed File System (HDFS) (Shvachko et al. 2010), are designed to store and manage data across multiple machines or nodes in a cluster. This enables parallel processing of data by dividing it into smaller blocks that can be processed simultaneously. Batch processing frameworks, like Apache Hadoop MapReduce (Ghazi et al. 2015), allow for large-scale data processing by breaking down tasks into smaller subtasks that can be executed in parallel across a cluster. This approach is suitable for processing historical data or running periodic analyses. The data is divided into smaller subtasks that can be executed in parallel across a cluster. For instance, in environmental monitoring, historical climate data from multiple sources can be processed using MapReduce to identify long-term trends or anomalies. The data is split into chunks, and each chunk is processed independently on different nodes, and the results are combined to generate insights.

Technologies like Apache Spark leverage in-memory processing and distributed computing to perform fast and efficient data processing. By distributing data and computation across multiple nodes, parallel computing frameworks can significantly speed up analysis tasks. Data partitioning techniques divide the data into smaller partitions based on specific criteria (e.g., geographic location or time) to distribute the workload evenly across the processing nodes. Data partitioning involves dividing the data into smaller partitions based on specific criteria such as geographic location or time. Each partition contains a subset of the data that can be processed independently on different nodes. This approach ensures that the workload is evenly distributed across the processing nodes, maximizing the utilization of resources and enabling efficient parallel processing. For instance, in climate research, if you have a dataset of temperature measurements from different regions, you can partition the data based on geographic location. Each node can then process the temperature data for its assigned partition, and the results can be combined later for analysis. Distributed computing refers to the concept of performing computation across multiple nodes in a cluster. In the context of technologies like Apache Spark, distributed computing allows for parallel processing of data by dividing it into smaller partitions that can be processed independently on different nodes (Diaz et al. 2012). This parallelism significantly speeds up data processing tasks. For example, in climate modeling, if you need to run complex simulations on large datasets, Spark can distribute the computation across multiple nodes, allowing for faster model execution and analysis. Sharding, on the other hand, involves horizontally splitting a dataset into subsets that can be processed independently. Sharding is a technique where a large dataset is horizontally split into subsets or shards, which can be processed independently. Each shard represents a portion of the dataset that can be analyzed separately. For example, in analyzing satellite imagery for land cover classification, the image dataset can be shared into smaller subsets based on spatial regions. Each shard can be processed on a different node, performing image analysis and classification independently. This parallel processing approach speeds up the overall analysis and enables quicker insights into land cover patterns.

To optimize storage and processing efficiency, compression techniques are used to reduce the size of data without significant loss of information. Compression techniques are used to reduce the size of data without significant loss of information. There are various compression algorithms available, such as gzip, zlib, or Snappy, that can be applied to compress different types of data. For example, in climate science, if you have a large dataset of weather observations with redundant information, compression can be applied to reduce the storage space required. By eliminating data redundancy and representing the information more efficiently, the dataset can be stored in a compressed format, saving storage costs and improving data transfer times.

Additionally, data serialization formats like Apache Avro, Apache Parquet, and Protocol Buffers are employed to efficiently store and exchange structured data. Data serialization formats are used to efficiently store and exchange structured data. These formats represent data in a compact and optimized manner, allowing for faster data processing and analysis.

2.4 *Data Visualization Tools*

Data visualization tools transform complex data into visually appealing and easily understandable representations. There are many tools that can accomplish the transformation. For example, Geographic Information Systems (GIS) tools like ArcGIS and QGIS (Steiniger et al. 2011) are commonly used to visualize geospatial data. They enable the mapping of climate and environmental variables onto geographic locations, allowing researchers and decision-makers to analyze spatial patterns and relationships. For example, GIS can be used to create maps displaying temperature variations across different regions or the distribution of pollutant emissions. Data dashboard tools, such as Tableau (Batt et al. 2020) and D3.js (Zhu 2013), provide interactive and customizable visualizations that consolidate multiple data sources into a single interface. These dashboards allow users to explore data, uncover insights, and track key metrics. In the context of climate and environmental science, a data dashboard can display real-time weather conditions, air quality indices, or renewable energy production trends. Time series visualization tools like Grafana and Plotly (Sievert 2020) are used to plot and analyze data over time. They facilitate the visualization of temporal trends, patterns, and anomalies in climate and environmental variables. For instance, time-series visualizations can illustrate long-term changes in global temperature or fluctuations in greenhouse gas emissions. Heatmap and choropleth map visualizations are effective for illustrating density, intensity, or concentration of a particular climate or environmental variable. Tools like Matplotlib, seaborn, and Leaflet enable the creation of such visualizations. For example, a heatmap can display the intensity of rainfall across different regions, while a choropleth map can represent the air pollution levels in various cities. Network visualizations, created using tools like Gephi and Cytoscape, represent relationships and interactions among elements in a system. In the context of climate and environmental science, network visualizations can illustrate the flow of energy within an ecosystem or the interconnectedness of climate-related factors. Tools such as Blender and Unity allow the creation of 3D visualizations that enhance the understanding of complex climate and environmental concepts. For instance, 3D visualizations can be used to represent the movement of ocean currents, the impact of sea level rise on coastal areas, or the spatial distribution of wildlife habitats.

These data visualization tools enable scientists, policymakers, and the general public to gain insights and make informed decisions based on the analysis and interpretation of climate and environmental data. By presenting data in visually compelling and intuitive formats, these tools enhance communication, facilitate data-driven decision-making, and promote wider understanding of climate and environmental issues.

3 Modeling and Simulation Technologies

3.1 Climate and Environmental Modeling Tools

Climate and environmental modeling tools are essential for understanding and predicting the behavior of Earth's climate system and the impact of various environmental factors. General Circulation Models (GCMs), such as the Community Earth System Model (CESM) (Kay et al. 2015) and the Hadley Centre Global Environment Model (HadGEM) (Bellouin et al. 2013), simulate the interactions between the atmosphere, oceans, land surface, and sea ice. They use mathematical equations to represent the physical processes and dynamics of the climate system. GCMs are employed to study long-term climate patterns, project future climate scenarios, and investigate the effects of greenhouse gas emissions. Regional Climate Models (RCMs) (Giorgi 2019) are higher-resolution models that focus on specific geographic regions. They downscale the output of global models to provide more detailed information on regional climate characteristics. Examples of RCMs include the Weather Research and Forecasting (WRF) model (Powers et al. 2017) and the Regional Climate Model (RegCM) (Ozturk et al. 2012). RCMs are particularly useful for studying localized climate phenomena, assessing regional impacts, and informing regional-scale decision-making. Ecological models, such as the Dynamic Global Vegetation Model (DGVM) (Sitch et al. 2008) and the Terrestrial Ecosystem Model (TEM) (Li et al. 2020), simulate the interactions between the climate system and ecosystems. These models represent vegetation dynamics, carbon cycling, and species distribution patterns. Ecological models help assess the impacts of climate change on biodiversity, land use, and ecosystem services. Air quality models, like the Community Multiscale Air Quality (CMAQ) model (Appel et al. 2021; Alnuaim et al. 2023) and the System for Integrated Modeling of Atmospheric Composition (SIM-air) (Guttikunda et al. 2012), simulate the dispersion and transformation of pollutants in the atmosphere. They consider emission sources, meteorological conditions, and chemical reactions to assess air pollution levels, understand pollutant transport, and inform air quality management strategies. Hydrological models, such as the Soil and Water Assessment Tool (SWAT) (Douglas-Mankin et al. 2010) and the Variable Infiltration Capacity (VIC) model (Gao et al. 2010), simulate the movement of water within the Earth's hydrological cycle. These models consider factors such as precipitation, evapotranspiration, and runoff to simulate river flows, groundwater levels, and water availability. Hydrological models support water resource management, flood forecasting, and watershed assessments. IAMs, such as the Integrated Assessment Modeling Framework (IAMF) (Zeng et al. 2020), combine climate, economic, and social factors to assess the impacts of climate policies and alternative development pathways. IAMs consider the interactions between climate change, energy systems, land use, and socioeconomic factors. They provide a holistic approach to understanding the long-term implications of climate action and policy decisions. These modeling tools help scientists and policymakers simulate and analyze the complex interactions within the climate system and the environment. They provide insights into future scenarios, aid in decision-making, and contribute to our understanding of the challenges posed by climate change and environmental degradation.

3.2 *Artificial Intelligence (AI) and Machine Learning*

Artificial intelligence (AI) techniques have a wide range of applications in actionable climate science (Sun et al. 2022). We will dive into the current AI techniques in Chap. 12. Here we will briefly introduce some examples to help understand the current application scope.

(a) Extreme Weather Prediction

AI and machine learning algorithms can analyze large datasets of historical weather patterns, atmospheric conditions, and climate variables to develop models that might be able to predict extreme weather events such as hurricanes, droughts, or heatwaves (Karpatne et al. 2018; Bennett et al. 2023). By identifying patterns and correlations, these models can improve early warning systems and assist in implementing proactive measures to mitigate the impacts of extreme weather (Sun et al. 2023).

(b) Renewable Energy Optimization

AI can be used to optimize the integration and management of renewable energy sources, such as solar and wind, into the power grid (Hu et al. 2023). Machine learning algorithms can analyze weather data, electricity demand, and energy generation patterns to predict the optimal scheduling and control of renewable energy resources. This helps maximize renewable energy utilization, reduce reliance on fossil fuels, and enhance grid stability.

(c) Climate Data Analysis

AI techniques can process large volumes of climate data, including satellite imagery, sensor data, and climate model outputs (Yang et al. 2023). Machine learning algorithms can identify trends, detect anomalies, and extract valuable insights from complex datasets. For example, AI can analyze satellite data to monitor deforestation rates, track changes in land cover (Mahoney et al. 2023; Ganji et al. 2023), or assess the health of coral reefs, enabling timely conservation efforts.

(d) Climate Change Impact Assessment

AI and machine learning can aid in assessing the impacts of climate change on various sectors, such as agriculture, water resources, and public health (Rolnick et al. 2022). By analyzing historical climate data and socioeconomic variables, AI models can predict the vulnerabilities and risks associated with climate change. This information helps policymakers and stakeholders develop adaptation strategies and allocate resources effectively.

(e) Climate Finance and Policy

AI can assist in evaluating the effectiveness and efficiency of climate finance and policy measures (Bhandary et al. 2021). Machine learning algorithms can analyze financial data, investment patterns, and policy frameworks to identify the areas for improvement and optimize resource allocation. AI can also help track and verify carbon emissions and support the development of emission reduction strategies.

(f) Environmental Monitoring and Conservation

AI-powered technologies, such as remote sensing and image recognition, can contribute to environmental monitoring and conservation efforts. For example, machine learning algorithms can analyze satellite imagery to detect deforestation or monitor wildlife habitats (Masrur et al. 2023). AI can also assist in identifying invasive species, monitoring air and water quality, detecting diseases in vegetation (Barbedo 2023), and managing protected areas more effectively.

These examples highlight how AI and machine learning can enhance actionable climate science by providing more accurate predictions, improving resource management, and enabling informed decision-making. The ability of AI to process vast amounts of data, identify patterns, and make predictions offers valuable insights and tools for addressing climate change and promoting sustainable practices.

4 Communication and Collaboration Technologies

Revolutionary technologies centered on the Internet provide efficient and convenient ways for scientists, policymakers, and stakeholders to communicate, collaborate, and share information in actionable climate and environmental science (Podestá et al. 2013). These technologies enable real-time interactions, facilitate remote collaboration, streamline data sharing, and foster engagement among diverse stakeholders, ultimately supporting collective efforts to address climate and environmental challenges.

4.1 Social Media and Online Communities

Social media platforms like Twitter, Facebook, or LinkedIn, along with online communities and forums, provide channels for scientists and stakeholders to engage in discussions, share knowledge, and promote awareness on climate and environmental issues (Heavey et al. 2020). These platforms enable the dissemination of research findings, crowdsource ideas, and engage with a broader audience.

4.2 Collaboration Tools and Platforms

Platforms such as Slack provide centralized spaces for teams to communicate, share files, and collaborate on projects (Marion et al. 2021). These platforms offer features like group chats, file sharing, and task management, fostering effective teamwork and coordination among researchers and stakeholders.

4.3 Video Conferencing and Remote Collaboration Tools

Video conferencing platforms like Zoom or Microsoft Teams enable real-time communication and collaboration regardless of geographical locations (Correia et al. 2020). Scientists and experts can conduct virtual meetings, share presentations, and discuss research findings. This technology allows for efficient knowledge exchange and collaboration without the need for physical travel.

5 Emerging Technologies for Making Science Actionable

5.1 Blockchain and Distributed Ledger Technology

Blockchain is a decentralized and secure digital ledger that records and verifies transactions across multiple computers. In the context of climate science, blockchain can be used for transparent and secure management of climate-related data, such as carbon credits, emissions trading, or climate finance (Puschmann et al. 2020). For example, blockchain can ensure the traceability and integrity of carbon offset projects by securely recording the carbon credits generated and verifying their authenticity.

5.2 *Quantum Computing*

Quantum computing is a revolutionary technology that leverages the principles of quantum mechanics to perform computations in a fundamentally different way than traditional computers. In the context of actionable climate science, quantum computing has the potential to greatly impact various areas of research and decision-making. Here's a simple explanation. Traditional computers, known as classical computers, process information using bits that represent either a 0 or a 1. Quantum computers, on the other hand, use quantum bits or qubits, which can represent a 0, a 1, or both simultaneously due to a property called superposition. This allows quantum computers to perform multiple computations in parallel, enabling them to solve complex problems more efficiently.

In the realm of climate science, quantum computing can contribute in several ways. One key application is in the optimization of energy systems (Ajagekar et al. 2019). Climate models involve vast amounts of data and complex calculations to identify the most effective strategies for mitigating greenhouse gas emissions, optimizing renewable energy deployment, or designing efficient energy networks. Quantum computing can enhance the speed and accuracy of these calculations, enabling researchers and policymakers to identify optimal solutions more quickly.

Quantum computing can also help simulate and understand molecular systems (McArdle et al. 2020). Climate scientists often need to analyze the behavior of molecules and chemical reactions to study atmospheric composition, air pollution, or climate processes. Quantum computers excel at simulating quantum systems, enabling more detailed and accurate molecular simulations. This can lead to a better understanding of climate-related phenomena and help guide policy decisions.

Besides, quantum computing can assist in data analysis tasks. Climate science generates vast amounts of data, including satellite observations, climate model outputs, and sensor readings. Quantum algorithms can analyze these large datasets more efficiently, identifying patterns and extracting insights that may be challenging for classical computers (Acharjya & Ahmed 2016; Ayoade et al. 2023). This can lead to improved climate predictions, more accurate impact assessments, and better-informed decisions.

However, it' is important to note that quantum computing is still in its early stages of development, and practical quantum computers with sufficient computational power for climate-related applications are not yet widely available (Berger et al. 2021). Nonetheless, ongoing research and development in quantum computing hold promise for transformative applications in climate science, enhancing our understanding of the Earth's climate system and aiding in the development of effective climate strategies.

5.3 5G and Other Advanced Network Technologies

5G, the fifth generation of wireless communication technology, offers significant improvements over previous generations. It provides faster data transmission speeds, lower latency (delay), and higher network capacity. In the context of climate science, these features are crucial for real-time monitoring, data collection, and analysis. For example, 5G can enable seamless communication between remote sensors, monitoring stations, and data centers, allowing for immediate access to environmental data (Agyapong et al. 2014). This enables scientists and decision-makers to respond rapidly to climate events, such as extreme weather events or environmental disasters.

Other advanced network technologies like satellite communications and Internet of Things (IoT) networks are key infrastructure in actionable climate science (Vermesan et al. 2013). Satellite communications provide global coverage, allowing data to be transmitted from remote locations, such as polar regions or remote wilderness areas, where traditional infrastructure may be limited. This enables continuous monitoring of critical climate parameters like ice melting, sea surface temperatures, or vegetation health.

IoT networks, consisting of interconnected devices with sensors, enable the collection of real-time environmental data from various sources. These devices can be deployed in diverse locations, including cities, forests, or oceans, to gather data on air quality, temperature, humidity, soil conditions, and more. The IoT network facilitates the integration of these data points, providing a comprehensive view of environmental conditions. This rich dataset empowers climate scientists and policymakers to make informed decisions and take action toward mitigating climate change.

Advanced network technologies enhance collaboration and knowledge sharing among researchers, institutions, and communities. High-speed data connections enable seamless sharing of large datasets, models, and simulations. This facilitates collaboration on a global scale, allowing experts from different regions to work together, exchange insights, and collectively address climate-related challenges. Additionally, cloud computing platforms can leverage advanced networks to enable efficient storage, processing, and analysis of big data, providing scalable infrastructure for climate research.

5.4 Nanotechnology and Advanced Materials

Nanotechnology and advanced materials have significant implications for actionable climate science by offering innovative solutions to address environmental challenges. Here is an understandable explanation of these technologies in the context of climate science.

Nanotechnology usually works with materials at the nanoscale, which is the scale of atoms and molecules. It allows scientists to manipulate and engineer materials with unique properties and functionalities. In the context of climate science, nanotechnology can be used to develop advanced sensors and detectors that enable precise monitoring of environmental parameters (Dwivedi et al. 2022). For example, nanosensors can be designed to detect and measure pollutants in the air or water, providing real-time data for effective pollution management (Das et al. 2015). It also enables the development of advanced materials with improved performance and sustainability. For instance, nanomaterials can be used to enhance the efficiency of solar panels by improving light absorption and energy conversion. These materials can also be employed in energy storage technologies, such as batteries, to enhance their capacity and lifespan. By utilizing nanotechnology, renewable energy sources can be harnessed more effectively, contributing to the reduction of greenhouse gas emissions.

Advanced materials, including nanomaterials but extend beyond them, refer to materials engineered with specific properties to address specific challenges. They can be used to develop lightweight and durable components for renewable energy infrastructure, such as wind turbines or tidal energy devices (Yang & Sun 2013). These materials can also be applied in insulation and construction materials to improve energy efficiency in buildings, reducing the demand for heating and cooling.

5.5 *Edge Computing*

Edge computing involves processing data near the source or edge of the network rather than sending it to a centralized cloud server. This technology reduces latency and enables real-time analytics and decision-making. In climate science, edge computing can be applied to process data collected by IoT sensors in remote locations (Uddin et al. 2021). For instance, an edge computing device installed on a satellite could process satellite imagery data onboard and transmit only relevant information to Earth, reducing the bandwidth requirements and enabling more efficient data analysis.

6 Case Studies of Technology Applications in Making Science Actionable

6.1 *Flood Monitoring and Early Warning Systems*

In flood-prone areas, sensors are deployed in rivers and water bodies to monitor water levels and provide early warning systems (Arshad et al. 2019). These sensors measure water levels and transmit the data to a centralized system. Through data analysis and modeling, thresholds can be set to trigger alerts when water levels rise rapidly, indicating a potential flood. This information enables authorities to take timely actions, such as evacuations or deploying flood barriers, to minimize damages and ensure public safety.

6.2 Forest Fire Detection and Monitoring

Detecting forest fires in remote areas to enable rapid response and containment. Satellite-based sensors capture images of forests, and advanced algorithms analyze the data to identify signs of smoke and heat signatures associated with wildfires (Barmpoutis et al. 2020). Real-time monitoring allows authorities to quickly locate and respond to fire outbreaks, dispatching firefighting teams and resources to contain the fire and prevent further damage to ecosystems and nearby communities.

6.3 Precision Agriculture and Irrigation Management

Optimizing agricultural practices to enhance crop yield while minimizing water usage. Soil moisture sensors installed in agricultural fields measure moisture levels, while aerial imagery captures high-resolution images of the crops (Scott 2003). These data sources, combined with weather data and advanced analytics, help farmers determine precise irrigation needs for different areas of their fields. Automated irrigation systems then deliver water based on the specific requirements of each crop, minimizing water waste and reducing the risk of over- or under-irrigation.

6.4 Biodiversity Monitoring and Conservation

Monitoring and protecting endangered species and their habitats. Satellite imagery and remote sensing technologies are used to assess changes in land cover and habitat fragmentation, aiding in the identification of critical habitats for endangered species (Pettorelli et al. 2014). GPS tracking devices attached to animals allow researchers to monitor their movements, migration patterns, and behavior. The data collected helps conservationists develop effective strategies to protect and restore habitats, mitigate human–wildlife conflicts, and support biodiversity conservation efforts.

6.5 Carbon Footprint Tracking and Emission Reduction

Monitoring and reducing greenhouse gas emissions from industrial processes and transportation. IoT sensors installed in manufacturing facilities or vehicles capture real-time data on energy consumption, fuel usage, and emissions (Maksimovic

2018). This data is analyzed to calculate carbon footprints and identify areas for improvement. Based on these insights, organizations can implement energy-efficient measures, optimize transportation routes, or adopt cleaner technologies to reduce emissions and contribute to climate change mitigation.

7 Challenges and Future Directions

Climate science relies on diverse datasets from various sources, which can be challenging to integrate and ensure interoperability. As the volume and complexity of climate data increase, there is a need for scalable technologies that can handle large-scale data processing and analysis. Climate data often involves uncertainties and errors due to measurement limitations, data gaps, or biases in collection methods (Wilby et al. 2017). Climate science requires collaboration among scientists, policymakers, and stakeholders to effectively address climate challenges and develop actionable solutions. Climate technologies raise ethical and social considerations regarding privacy, data ownership, and equitable access to information and resources.

7.1 Ethical Considerations in Technology Applications

Ensuring responsible and inclusive deployment of technologies by incorporating ethical frameworks, privacy protection mechanisms, and equitable data access policies. This involves engaging diverse stakeholders, considering social impacts, and addressing ethical concerns to ensure that climate technologies serve the common good and prioritize sustainability and equity (Luck et al. 2012). More discussion about ethics can be found in Chapter 14.

7.2 Interdisciplinary Collaboration Challenges

These challenges resides in encouraging open data initiatives, fostering interdisciplinary collaborations, and creating platforms for knowledge sharing and stakeholder engagement. It includes leveraging technologies like online forums, data repositories, and collaborative tools to facilitate information exchange, co-design of solutions, and collective decision-making (Randhawa et al. 2017).

7.3 *Future Technological Developments and Their Potential Impact*

This usually refers to developing standardized data formats and protocols that enable seamless data sharing and integration across different platforms and organizations (Giuliani et al. 2011). This would facilitate comprehensive analysis and decision-making based on a holistic view of climate and environmental data.

8 Conclusion

This chapter on the technology landscape for making climate and environmental science "actionable" provides a detailed overview of the technologies used in addressing climate challenges. It discusses key technologies such as remote sensing, big data analytics, artificial intelligence, communication tools, and emerging technologies like quantum computing and nanotechnology. These technologies enable the collection and analysis of large volumes of climate data, facilitating informed decision-making and policy formulation. The chapter highlights challenges such as data integration, scalability, technology limitation, collaboration, and ethical considerations.

Future directions for technology development include improving data integration and interoperability, enhancing processing power and scalability, addressing data quality and uncertainties, promoting collaboration and knowledge sharing, and ensuring ethical and social considerations. The chapter emphasizes the important role of technology in actionable climate and environmental science, acknowledging the challenges while highlighting the potential for future advancements to drive effective climate action and sustainable environmental management.

References

Acharjya, Debi Prasanna, and Kauser Ahmed. 2016. A survey on big data analytics: Challenges, open research issues and tools. *International Journal of Advanced Computer Science and Applications* 7 (2): 511–518.

Afghah, Fatemeh, Abolfazl Razi, Jacob Chakareski, and Jonathan Ashdown. 2019. Wildfire monitoring in remote areas using autonomous unmanned aerial vehicles. In *IEEE INFOCOM 2019-IEEE Conference on Computer Communications Workshops (INFOCOM WKSHPS)*, 835–840. IEEE.

Agyapong, Patrick Kwadwo, Mikio Iwamura, Dirk Staehle, Wolfgang Kiess, and Anass Benjebbour. 2014. Design considerations for a 5G network architecture. *IEEE Communications Magazine* 52 (11): 65–75.

Ahmad, Tanveer, Dongdong Zhang, Chao Huang, Hongcai Zhang, Ningyi Dai, Yonghua Song, and Huanxin Chen. 2021. Artificial intelligence in sustainable energy industry: Status Quo, challenges and opportunities. *Journal of Cleaner Production* 289: 125834.

Ajagekar, Akshay, and Fengqi You. 2019. Quantum computing for energy systems optimization: Challenges and opportunities. *Energy* 179: 76–89.

Ali, Syed Mohd, Noopur Gupta, Gopal Krishna Nayak, and Rakesh Kumar Lenka. 2016. Big data visualization: Tools and challenges. In *20162nd International Conference on Contemporary Computing and Informatics (IC3I)*, 656–660. IEEE.

Ali, Sahara, Yiyi Huang, and Jianwu Wang. 2023. AI for sea ice forecasting. In *Artificial intelligence in earth science*, 41–58. Elsevier.

Alnuaim, Ahmed, Ziheng Sun, and Didarul Islam. 2023. AI for improving ozone forecasting. In *Artificial intelligence in earth science*, 247–269. Elsevier.

Appel, K. Wyat, Jesse O. Bash, Kathleen M. Fahey, Kristen M. Foley, Robert C. Gilliam, Christian Hogrefe, William T. Hutzell, et al. 2021. The Community Multiscale Air Quality (CMAQ) model versions 5.3 and 5.3. 1: System updates and evaluation. *Geoscientific Model Development* 14 (5): 2867–2897.

Arshad, Bilal, Robert Ogie, Johan Barthelemy, Biswajeet Pradhan, Nicolas Verstaevel, and Pascal Perez. 2019. Computer vision and IoT-based sensors in flood monitoring and mapping: A systematic review. *Sensors* 19 (22): 5012.

Ayoade, O., P. Rivas, J. Orduz, and N. Rafi. 2023. Satellite image classification using quantum machine learning. In *Artificial intelligence in earth science*, 337–355. Elsevier.

Barbedo, Jayme Garcia Arnal. 2023. Artificial intelligence for plant disease recognition. In *Artificial intelligence in earth science*, 101–118. Elsevier.

Barmpoutis, Panagiotis, Periklis Papaioannou, Kosmas Dimitropoulos, and Nikos Grammalidis. 2020. A review on early forest fire detection systems using optical remote sensing. *Sensors* 20 (22): 6442.

Barton, Rick. 2013. *Talend open studio cookbook*. Packt Publishing Ltd.

Batt, Steven, Tara Grealis, Oskar Harmon, and Paul Tomolonis. 2020. Learning Tableau: A data visualization tool. *The Journal of Economic Education* 51 (3-4): 317–328.

Bellouin, Nicolas, G.W. Mann, M.T. Woodhouse, C. Johnson, K.S. Carslaw, and M. Dalvi. 2013. Impact of the modal aerosol scheme GLOMAP-mode on aerosol forcing in the Hadley Centre Global Environmental Model. *Atmospheric Chemistry and Physics* 13 (6): 3027–3044.

Bennett, Andrew. 2023. AI for physics-inspired hydrology modeling. In *Artificial intelligence in earth science*, 157–203. Elsevier.

Berger, Casey, Agustin Di Paolo, Tracey Forrest, Stuart Hadfield, Nicolas Sawaya, Michał Stęchły, and Karl Thibault. 2021. Quantum technologies for climate change: Preliminary assessment. *arXiv preprint arXiv:2107.05362*.

Bhandary, Rishikesh Ram, Kelly Sims Gallagher, and Fang Zhang. 2021. Climate finance policy in practice: A review of the evidence. *Climate Policy* 21 (4): 529–545.

Buytaert, Wouter, Zed Zulkafli, Sam Grainger, Luis Acosta, Tilashwork C. Alemie, Johan Bastiaensen, Bert De Bièvre, et al. 2014. Citizen science in hydrology and water resources: Opportunities for knowledge generation, ecosystem service management, and sustainable development. *Frontiers in Earth Science* 2: 26.

Campbell, James B., and Randolph H. Wynne. 2011. *Introduction to remote sensing*. Guilford Press.

Carbone, Paris, Asterios Katsifodimos, Stephan Ewen, Volker Markl, Seif Haridi, and Kostas Tzoumas. 2015. Apache flink: Stream and batch processing in a single engine. *The Bulletin of the Technical Committee on DataEngineering* 38 (4).

Chadwick, K. Dana, Philip G. Brodrick, Kathleen Grant, Tristan Goulden, Amanda Henderson, Nicola Falco, Haruko Wainwright, et al. 2020. Integrating airborne remote sensing and field campaigns for ecology and Earth system science. *Methods in Ecology and Evolution* 11 (11): 1492–1508.

Chi, Mingmin, Antonio Plaza, Jon Atli Benediktsson, Zhongyi Sun, Jinsheng Shen, and Yangyong Zhu. 2016. Big data for remote sensing: Challenges and opportunities. *Proceedings of the IEEE* 104 (11): 2207–2219.

Correia, Ana-Paula, Chenxi Liu, and Xu Fan. 2020. Evaluating videoconferencing systems for the quality of the educational experience. *Distance Education* 41 (4): 429–452.

Das, Sumistha, Biswarup Sen, and Nitai Debnath. 2015. Recent trends in nanomaterials applications in environmental monitoring and remediation. *Environmental Science and Pollution Research* 22: 18333–18344.

Delp, William W., and Brett C. Singer. 2020. Wildfire smoke adjustment factors for low-cost and professional PM2. 5 monitors with optical sensors. *Sensors* 20 (13): 3683.

Diaz, Javier, Camelia Munoz-Caro, and Alfonso Nino. 2012. A survey of parallel programming models and tools in the multi and many-core era. *IEEE Transactions on parallel and distributed systems* 23 (8): 1369–1386.

Douglas-Mankin, K.R., Raghavan Srinivasan, and J.G. Arnold. 2010. Soil and Water Assessment Tool (SWAT) model: Current developments and applications. *Transactions of the ASABE* 53 (5): 1423–1431.

Dwivedi, Yogesh K., Laurie Hughes, Arpan Kumar Kar, Abdullah M. Baabdullah, Purva Grover, Roba Abbas, Daniela Andreini, et al. 2022. Climate change and COP26: Are digital technologies and information management part of the problem or the solution? An editorial reflection and call to action. *International Journal of Information Management* 63: 102456.

Eranna, G., B.C. Joshi, D.P. Runthala, and R.P. Gupta. 2004. Oxide materials for development of integrated gas sensors—A comprehensive review. *Critical Reviews in Solid State and Materials Sciences* 29 (3-4): 111–188.

Ganji, Geetha Satya, Mounika, and L. Wai Hang Chow. 2023. Explainable AI for understanding ML-derived vegetation products. In *Artificial intelligence in earth science*, 317–335. Elsevier.

Gao, Jidong, Ming Xue, Alan Shapiro, and Kelvin K. Droegemeier. 1999. A variational method for the analysis of three-dimensional wind fields from two Doppler radars. *Monthly weather review* 127 (9): 2128–2142.

Gao, Huilin, Qiuhong Tang, Xiaogang Shi, Chunmei Zhu, Ted Bohn, Su Fengge, Ming Pan, Justin Sheffield, Dennis Lettenmaier, and Eric Wood. 2010. *Water budget record from variable infiltration capacity (VIC) model*, 120–173. University of Princeton.

Garg, Nishant. 2013. *Apache Kafka*. Birmingham: Packt Publishing.

Ghazi, MohdRehan, and Durgaprasad Gangodkar. 2015. Hadoop, MapReduce and HDFS: A developers perspective. *Procedia Computer Science* 48: 45–50.

Giorgi, Filippo. 2019. Thirty years of regional climate modeling: Where are we and where are we going next? *Journal of Geophysical Research: Atmospheres* 124 (11): 5696–5723.

Giuliani, Gregory, and Pascal Peduzzi. 2011. The PREVIEW Global Risk Data Platform: A geoportal to serve and share global data on risk to natural hazards. *Natural hazards and earth system sciences* 11 (1): 53–66.

Guttikunda, Sarath K., and Puja Jawahar. 2012. Application of SIM-air modeling tools to assess air quality in Indian cities. *Atmospheric Environment* 62: 551–561.

Hadoop, Apache. *Apache hadoop*.2011http://hadoop.apache.org.

Hagan, David H., and Jesse H. Kroll. 2020. Assessing the accuracy of low-cost optical particle sensors using a physics-based approach. *Atmospheric measurement techniques* 13 (11): 6343–6355.

Haggag, May, Ahmed Yosri, Wael El-Dakhakhni, and Elkafi Hassini. 2022. Interpretable data-driven model for Climate-Induced Disaster damage prediction: The first step in community resilience planning. *International Journal of Disaster Risk Reduction* 73: 102884.

Hariri, Reihaneh H., Erik M. Fredericks, and Kate M. Bowers. 2019. Uncertainty in big data analytics: Survey, opportunities, and challenges. *Journal of Big Data* 6 (1): 1–16.

Heavey, Ciaran, Zeki Simsek, Christina Kyprianou, and Marten Risius. 2020. How do strategic leaders engage with social media? A theoretical framework for research and practice. *Strategic Management Journal* 41 (8): 1490–1527.

Hu, Weiming, Guido Cervone, and George Young. 2023. Theory of spatiotemporal deep analogs and their application to solar forecasting. In *Artificial intelligence in earth science*, 205–246. Elsevier.

Kaack, Lynn H., Priya L. Donti, Emma Strubell, George Kamiya, Felix Creutzig, and David Rolnick. 2022. Aligning artificial intelligence with climate change mitigation. *Nature Climate Change* 12 (6): 518–527.

Karpatne, Anuj, Imme Ebert-Uphoff, Sai Ravela, Hassan Ali Babaie, and Vipin Kumar. 2018. Machine learning for the geosciences: Challenges and opportunities. *IEEE Transactions on Knowledge and Data Engineering* 31 (8): 1544–1554.

Kay, Jennifer E., A. Clara Deser, A. Mai Phillips, Cecile Hannay, Gary Strand, Julie Michelle Arblaster, et al. 2015. The Community Earth System Model (CESM) large ensemble project: A community resource for studying climate change in the presence of internal climate variability. *Bulletin of the American Meteorological Society* 96 (8): 1333–1349.

Kim, Sang-Su, Wang-Ro Lee, and Jun-Hui Go. 2019. A study on utilization of spatial information in heterogeneous system based on apache NiFi. In *2019 International Conference on Information and Communication Technology Convergence (ICTC)*, 1117–1119. IEEE.

Li, Tingting, Lu Yanyu, Yu Lingfei, Wenjuan Sun, Qing Zhang, Wen Zhang, Guocheng Wang, et al. 2020. Evaluation of CH4MOD wetland and Terrestrial Ecosystem Model (TEM) used to estimate global CH 4 emissions from natural wetlands. *Geoscientific Model Development* 13 (8): 3769–3788.

Luck, Gary W., Kai MA Chan, Uta Eser, Erik Gómez-Baggethun, Bettina Matzdorf, Bryan Norton, and Marion B. Potschin 2012. Ethical considerations in on-ground applications of the ecosystem services concept. *BioScience 62 (12): 1020–1029.*

Lynch, Peter. 2008. The origins of computer weather prediction and climate modeling. *Journal of computational physics* 227 (7): 3431–3444.

Ma, Yan, Wu Haiping, Lizhe Wang, Bormin Huang, Rajiv Ranjan, Albert Zomaya, and Wei Jie. 2015. Remote sensing big data computing: Challenges and opportunities. *Future Generation Computer Systems* 51: 47–60.

Madakam, Somayya, Vihar Lake, Vihar Lake, and Vihar Lake. 2015. Internet of Things (IoT): A literature review. *Journal of Computer and Communications* 3 (05): 164.

Mahoney, Michael J., Lucas K. Johnson, and Colin M. Beier. 2023. AI for shrubland identification and mapping. In *Artificial intelligence in earth science*, 295–316. Elsevier.

Maksimovic, M. 2018. Greening the future: Green Internet of Things (G-IoT) as a key technological enabler of sustainable development. *Internet of things and big data analytics toward next-generation intelligence*: 283–313.

Marion, Tucker J., and Sebastian K. Fixson. 2021. The transformation of the innovation process: How digital tools are changing work, collaboration, and organizations in new product development. *Journal of Product Innovation Management* 38: 192–215.

Masrur, Arif, and Yu. Manzhu. 2023. Spatiotemporal attention ConvLSTM networks for predicting and physically interpreting wildfire spread. In *Artificial intelligence in earth science*, 119–156. Elsevier.

McArdle, Sam, Suguru Endo, Alán Aspuru-Guzik, Simon C. Benjamin, and Xiao Yuan 2020. Quantum computational chemistry. *Reviews of Modern Physics* 92 (1): 015003.

Munawar, Hafiz Suliman, Ahmed W.A. Hammad, and S. Travis Waller. 2021. A review on flood management technologies related to image processing and machine learning. *Automation in Construction* 132: 103916.

Ozturk, Tugba, Hamza Altinsoy, Murat Türkeş, and M. LeventKurnaz. 2012. Simulation of temperature and precipitation climatology for the Central Asia CORDEX domain using RegCM4.0. *Climate Research* 52: 63–76.

Paul, Frank. 2015. Revealing glacier flow and surge dynamics from animated satellite image sequences: Examples from the Karakoram. *The Cryosphere* 9 (6): 2201–2214.

Pearce, Warren, Sabine Niederer, Suay Melisa Özkula, and Natalia Sánchez Querubín. 2019. The social media life of climate change: Platforms, publics, and future imaginaries. *Wiley interdisciplinary reviews: Climate change* 10 (2): e569.

Pettorelli, Nathalie, William F. Laurance, Timothy G. O'Brien, Martin Wegmann, Harini Nagendra, and Woody Turner. 2014. Satellite remote sensing for applied ecologists: opportunities and challenges. *Journal of Applied Ecology* 51 (4): 839–848.

Podestá, Guillermo P., Claudia E. Natenzon, Cecilia Hidalgo, and Fernando Ruiz Toranzo. 2013. Interdisciplinary production of knowledge with participation of stakeholders: A case study of a collaborative project on climate variability, human decisions and agricultural ecosystems in the Argentine Pampas. *Environmental science & policy* 26: 40–48.

Powers, Jordan G., Joseph B. Klemp, William C. Skamarock, Christopher A. Davis, Jimy Dudhia, David O. Gill, Janice L. Coen et al. 2017. The weather research and forecasting model: Overview, system efforts, and future directions. *Bulletin of the American Meteorological Society 98 (8): 1717–1737.*

Puschmann, Thomas, Christian Hugo Hoffmann, and Valentyn Khmarskyi. 2020. How green FinTech can alleviate the impact of climate change—The case of Switzerland. *Sustainability* 12 (24): 10691.

Randhawa, Krithika, Emmanuel Josserand, Jochen Schweitzer, and Danielle Logue. 2017. Knowledge collaboration between organizations and online communities: the role of open innovation intermediaries. *Journal of Knowledge Management.*

Riediker, Michael, Ronald Williams, Robert Devlin, Thomas Griggs, and Philip Bromberg. 2003. Exposure to particulate matter, volatile organic compounds, and other air pollutants inside patrol cars. *Environmental science & technology* 37 (10): 2084–2093.

Rolnick, David, Priya L. Donti, Lynn H. Kaack, Kelly Kochanski, Alexandre Lacoste, Kris Sankaran, Andrew Slavin Ross, et al. 2022. Tackling climate change with machine learning. *ACM Computing Surveys (CSUR)* 55 (2): 1–96.

Salloum, Salman, Ruslan Dautov, Xiaojun Chen, Patrick Xiaogang Peng, and Joshua Zhexue Huang. 2016. Big data analytics on Apache Spark. *International Journal of Data Science and Analytics* 1: 145–164.

Schaefer, Garry L., and Ron F. Paetzold. 2001. SNOTEL (SNOwpackTELemetry) and SCAN (soil climate analysis network). *Automated Weather Stations for Applications in Agriculture and Water Resources Management: Current Use and Future Perspectives* 1074: 187–194.

Scott, Christopher A., Wim G.M. Bastiaanssen, and Mobin-ud-Din Ahmad. 2003. Mapping root zone soil moisture using remotely sensed optical imagery. *Journal of Irrigation and Drainage Engineering* 129 (5): 326–335.

Shvachko, Konstantin, Hairong Kuang, Sanjay Radia, and Robert Chansler. 2010. The hadoop distributed file system. In *2010 IEEE 26th Symposium on Mass Storage Systems and Technologies (MSST)*, 1–10. IEEE.

Sievert, Carson. 2020. *Interactive web-based data visualization with R, Plotly, and shiny*. CRC Press.

Sitch, Stephen, C. Huntingford, N. Gedney, P.E. Levy, M. Lomas, S.L. Piao, R. Betts, et al. 2008. Evaluation of the terrestrial carbon cycle, future plant geography and climate-carbon cycle feedbacks using five Dynamic Global Vegetation Models (DGVMs). *Global Change Biology* 14 (9): 2015–2039.

Steiniger, Stefan, and Andrew J.S. Hunter. 2011. Free and open source GIS software for building a spatial data infrastructure. In *Geospatial Free and Open Source Software in the 21st Century: Proceedings of the First Open Source Geospatial Research Symposium, OGRS 2009*, 247–261. Berlin, Heidelberg: Springer.

Sun, Ziheng, and Nicoleta Cristea. 2023. Introduction of artificial intelligence in Earth sciences. In *Artificial intelligence in earth science*, 1–15. Elsevier.

Sun, Ziheng, Laura Sandoval, S. Robert Crystal-Ornelas, Mostafa Mousavi, Jinbo Wang, Cindy Lin, Nicoleta Cristea, et al. 2022. A review of earth artificial intelligence. *Computers & Geosciences*: 105034.

Sungheetha, Akey, and Rajesh Sharma. 2020. Real time monitoring and fire detection using internet of things and cloud based drones. *Journal of Soft Computing Paradigm (JSCP)* 2 (03): 168–174.

Tantalaki, Nicoleta, Stavros Souravlas, and Manos Roumeliotis. 2020. A review on big data real-time stream processing and its scheduling techniques. *International Journal of Parallel, Emergent and Distributed Systems* 35 (5): 571–601.

Uddin, M. Ammad, Muhammad Ayaz, Ali Mansour, El-Hadi M. Aggoune, Zubair Sharif, and Imran Razzak. 2021. Cloud-connected flying edge computing for smart agriculture. *Peer-to-Peer Networking and Applications* 14 (6): 3405–3415.

Vassiliadis, Panos, Alkis Simitsis, and Spiros Skiadopoulos. 2002. *Conceptual modeling for ETL processes*, 14–21. Proceedings of the 5th ACM international workshop on Data Warehousing and OLAP.

Vermesan, Ovidiu, and Peter Friess, eds. 2013. *Internet of things: Converging technologies for smart environments and integrated ecosystems*. River Publishers.

Walden, Robert H. 1999. Analog-to-digital converter survey and analysis. *IEEE Journal on selected areas in communications* 17 (4): 539–550.

Wilby, Robert L., Nicholas J. Clifford, Paolo De Luca, Shaun Harrigan, John K. Hillier, Richard Hodgkins, Matthew F. Johnson et al. 2017. The 'dirty dozen'of freshwater science: detecting then reconciling hydrological data biases and errors. Wiley Interdisciplinary Reviews: Water 4 (3): e1209.

Yang, Bin, and Dongbai Sun. 2013. Testing, inspecting and monitoring technologies for wind turbine blades: A survey. *Renewable and Sustainable Energy Reviews* 22: 515–526.

Yang, Jun, Peng Gong, Fu Rong, Minghua Zhang, Jingming Chen, Shunlin Liang, Xu Bing, Jiancheng Shi, and Robert Dickinson. 2013. The role of satellite remote sensing in climate change studies. *Nature climate change* 3 (10): 875–883.

Yang, Kehan, Aji John, Ziheng Sun, and Nicoleta Cristea. 2023. Machine learning for snow cover mapping. In *Artificial intelligence in earth science*, 17–39. Elsevier.

Zeng, Sidong, Du Hong, and Jun Xia. 2020. Development of an interface-oriented add-in modeling framework for integrated water system simulation and its application. *Environmental Modelling & Software* 134: 104840.

Zhu, Nick Qi. 2013. *Data visualization with D3.js cookbook*. Packt Publishing Ltd.

Chapter 4
Actionable Science for Greenhouse Gas Emission Reduction

Bhargavi Janga, Ziheng Sun, and Gokul Prathin Asamani

Contents

1 Introduction

Greenhouse gases, such as carbon dioxide (CO_2), methane (CH_4), and nitrous oxide (N_2O), have the ability to trap heat in the Earth's atmosphere (Kweku et al. 2018). They absorb and release infrared radiation, a form of heat energy, causing the molecules to vibrate and absorb energy. This leads to a radiative forcing effect, which describes the imbalance between incoming solar radiation and outgoing radiation

B. Janga · Z. Sun (✉) · G. P. Asamani
Center for Spatial Information Science and Systems, Department of Geography and Geoinformation Science, George Mason University, Fairfax, VA, USA
e-mail: bjanga@gmu.edu; zsun@gmu.edu

Z. Sun (ed.), *Actionable Science of Global Environment Change*,
https://doi.org/10.1007/978-3-031-41758-0_4

from the atmosphere (Thomas et al. 2002). As a result, some heat becomes trapped. The re-emitted radiation further contributes to warming the Earth's surface and lower atmosphere. Human activities, such as deforestation and burning fossil fuels, have significantly increased greenhouse gas concentrations in the atmosphere (Houghton 1995). This amplifies the greenhouse effect, trapping more heat and leading to global warming and climate change. The heightened greenhouse gas levels can also trigger feedback loops that intensify the warming effect (Kweku et al. 2018). For example, as temperatures rise, natural sources like melting permafrost or oceanic methane hydrates release additional greenhouse gases, further enhancing the greenhouse effect and raising global temperatures (O'Connor et al. 2010).

Greenhouse gases also contribute to air pollution and the formation of harmful pollutants, such as ground-level ozone and particulate matter (Amann et al. 2008). These pollutants have severe health implications, including respiratory problems, cardiovascular diseases, and premature deaths. Rising temperatures, altered precipitation patterns, and extreme weather events can disrupt ecosystems, alter habitats, and threaten species' survival, including humans (Sintayehu et al. 2018). Changes in temperature and precipitation can also impact crop yields, water availability, and the dynamics of pests and diseases, posing challenges to food production (Skendžić et al. 2021). Reducing greenhouse gas emissions is necessary to mitigate its effects. The benefits of emission reduction extend far beyond environmental concerns, encompassing improvements in public health and a decreased dependence on fossil fuels (Fig. 4.1).

Scientific research establishes a solid foundation of knowledge about the sources, impacts, and potential solutions to greenhouse gas emissions (Rogelj et al. 2018). Interdisciplinary researchers are collaborating to explore various factors contributing to climate change, such as fossil fuel combustion, deforestation, industrial

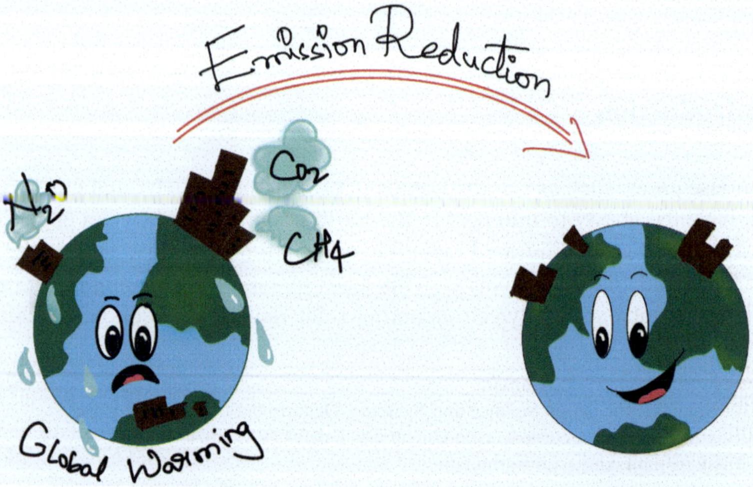

Fig. 4.1 Emission reduction cartoon

activities, and transportation. Their findings serve as a basis for policymakers to formulate strategies for mitigation and adaptation.

Meanwhile, governments and other decision-making entities are actively developing and implementing policies, laws, regulations, and programs in response to climate challenges (Head 2016). The Paris Agreement of 2015 establishes a global framework to limit global warming to below 2 °C and pursue efforts to restrict it to 1.5 °C (Rogelj et al. 2016). It also aims to enhance countries' resilience to climate change impacts and support their initiatives. At the local level, communities can take actions such as reducing driving, recycling, and choosing renewable energy sources to minimize their carbon footprint. Robust policy and regulatory frameworks are also necessary to hold companies and industries accountable for their emissions. Researchers can pinpoint effective strategies to combat climate change through rigorous scientific studies. These scientific findings serve as the basis for evidence-based policy recommendations, ensuring decisions align with environmental and societal interests. Local communities can access research findings to make informed choices, considering the environmental impact and striving to reduce emissions. By leveraging science, policymakers and local communities can collaborate toward an environmentally sustainable and economically feasible future.

2 Current Practice in the Industry, Government, and Local Community

In this section, we will delve into the industries, local communities, and governments that are dedicated to taking action on a global scale to reduce emissions. Our goal is to comprehensively explore the potential and challenges associated with implementing emission reduction measures. By doing so, we aim to offer a deeper understanding of the complexities involved in this process.

2.1 Emission Reduction in Various Industries

Across diverse industries, local communities, and organizations, there is a strong commitment to taking proactive measures for emission reduction and creating a sustainable future. Let us explore some key initiatives currently underway.

(1) Energy Industry

Coal-fired power plants have been constantly releasing greenhouse gases into the atmosphere. As our understanding of climate change and its consequences grew, the environmental impact of these power plants became a pressing concern (Fatih Birol et al. 2021). With increased awareness of these environmental and health impacts, a heightened focus has been on mitigating power plant emissions and transitioning to cleaner energy sources. Industries have taken steps such as implementing regulations, adopting cleaner technologies, and promoting renewable energy (Sinha et al. 2019). This shift aims to reduce the harmful effects of power plant emissions on the Earth's climate and ecosystems (Sinha et al. 2019). As part of this transition, power plants are moving away from high-emission fossil fuels and embracing cleaner alternatives. Renewable energy sources like solar, wind, and hydropower are being utilized to generate electricity (Sinha et al. 2019).

Power plants are employing Carbon Capture and Storage (CCS) technologies to capture CO_2 emissions from fossil fuel combustion and store them underground, preventing their release into the atmosphere (Sinha et al. 2019). These efforts reflect the power plant industry's commitment to achieving sustainable power generation and contributing to global emission reduction goals. By embracing cleaner energy sources and implementing CCS technologies, they strive to mitigate the adverse effects of power plant emissions on our environment (Fig. 4.2).

Oil and gas companies have historically been significant contributors to greenhouse gas emissions and climate change. However, in recent years, many of these companies have taken steps toward a low-carbon future by actively working to reduce carbon emissions in their operations and production processes. They employ various practices to achieve this goal, such as investing in low-carbon technologies and products. They utilize monitoring systems to detect and repair leaks along their operations, minimizing flaring and venting. They effectively reduce emissions by implementing technologies and practices that capture, utilize, or reinject the gas.

ExxonMobil, for example, employed computational models and technologies for emission reduction attempt. They utilize the fugitive emissions abatement simulation toolkit (FEAST) and Leak Detection and Repair programs simulation (LDAR-Sim) models to compare emissions reductions in different Leak Detection and Repair (LDAR) programs (Kemp et al. 2021). These programs incorporate various technologies, including aerial surveys, satellites, drones, trucks, and fixed monitors.

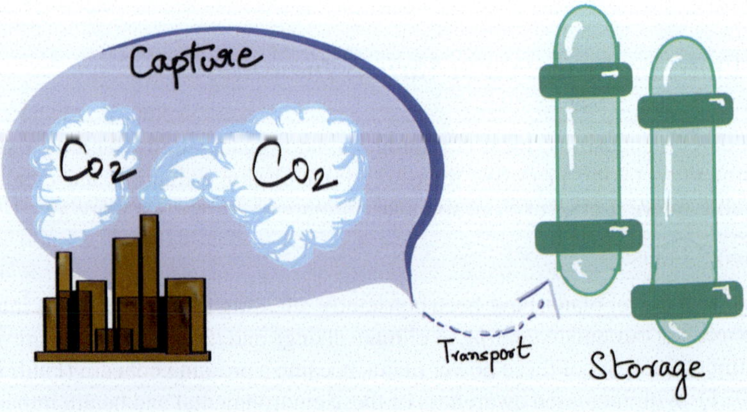

Fig. 4.2 Carbon capture and storage basic flow

Their purpose is to detect emissions, especially larger ones that go unnoticed in routine inspections, particularly in the Permian region. They rely on the Methane Emission Estimation Tool (MEET) to assess emissions from well sites with detailed temporal and spatial resolution (Allen et al. 2022). MEET simulates transitions between leaking and nonleaking states, estimates equipment failure rates and repair times, and assesses the effectiveness of LDAR programs in reducing emissions through stochastic simulations. The simulations span 5 years with a daily resolution, focusing on tank batteries and other facilities. The results highlight the efficacy of combining technologies, demonstrating that they can achieve greater reductions than regular inspections alone (Frate et al. 2021). These simulation outcomes are highly valuable in shaping policy decisions and industry practices to reduce oil and gas production emissions (Cardoso-Saldaña et al. 2023).

Another example of CCS technology in action is the Boundary Dam Carbon Capture and Storage Facility at the Boundary Dam power station. Located in Estevan, Saskatchewan, this commercial-scale CCS project has been at the forefront of carbon dioxide capture and storage from a coal-fired power plant since 2014. SaskPower implemented this pioneering solution, resulting in capturing 2,000,000 tons of CO_2 by March 2018 (Preston et al. 2018). This project not only demonstrates the potential for emission reduction in the power sector but also represents a significant advancement in CCS technology.

They used Shell Cansolv's post-combustion technology to effectively capture CO_2 emissions. This technology utilizes regenerable amines to capture both sulfur dioxide (SO_2) and CO_2 (Stéphenne 2014). In this process, the flue gas from the coal-fired unit is directed through a solution of amines, which selectively absorb the CO_2. Once captured, the CO_2 emissions are compressed and transported via pipelines to Cenovus Energy. Cenovus utilizes the CO_2 for Enhanced Oil Recovery (EOR) in the Weyburn oil field, where the captured CO_2 emissions are stored underground. The injection of CO_2 into the oil reservoirs reduces oil viscosity, increasing the flow rate and facilitating extraction. This technique, called CO_2-EOR, enhances oil production in mature oil fields.

The project encountered several challenges during the CO_2 capture process at SaskPower Boundary Dam Power Station. One significant challenge involved preventing cross-contamination among the prescrubber section, SO_2 absorption section, and caustic polisher section of the SO_2 Absorber. To address this, the chimney tray design was enhanced. The focus was particularly on the SO_2-Rich Amine Chimney Tray as any amine leakage into the prescrubber would result in amine traces in the prescrubber purge stream, eventually released into the ash ponds. Maintaining amine levels below 1 part per million by weight (ppmv) in the prescrubber purge was crucial to adhere to water release permits for the ash ponds. Shell Cansolv recognized the difficulty in achieving this condition and made chimney tray leak testing an integral part of the Water Commissioning activities. Despite the initial fabrication quality not meeting requirements, multiple test and repair sessions were conducted in a confined area until the weld quality met specifications. As a result, no leaks were detected under both static and water circulation modes.

Another challenge revolved around the delivery of the CO_2 capture absorbent inventory. To simplify storage and transportation logistics and reduce costs, the absorbent was shipped as 99 wt% amine (Stéphenne 2014). Dilution with demineralized water would occur on-site. The Lean Amine Tank on-site needed to be prepared to receive the complete absorbent inventory. To address this, temporary tanks need to be installed to store the amine until water commissioning activities concluded, enabling the transfer to the Lean Amine Tank. Despite these obstacles, the amine was delivered punctually, meeting specifications, and successfully loaded into the CO_2 Capture Plant. Using captured CO_2 emissions for EOR activities not only offers a sustainable solution to reduce carbon emissions but also promotes environmentally friendly practices in the pursuit of sustainable energy (Stéphenne 2014).

(2) Transportation Industry

The transportation sector has long been acknowledged as a significant contributor to emissions, prompting companies in the industry to take active measures to reduce their environmental impact. Transportation companies have adopted various strategies and practices to mitigate emissions to address this challenge. One effective approach they have embraced is the transition to battery electric vehicles (BEVs) or plug-in hybrid vehicles (PHEV) (Bibra et al. 2021). This shift involves replacing conventional internal combustion engine vehicles with cleaner alternatives that run on electricity or a combination of electricity and fuel. In parallel with vehicle electrification, companies also invest in the necessary infrastructure to support these alternative fuel types. This includes establishing charging stations and collaborating with fuel suppliers to ensure a reliable and accessible supply of electricity or alternative fuels (Bibra et al. 2021).

Encouraging a shift from individual car usage to shared transportation modes and promoting intermodal transportation is important in emission reduction efforts. Transportation companies are actively promoting ride-sharing, carpooling, and bike-sharing initiatives, aiming to reduce the number of vehicles on the road and maximize the utilization of existing resources (Nikitas et al. 2017). Companies are developing efficient intermodal transportation networks that seamlessly integrate different modes of transport, such as rail, road, and sea. This integration optimizes freight movement, minimizing emissions associated with transportation activities. Companies can use route optimization software to identify the most efficient routes, considering factors like traffic conditions and distance (Mngomezulu et al. 2023). This approach minimizes unnecessary mileage and the emissions associated with it. Real-time information and GPS tracking enable companies to adjust dynamically, ensuring continuous route optimization and enhanced fuel efficiency (White et al. 2023). By implementing these strategies and leveraging technological advancements, the transportation industry is actively working toward reducing its environmental footprint.

One effort by Toyota in emission reduction is the development of hybrid vehicles (Pratt et al. 2021). These hybrid vehicles combine a gasoline engine with an electric motor, improving power and efficiency. Plug-in hybrid vehicles, on the other hand, can operate on electricity for a limited distance before the gasoline engine engages, extending the driving range. Battery electric vehicles solely rely on externally charged batteries as their primary power source, eliminating tailpipe emissions (Poullikkas 2015).

Toyota has implemented a manufacturing practice known as the chemical recycling process to minimize emissions during the production phase. This method focuses on producing high-quality rare-earth oxides for driving motors and vehicle exhaust gas filters. The process commences by dissolving magnets, such as electric motors and driver motors, in nitric acid, specifically selected as the most suitable chemical. To ensure efficiency, the concentration of nitric acid in the solution is carefully regulated within the range of 5–20% while maintaining the temperature between 50 °C and 95 °C (Isomura et al. 2017). The utilization of waste acid in this step enhances its cost-effectiveness. Next, the solution undergoes a reduced heat energy requirement process to eliminate iron content to prevent any interference with the rare-earth separation process, ultimately yielding optimal outcomes. The ideal values for achieving such a high level of separation include an acid-extractant ratio of 1, an acid concentration of 6.39%, and a minimum of 10 separation steps (Isomura et al. 2017). Finally, the remaining trace elements are eliminated via oxalic acid crystallization. This meticulous procedure ensures that the recovered rare-earth oxides reach the required purity level, making them suitable catalysts for exhaust gas filters in automobiles. The recovered rare-earth oxides demonstrated sufficient purity, establishing their suitability as catalysts for automobile exhaust gas filters (Isomura et al. 2017).

(3) Chemical Industry

The chemical industry, however, also emits greater than two gigatons of greenhouse gases per annum globally (Yankovitz et al. 2022). As the demand for chemicals grew, so did GHG emissions, posing a significant environmental hazard. Fortunately, chemical factories have recognized the urgency and have implemented measures to reduce emissions and mitigate their impact on the planet. These practices include improving energy efficiency measures and implementing proper water management, and recycling that can help reduce emissions and minimize environmental impact. Effective waste segregation, treatment, and disposal methods prevent the release of hazardous substances into the environment. Chemical factories are increasingly shifting toward low-carbon investments by adopting renewable energy generation, such as solar panels or wind turbines (Nalini 2023).

Daikin Industries, Ltd. is a manufacturing company specializing in heating, ventilation, and air conditioning (HVAC) systems (Daikin Global 2023). Daikin has faced scrutiny due to the release of fluorochemicals, particularly fluorocarbons. Recognizing the importance of energy efficiency, they conducted a comprehensive life-cycle assessment (LCA) to compare the environmental impacts of different treatments for recovered refrigerants from air-conditioning equipment. When air-conditioning units are replaced or undergo maintenance, the refrigerants can be reclaimed by removing impurities. LCA compares the environmental impact of destroying and reclaiming recovered refrigerants and assesses the environmental effects of a product throughout its entire life cycle, from raw material extraction to waste disposal. Data was collected from surveys in Japan and Europe and focused on R410A and R22 in Japan and R410A and R134a in Europe. GHG emissions were calculated using the Global Warming Potential (GWP) AR5 values developed by the IPCC. Energy consumption was evaluated by considering the resource consumption of coal, oil, and natural gas. LIME3 impacts were assessed using consumption-based integration factors for Japan, Germany, and China, with a G20 population weighting average and a discount rate of 5% (Yasaka et al. 2022).

The assessment results demonstrated that the reclamation process significantly reduces GHG emissions and energy consumption compared to destructive treatment. This is primarily because the reclamation process replaces the production of new refrigerant, which has a high environmental burden, while the destruction process does not. The reclamation process allows for the recovery of valuable materials, such as hydrofluoric and hydrochloric acids and fluorite, which can be reused in other processes, further reducing the environmental impact. The primary data collected from surveys of facilities involved in the destruction and reclamation processes to obtain the values for GHG emissions, energy consumption, and LIME3 impacts. Secondary data from the AIST IDEA v2.3 database, contained within the LCA software Simapro 9.3, was also used for background processes. Specifically, the study found that reclamation can reduce GHG emissions by 5.7–15.9 kgCO2eq per kg of used refrigerant compared to destructive treatment in Japan and approximately 7.8–9.9 kgCO2eq less than destructive treatment in Europe. The reclamation process can reduce energy consumption by an average of 155 MJ per 1 kg of used refrigerant compared to the destruction process. These findings strongly advocate for the reclamation of used refrigerants and can contribute to reducing the environmental impact of air-conditioning equipment while aiding in emission reduction efforts (Yasaka et al. 2022).

Honeywell also launched an Emissions Control & Reduction Initiative to support customers in their carbon neutrality goals (Benjaafar et al. 2012). This initiative combines wireless gas detector technology with enterprise-wide data management solutions, providing an innovative approach to enhance reporting accuracy and increase productivity in conjunction with existing Leak Detection and Repair (LDAR) testing methods. By utilizing various gas detection solutions, such as gas cloud imaging cameras, they used early detection of leaks, allowing customers to swiftly identify leak locations. Integrating data analytics further improves leak management, minimizing production losses and ensuring compliance with legislation.

(4) Agriculture Sector

Agriculture is the leading source of pollution in many countries. Pesticides, fertilizers, and other toxic farm chemicals can poison fresh water, marine ecosystems, air, and soil. They also can remain in the environment for generations (Anju et al. 2010). As the demand for food continues to rise, agriculture has undergone significant changes to meet these demands. As awareness of these environmental challenges has grown, there has been a surge in efforts to address them through sustainable agricultural practices and emission reduction strategies. Various practices are being employed in the agricultural sector to tackle these emissions and work toward reduction. One effective strategy to combat the adverse effects of traditional agriculture is the adoption of precision agriculture techniques. This approach utilizes advanced technologies like soil testing, satellite imagery, and crop sensors to accurately apply fertilizers based on the specific needs of individual crops (Shafi et al. 2019). Farmers can reduce excess fertilizer application, thereby decreasing nutrient runoff and nitrous oxide emissions. However, despite these advances, traditional tillage methods still contribute to increased carbon dioxide emissions. This is where conservation tillage practices, such as minimum or no-till farming, come into play. These practices minimize soil disturbance and allow crop residues to remain on the surface, aiding in carbon sequestration (Lal et al. 1997). Farmers can promote soil health and reduce erosion by planting cover crops during fallow periods or between main crops. Crop rotation is also an effective strategy, breaking pest and disease cycles while improving soil fertility.

Methane emissions from livestock manure significantly contribute to greenhouse gas emissions (Shakoor et al. 2021). However, farmers can convert livestock waste into a renewable energy source through anaerobic digestion systems. The captured methane can then be utilized on-farm or fed into the grid, reducing greenhouse gas emissions and enabling farmers to become energy self-sufficient.

Indigo Ag, a recent player in the environment-friendly agriculture industry, focuses on reducing emissions through regenerative agricultural practices such as using cover crops (noncommercial plants that hold carbon in the soil and prevent erosion over the winter); crop rotation; and reducing tillage help to keep carbon out of the atmosphere, making farms stronger while reducing reliance on fertilizer and pesticides (Ahn et al. 2023). By integrating cover crops into their farming systems, farmers can capture carbon dioxide through photosynthesis and convert it into organic matter, which is then stored in the soil. This not only acts as a long-term carbon storage reservoir but also improves soil health by enhancing structure, increasing organic matter, and improving nutrient cycling (McDowell et al. 2019). As a result, soil fertility improves, reducing the need for synthetic fertilizers and decreasing the release of associated greenhouse gases.

Indigo Ag works in supporting farmers in adopting cover crops through tools like the Cover Crop Recommendation Tool, tailored recommendations, and the assistance of agronomists (Bruno et al. 2022). This empowers farmers to implement regenerative practices effectively, enhancing soil health, productivity, and emissions reduction. Indigo Ag collaborates with the USDA to develop the Cover Crop Recommendation Tool, simplifying the selection process. Agronomists provide guidance based on location and specific challenges, recommending appropriate cover crops. Indigo Ag utilizes technologies like geographic information systems (GIS) and remote sensing for data analysis to optimize resource utilization and support the transition to regenerative practices. Recognizing the environmental benefits of cover crops, Indigo Ag offers programs enabling farmers to earn carbon credits, further incentivizing adopting these practices and promoting sustainable agriculture.

2.2 How Is Policy Made?

The policymaking process for emission reduction is a complex and multilevel approach involving various actors at national and international levels. It usually begins by identifying a problem, such as climate change or air pollution, and assessing its sources and impacts. Once the problem is evaluated, policies are formulated and implemented at different levels, including national, regional, and local.

At the national level, the process starts with establishing a national climate change strategy (Kok et al. 2008). This strategy sets emission reduction goals and outlines the measures to be taken. Government agencies and departments responsible for environmental protection and climate change mitigation develop this strategy. The government then implements it through legislation, rules, and financial incentives, often collaborating with businesses and organizations. The policymaking process becomes more complex on the international level and involves multiple actors, such as nations, international organizations, and nongovernmental organizations. It begins with negotiations to develop a global climate change accord, usually led by groups like the United Nations Framework Convention on Climate Change (UNFCCC) (Kuyper et al. 2018) and IPCC (Intergovernmental Panel on Climate Change 2022). This accord defines carbon reduction targets and outlines nations' obligations. Countries ratify the agreement to formally accept its provisions, and once enough ratification is achieved, it becomes effective. Implementation of the accord involves enacting laws, developing regulations, providing financial incentives, and fostering collaboration between countries.

Political factors are prominent and decisive throughout the policymaking process. Perspectives of elected officials, the influence of special interest groups, and public opinion all shape the development of policies. The process must consider political, economic, and social factors, making consensus on a policy challenging. A lack of effective policy implementation can hinder progress in reducing emissions. Scientific research is essential in understanding the causes and effects of climate change and developing effective emission reduction policies. It provides decision-makers valuable information based on rigorous study, data collection, and analysis. Science helps identify emission sources, assess their consequences, and estimate future climate scenarios, and also evaluates the success of emission reduction strategies and considers the feasibility, potential side effects, risks, and benefits of different approaches (Dilling et al. 2011). Scientific assessments enable policymakers to make informed decisions, anticipate challenges, and mitigate uncertainties associated with emission reduction policies.

(1) Carbon Prices

Countries around the world are exploring policy solutions to encourage emission reduction, control polluting activities, and facilitate the transition to greener energy sources. One effective solution gaining momentum is carbon pricing (Baranzini et al. 2017). Carbon pricing is an economic strategy that involves charging for carbon emissions, providing businesses and communities with a financial incentive to reduce their emissions. This can be implemented through carbon taxes or cap-and-trade programs.

For instance, carbon taxes on residential CO_2 emissions in Sweden were implemented to address the problem of carbon emissions. These taxes applied to energy consumption in residential areas, encompassing heating and electricity. Initially set at 26 euros per ton of emitted CO_2 in 1991, the tax has gradually increased over the years, reaching 120 euros by 2018. The implementation of this carbon tax has yielded positive results, notably in reducing the consumption of fossil fuels and consequently leading to a decline in carbon emissions. It is worth noting that this decline occurred simultaneously with the increase in the carbon tax and the growing adoption of alternative energy sources like heat pumps. These findings support the effectiveness of the carbon tax in curbing residential carbon emissions. To study the impact of the carbon tax on residential CO_2 emissions, various methodologies such as difference-in-differences (DiD) and synthetic control methods (SCM) were employed. Through these analyses, the results consistently demonstrated a strong and robust negative causal relationship between the augmentation of the carbon tax and residential carbon emissions (Runst et al. 2020).

(2) Emission Standards

Emission standards serve as another effective solution to limit pollutants emitted from different sources such as cars, power plants, and industrial sites. These regulations ensure that industries operate within specified emission limits and promote the utilization of cleaner technologies. In the transportation sector, the burning of fossil fuels in cars significantly contributes to greenhouse gas emissions. The combination of high fuel consumption and low fuel efficiency further exacerbates climate change and air pollution. To ensure progress in reducing emissions and encourage automakers to prioritize fuel efficiency, regulations and incentives play an important role. The United States Congress recognized this need and developed the Corporate Average Fuel Economy (CAFE) standards as part of the Energy Policy and Conservation Act (Shiau et al. 2009). This comprehensive program has proven successful in enhancing fuel economy and reducing emissions from new cars and light trucks. Over time, it has adapted to changing policy objectives, technological advancements, and insights gained from real-world implementation. One of the key outcomes of the CAFE standards is the steady increase in on-road fuel economy as older, less efficient vehicles are phased out and replaced by newer, more fuel-efficient models. While the CAFE program focuses on fuel efficiency, it does not compromise on other aspects of vehicle quality. As automakers strive to meet the standards, they are incentivized to invest in technological advancements and innovative designs that enhance fuel efficiency. This encourages the development of more sustainable transportation options and fosters a positive impact on both the environment and the economy (Greene et al. 2020).

By gradually raising the fuel efficiency standards over time, the program encourages automakers to prioritize fuel efficiency in their designs and production processes. As a result, the CAFE program leads to increased average fuel efficiency of vehicles in the market, thereby reducing emissions from the transportation sector.

This regulatory framework aligns industry practices with emission reduction goals, promotes sustainable practices, and drives progress toward a more fuel-efficient and environmentally friendly transportation system. Through the implementation of the CAFE program, the United States can work toward mitigating climate change, improving air quality, and achieving energy efficiency in the transportation sector.

(3) Renewable Energy Targets

To promote the use of renewable energy technology to transition away from fossil fuels, policy initiatives are being developed to set goals for the proportion of energy generated from renewable sources within a specified time frame.

Rosa et al. (2021) have highlighted the significant achievements of the European Union (EU) in meeting its renewable energy targets. The EU's heavy reliance on traditional energy generation methods contributes to emissions and worsens climate change's impact. The EU has implemented renewable energy targets to address these challenges and promote the adoption of cleaner energy sources as part of its comprehensive climate and energy framework. These targets provide a clear direction and demonstrate a strong commitment to transitioning to renewable energy, which is critical in advancing sustainability and mitigating climate change.

Over the years, the share of renewable energy sources (RES) in total energy consumption has steadily increased. In 2019, the EU came close to achieving its target of 20% RES in gross final energy consumption, set for 2020 according to the RES Directive 2009/28/EC (Simionescu et al. 2020). Although slightly below the more ambitious trajectory defined by Member States (MS) in their National Renewable Energy Action Plans (NREAPs) in 2018, the overall progress was still significant. The RES-E (renewable electricity) and RES-H&C (renewable heating and cooling) sectors showed promising advancements (Steinhilber et al. 2016). Photovoltaics made substantial contributions to the RES-E sector, while the RES-H&C sector witnessed increased utilization of heat pumps. Examining the various support schemes employed by Member States to encourage the deployment of renewable energy, we find that the RES-E sector predominantly relied on premium and feed-in tariffs, often complemented by tendering systems (auctions). The RES-T sector heavily depended on biofuel quota obligations, although some countries also implemented tax incentives and subsidies. The EU has made remarkable strides toward a sustainable future by implementing these renewable energy targets. The increasing share of renewable energy sources in overall energy consumption demonstrates a clear shift away from traditional methods, thereby reducing emissions and combating climate change. The EU's commitment to supporting renewable energy deployment through various support schemes contributes to successfully attaining these targets.

2.3 Community Decision-Making for Emission Reduction

As we become more conscious of environmental issues, it is evident that our daily activities and lifestyle choices have a significant impact on the planet. These activities often require energy, primarily derived from fossil fuels, resulting in carbon dioxide (CO_2) release and contributing to climate change. Therefore, local communities strive to reduce energy consumption and adopt sustainable practices.

Various aspects of our daily lives contribute to energy consumption and emissions. For instance, heating, water overuse, appliances and electronics, and transportation play significant roles (Driga et al. 2019). Private vehicles, especially those running on fossil fuels, and waste disposal, particularly organic waste decomposition in landfills. Food production, including livestock farming, generates substantial greenhouse gas emissions, while water consumption indirectly affects greenhouse gas emissions due to energy requirements for water treatment, distribution, and heating. Energy-intensive processes such as water pumping and heating also contribute to emissions. It is important to recognize that the impact of human activity on the environment extends beyond food and water. Buildings, for example, have a notable environmental impact, with heating, cooling, and lighting consuming significant amounts of energy. The materials used in construction have implications for the environment.

Considering these factors, it is evident that we need to be mindful of how our actions impact the planet. Communities are taking steps to mitigate this impact through sustainable practices. They are installing LED light bulbs that consume up to 75% (John et al. 2023) less energy and using programmable thermostats to optimize home heating and cooling mechanisms. They are also adopting renewable energy sources such as solar panels and wind power for electricity generation, which reduces greenhouse gas emissions. Another approach is to choose public transportation, carpooling, or ridesharing to reduce individual vehicle emissions. Electric vehicles (E.V.s) and hybrid vehicles are also becoming popular alternatives to conventional gasoline or diesel cars (Penney 2021; Ice 2023). Waste reduction, recycling, and reusing are encouraged, along with participating in community recycling programs and raising awareness about waste reduction. Examples include reprocessing single-use N95 masks without compromising their integrity and protective ability and diverting a significant amount of waste from landfills (Silva et al. 2021). Installing low-flow fixtures, water-efficient appliances, and collecting rainwater for irrigation also reduce water-related emissions.

Engaging in activities that promote awareness and influence policy can make a significant difference in promoting global emission reduction efforts. Joining local environmental groups, participating in community events and workshops, and communicating with elected representatives to advocate for sustainable policies are impactful actions.

For example, the Smart Energy G.B campaign advocates adopting smart meters, which provide real-time energy usage data (Sovacool et al. 2017). These meters have proven scientific benefits in reducing energy consumption and carbon emissions. The campaign effectively communicates the advantages of smart meters to the public, raising awareness and encouraging adoption. It educates people about the benefits, helps decrease energy usage, aims to reduce carbon emissions, and enhances customer service by providing reliable energy usage information to suppliers (Bbdo et al. 2023).

3 State-of-the-Art Research for Emission Reduction

This section will provide an overview of numerous ongoing research subjects and their present experimental status with an overview of research outputs related to emission reduction.

3.1 Carbon Capture and Storage (CCS)

CCS is a promising solution for providing low-carbon products, aiding industries in decarbonization, and meeting climate targets (Townsend et al. 2020). By capturing and storing CO_2 emissions, CCS effectively reduces atmospheric CO_2 concentrations, mitigating the impacts of climate change. It has the potential to significantly decrease greenhouse gas emissions significantly, resulting in cleaner industries and a slowdown in the effects of climate change.

The concept of CCS was introduced in 1977 to capture CO_2 from power plants and inject it into suitable geological formations (Raza et al. 2019). Numerous CCS projects have been initiated worldwide, such as Sleipner, In-Salah, CO2SINK, and RECOPOL, to test the feasibility and effectiveness of CO_2 storage in different geological sites (Wildbolz et al. 2007). Ongoing large-scale CCS projects like the Tomakomai CCS Demonstration Project, the Illinois Industrial CCS Project, and the Petra Nova Carbon Capture Project demonstrate the increasing interest in implementing CCS on a larger scale. Successful CCS implementation requires careful consideration of various factors, including the safe capture, transportation, injection, and storage of CO_2 in subsurface geological formations. This involves understanding the properties and behavior of CO_2 under different pressure and temperature conditions. During injection into geological formations at depths greater than 800 meters, CO_2 often appears as a supercritical fluid. Therefore, understanding the physical properties and phase change of CO_2 is crucial for site selection and monitoring processes (Raza et al. 2019).

Gabrielli et al. (2020) explored three technology chains. These chains involve the use of fossil fuels with CCS (CCS route), captured CO_2 as feedstock with "green" hydrogen in new chemical processes (CCU route), and biomass dedicated to chemical production (BIO route). They conducted a quantitative comparative assessment of methanol production and found that defossilization of the chemical industry is feasible through any of the three approaches, each with its advantages and challenges. They emphasized the need for systemic analysis and identified key hurdles for each route. The CCS route offers the advantage of leveraging existing infrastructure and contributes to negative emissions technologies important for climate targets.

Yan et al. (2021) utilized machine learning (ML) in CO_2 capture, transport, storage, and utilization. ML has been extensively applied in absorbent- and adsorbent-based CO_2 capture, facilitating a simulation, optimization, thermodynamic analysis, solvent selection, and design. ML has also been used to predict combustion characteristics and monitor oxyfuel combustion processes for CO_2 capture. ML shows promise in calcium looping and chemical looping combustion for CO_2 capture. In CO_2 transportation and storage, ML enables accurate measurement of CO_2 flow and leak detection. Machine learning algorithms study trapping mechanisms, predict and monitor CO_2 leakage, and optimize CO_2 CCS-EOR processes. They explained the advantages of ML in CCUS, including identifying hidden data relationships, cost-effective computing, and expediting material design and process optimization.

Bui et al. (2018) highlighted the research challenges and nontechnical barriers that must be addressed for the widespread deployment of CCS in the next decade. Despite the ambitions set a decade ago, CCS has yet to be implemented at the necessary scale. They also explored the potential of negative emissions technologies like bioenergy with CCS (BECCS) and direct air capture (DAC). Implementing CCUS involves capturing and transporting CO_2 emissions for underground storage, enabling enhanced oil recovery and conversion into valuable products. However, it requires supportive regulatory frameworks, infrastructure investments, and public acceptance. Successful deployment of CCUS has the potential to reduce CO_2 emissions and enable low-carbon operations, and foster a circular economy. Regarding the costs of CCS, Irlam et al. (2017) investigated and found that deploying CCS can result in significant cost savings for achieving emission reduction targets compared to other renewable energy sources. Considering the value for money and long-term cost reductions achieved through CCS deployment is essential.

Another study by Tcvetkov et al. (2019) showed the need to better understand the public perception of CO_2 storage, capture, and transportation. They noted that while CO_2 storage receives the most attention in public perception studies, capture and transportation aspects need to be better studied for optimal database establishment and public understanding. Smit et al. (2014) addressed concerns about capturing CO_2 from power plants and explored alternative methods like solid adsorption and membranes to improve efficiency. They also discussed applications such as enhanced oil recovery and using CO_2 as a feedstock for the chemical industry. CO_2 storage involves injecting it into geological formations, and this research focuses on safety monitoring and enhancing mineralization processes. They explained carbon capture technologies for stationary sources like coal-fired power plants and explored alternatives to increase efficiency in oil recovery, chemicals, fuels, and construction materials. Martin-Roberts et al. (2021) proposed a direct approach that involves implementing regulations requiring companies to include CO_2 storage plans post-extraction and directing funding toward challenging sectors.

3.2 *Renewable Energy Technologies*

Transitioning from fossil fuels to renewable energy sources is essential for generating clean and sustainable energy. Renewable technologies offer numerous benefits, including abundant potential, improved efficiency and cost-effectiveness, reliable power generation, and a lower carbon footprint throughout their lifecycle. Embracing renewables enables us to transition to a low-carbon economy and mitigate greenhouse gas emissions.

Østergaard et al. (2020) explored a range of renewable energy technologies within a broader context, including wind, wave, geothermal, solar, and salinity gradients. They evaluate the economics of renewable energy in different scenarios, such as assessing wave power potential, applications of wave energy converters, wind energy modeling, and repowering options for wind farms. They focus on energy system analyses, decarbonizing transportation, and socioeconomic aspects of renewable energy transitions. The study reveals significant advancements in renewable energy systems from technological, resource assessment, and systems design perspectives. Rahman et al. (2022) highlighted the importance of renewable energy (RE) in agricultural applications to mitigate climate change, especially in developing countries where agriculture contributes a higher share of greenhouse gas emissions (35%) compared to developed nations (12%). They propose utilizing solar, wind, and hydro-powered systems for water pumps, greenhouse heating, cooling, post-harvest processing, and lighting. Agro-photovoltaic systems and solar irrigation pumps are identified as beneficial for developing countries, while developed nations can leverage various renewable energy technologies based on their specific needs and environmental considerations.

Let us explore renewable energy supply by integrating the power, heat, transport, industry sectors, and even desalination. Bogdanov et al. (2021) demonstrate the feasibility of such a transition by modeling five scenarios that consider different energy sectors and additional demands like international transport, chemical industry feedstock, and desalination. The results indicate that achieving a 100% renewable energy system is feasible in Kazakhstan, even with its harsh climate and energy-intensive economy. Integrating different sectors allows for low-cost flexibility through power-to-heat, heat storage, hydrogen, and synthetic fuel storage. Similar transitions may be possible in regions with comparable conditions. González-Arias et al. (2022) analyze the economic feasibility of utilizing locally produced biomethane as a renewable fuel for light marine transport in Cornwall, UK, to reduce CO_2 emissions. They provide valuable insights into using biomethane as a green fuel for shipping, recognizing its importance for local economies.

Accelerating the deployment of renewable energy technologies, particularly in households, is crucial for reducing energy consumption and greenhouse gas emissions. Siksnelyte-Butkiene et al. (2020) employ multiple-criteria decision-making (MCDM) methods to evaluate household renewable energy technologies, considering economic, social, technological, and environmental criteria. They discuss various MCDM methods, including the analytical hierarchy process (AHP), the technique for order of preference by similarity to the ideal solution (TOPSIS), and the preference ranking organization method for enriching evaluation (PROMETHEE), as suitable approaches for assessment. Qazi et al. conduct a systematic review of renewable energy resources, highlighting the world's energy needs, technologies for domestic use, and public opinions. They find that fossil fuels still contribute 73.5% to global electricity production, while renewable sources contribute only 26.5% (Qazi et al. 2019). The lack of public awareness is identified as a significant barrier to the acceptance of renewable energy technologies. The study emphasizes integrating renewable energy sources into power generation to manage global energy crises effectively. They explore the significance of public opinion and include real-time analysis of public tweets as a novel initiative to guide future researchers and policymakers.

3.3 *Why Do Most of These Emission Studies Have Low Actionableness?*

The effectiveness of emission studies in mitigating climate change has been questioned due to their limited actionable nature. This discrepancy between research and real-world impact is a multifaceted challenge that hinders the translation of findings into practical actions. This section explores the reasons behind the low actionability observed in most emission studies, shedding light on the barriers that impede the implementation of research findings. It is key to prioritize climate research and catalyze bold action to address the urgent and complex challenges of greenhouse gas emissions.

For example, Rahman et al. (2022) provide valuable insights on using renewable energy (RE) in agriculture to combat climate change. However, several factors hinder the actionability of this research. High installation costs pose a significant barrier for small farmers without substantial government support or direct investment. Specialized equipment and long payback periods increase financial risks for investors. In developing countries, energy sources for rural areas are often distant from agricultural regions, resulting in limited incentives for energy-resource investors and difficulties in financing renewable energy-based agricultural initiatives. Limited technical expertise and resources in the developing world make it challenging to design reliable agro-photovoltaic systems with stable energy output. Technical standards and effective quality control processes are necessary for adopting renewable energy solutions in agriculture, but they need to improve system quality assurance and issue resolution. The lack of skilled personnel and training programs further contributes to the prevalence of low-quality solar systems and services, making installation and maintenance difficult and limiting farmers' access to necessary technical skills.

Similarly, according to González-Arias et al. (2022), using biomethane as a renewable fuel for light shipping is not economically viable under current market conditions. Subsidies alone are insufficient to make this approach profitable, and expanding the plant size is impractical due to the negative impact on profitability prospects. The search for cheaper technical solutions for biogas upgrading to biomethane is unrealistic in the near future, given the limitations of existing commercial-scale technologies. Similarly, despite extensive research and development, Martin-Roberts et al. (2021) highlight the ineffective functioning of many CCS initiatives. Limited progress in utilizing available storage resources delays verifying bankable storage options. The success of CCS projects in different countries, such as China and India, depends on their ability to progress from initial planning to actual implementation. Negative perceptions also arise due to limited understanding or a "not in my backyard" attitude. The absence of specific legislation and regulatory frameworks in certain countries further presents a significant barrier to CCS implementation.

Additionally, the machine learning studies like Yan et al. (2021) have challenges like ethical concerns, data availability, quality, misapplication, and interpretability. ML models heavily rely on data, which can introduce biases and prejudices. Testing and investigations are necessary to ensure the effectiveness of models at scale, questioning the applicability of lab and pilot-scale optimizations to industrial scales. Robust models require sufficient data and careful consideration of training dataset quality, feature selection, and algorithm choice.

4 Success Use Case and Analyzing Actionableness

By examining several recent studies on emission reduction and identifying effective instances of science applications, in this section we demonstrate practical science highlighting the essential factors that contribute to making science more relevant and actionable.

4.1 Solar PV-Diesel Hybrid Mini Cold Storage System

Infrastructure Development Company Limited (IDCOL) of Bangladesh has funded research teams to develop a hybrid mini cold storage and study the financial feasibility of establishing cold storage facilities for vegetables and fruits to prevent spoilage (Hossain et al. 2021).

The cold storage operates primarily during daylight hours, utilizing power generated from rooftop solar PV panels. They ensure that the interior temperature maintains a minimum of 2–3 degrees Celsius during the day and gradually increases to a maximum of 7–8 degrees Celsius at night when solar energy is unavailable. The researchers took specific temperature requirements for different crops during the design process. In order to ensure an uninterrupted power supply, a backup diesel generator was made available for emergencies. The cold storage unit encompasses

1000 m^3 of space and has a roof area that accommodates 20 kWp of solar panels to accommodate the required equipment and maximize the use of solar energy. The maximum cooling load is estimated to be 24 kW, divided into four 6 kW units. Under normal conditions, the solar panels power the compressors, while the diesel generator supplies energy during full-load operations, emergencies at night, or periods of low sunshine. The research team conducted a comprehensive financial analysis, considering the capital investment required to establish the cold storage, which accounts for over 70% of the total cost. The breakdown includes costs for civil engineering, solar PV panels with installation, compressors, inverters, a diesel generator, land expenses, and miscellaneous items. The total capital investment is determined to be Tk. 8,090,000. It was clear from the cost structure that capital investment primarily drives the storage cost. Although the cost of PV energy (without battery backup) is relatively higher than conventional grid power, it minimizes the overall storage cost. Later, their calculations indicate that an average increase of Tk. 0.10 per kg/week in product price can sustain the scheme's viability. In conclusion, the hybrid mini cold storage maintains appropriate temperature levels, effectively reduces market surpluses, and stabilizes prices. It presents an attractive solution for short-term storage needs in rural areas of Bangladesh, with a cost per kilogram of stored produce significantly lower than Tk. 1.00, making it an affordable option for farmers (Khan et al. 2014).

4.2 *Conservation Agriculture and Carbon Sequestration*

Scientists and farmers have embraced conservation agriculture as a means to bolster carbon sequestration within agricultural systems. To encourage its adoption, agricultural extension services, and organizations have taken on the responsibility of offering training and technical support to farmers. They have conducted practical demonstrations and on-farm trials to showcase the benefits of conservation agriculture, including carbon sequestration and improved soil health. Farmers have been educated about the significance of soil organic matter and how conservation agriculture practices enhance soil fertility, water retention, and erosion control.

Researchers have studied the role of organic matter in soil fertility and carbon sequestration, aiming to identify effective management practices that improve soil health and productivity. Through practices like reduced tillage, crop rotation, cover cropping, and intercropping, the input of organic matter into the soil has increased, leading to enhanced sequestration of soil organic carbon (SOC). Long-term carbon sequestration depends on maintaining these practices to promote organic matter accumulation in the soil. Financial incentives and policy support have also been provided through programs like the Clean Development Mechanism (CDM) of the Kyoto Protocol, which enables farmers to earn credits for the carbon they sequester. These initiatives encourage the adoption of conservation agriculture practices by offering economic benefits to farmers. As a result, carbon sequestration in agricultural systems has increased (Govaerts et al. 2009).

5 Suggestions for Improving Actionableness of Emission Control Research

We have observed that the utilization of science has resulted in reducing emissions. This section will discuss how to make emission control research more applicable to the real world.

5.1 Suggestions for Scientists

Scientific progress has yielded valuable insights into practical approaches for reducing emissions. We propose leveraging AI and machine learning techniques to enhance the actionability and effectiveness of emission reduction research. These advanced technologies can significantly improve the accuracy and efficiency of monitoring, predicting, and controlling emissions. Scientists can identify precise patterns and develop targeted strategies by analyzing extensive datasets encompassing historical information and meteorological factors. AI-driven virtual simulations and scenario analysis tools offer valuable insights for effective emission reduction planning (Degot et al. 2022). By harnessing the power of AI, researchers can optimize their approaches and make well-informed decisions based on real-time data, leading to a more practical and effective process.

In addition, exploring the potential of nanomaterials in emission control research can unlock innovative solutions. Nanocatalysts and nano sorbents promise to capture and convert harmful pollutants from industrial emissions efficiently (Ningthoujam et al. 2022). Nanotechnology, with its unique properties and high surface area-to-volume ratios, holds great potential for enhancing emission control efficiency. Nature-inspired approaches, such as studying plants and microorganisms to uncover natural mechanisms for pollutant capture, conversion, and remediation, are also worth exploring. Developing bio-inspired technologies like artificial photosynthesis, microbial fuel cells, and bio-inspired heat transfer systems based on these natural processes can contribute to effective emission reduction (Tschörtner et al. 2019).

Research should investigate new catalyst materials and electrode designs to improve the efficiency and selectivity of electrochemical conversion processes, providing a sustainable pathway for emission control and utilization (American Association for the Advancement of Science 2023). Real-time emissions monitoring is crucial, necessitating the development of compact and affordable sensors capable of collecting accurate and high-resolution emission data in various environments. Integrating these sensors into a network for comprehensive and continuous monitoring of pollutant sources will significantly enhance our understanding of emissions and enable effective control measures.

5.2 Suggestions for Policymakers

Policymakers are usually the final action-taker of proposing and creating an enabling environment and implementing effective measures to combat climate change and reduce emissions. To achieve substantial and sustainable progress, policymakers can employ various strategies. One approach is to initiate innovation challenges and competitions that specifically target emission reduction across different sectors (Williamson et al. 2018). By doing so, policymakers can inspire individuals and organizations to develop groundbreaking technologies, business models, and strategies that significantly decrease emissions and contribute to long-term sustainability. Policymakers should actively engage with scientists and researchers to enhance their understanding of scientific concepts and research findings. By engaging in dialogue, policymakers can clarify their information needs and explore effective ways to communicate research outcomes to policymakers and the public. They can also recognize the value of systematic reviews in synthesizing diverse evidence and providing comprehensive assessments of research findings. Collaborating with experts in conducting systematic reviews ensures a robust evidence base for decision-making.

Active stakeholder engagement throughout the research process is essential. Policymakers should involve stakeholders such as researchers, industry representatives, and affected communities in the design, implementation, and evaluation of research projects. This collaboration promotes cooperation, ensures relevance to real-world challenges, and increases the likelihood of research uptake in policy and practice. Building and nurturing relationships and networks between researchers, policymakers, and academic institutions with policy actors, including government agencies, nonprofit organizations is essential. Regular interaction, collaborative initiatives, and knowledge-sharing platforms facilitate the exchange of insights and the uptake of research findings in policy formulation. Evidence-based policymaking uses the best available research and information on program results to guide decisions at all stages of the policy process and in each branch of government (Vanlandingham et al. 2014). By promoting systematic reviews, rapid synthesis of evidence, and the application of research findings, policymakers can enhance the effectiveness and efficiency of policy interventions. One framework that policymakers can explore is the Model for Dissemination of Research (Tabak et al. 2012), which provides a framework for translating research into actionable insights for policymakers. By applying this model, policymakers can enhance their understanding of the research-to-policy process and identify effective strategies for research adoption.

5.3 Suggestions for Industries

Industries have a significant impact on climate change, so it is essential for them to actively seek partnerships with academic and research institutions to reduce emissions. These partnerships can involve joint funding projects, knowledge exchange programs, and dedicated research facilities. By collaborating with the research community, industries can access ongoing research, leverage educational resources, and contribute to developing practical solutions for controlling greenhouse gas emissions (Center for Climate and Energy Solutions 2020). Investing in technology transfer programs can facilitate translating experimental research results into tangible applications, such as licensing agreements, spin-off businesses, or technology incubators.

To adopt and implement research-based technologies, industries can promote the transition from research to comprehensive technology solutions. Conducting pilot studies based on experimental research findings allows industries to assess the practicality and effectiveness of these solutions in real-world scenarios (Liu et al. 2015). These pilot studies serve as testbeds for evaluating the applicability of research findings in industry-specific contexts. By gathering data and analyzing the results, industries can fine-tune their strategies and identify potential challenges or limitations.

Establishing partnerships between industries and academic institutions is crucial to bridge the gap between scientific research and practical implementation. These partnerships facilitate knowledge transfer, technology transfer, and collaborative problem-solving. Engaging with industry networks, associations, and platforms facilitating knowledge exchange and collaboration among industry peers is also valuable. These networks provide opportunities to learn from best practices, share experiences, and collectively work toward implementing research findings. For instance, a construction industry association can establish a platform where companies share experiences and lessons from implementing energy-efficient building designs and materials based on experimental research findings.

Industries can benefit from expertise in experimental design, data collection techniques, and statistical analysis employed in research. Collaborative efforts can involve regular interaction, knowledge sharing, and access to experimental data. By closely working with researchers, staying up to date with the latest research publications, engaging with research institutions, and participating in conferences or workshops related to emissions reduction, industries can translate scientific findings into actionable strategies and technologies. By actively collaborating with academic and research institutions, conducting pilot studies, establishing partnerships, and engaging with industry networks, industries can effectively adopt and implement scientific research results in the experimental stage to tackle greenhouse gas emissions.

5.4 Suggestions for Local Communities

Local communities can help in emission control by understanding and addressing climate change. They can collaborate with scientists to monitor air quality and greenhouse gas emissions in their area using portable monitoring devices. By collecting data on emissions from local sources such as transportation and industry, communities can work alongside researchers to analyze the data and identify effective strategies for reducing emissions. This collaboration allows communities to contribute valuable data that supplements scientific research and enhances understanding of local emission sources.

In addition to monitoring, communities can actively participate in field trials and pilot projects that test new technologies, practices, or policies to reduce greenhouse gas emissions. By volunteering for these initiatives, communities provide valuable feedback, assess the feasibility and effectiveness of interventions, and help refine them for broader implementation. To directly impact decision-making processes, community representatives can participate in local government meetings or advisory panels focused on climate change and emission reduction. Their input on the feasibility and acceptability of proposed interventions is crucial as they advocate for measures that align with the community's needs and aspirations. Implementing specific initiatives based on scientific research findings can also drive emission reduction. For example, a community can establish a neighborhood composting program based on research demonstrating the benefits of organic waste management in reducing methane emissions. By coordinating waste collection, educating residents on composting practices, and measuring emissions reduction, communities actively contribute to addressing climate change.

Creating living labs focused on sustainability is another approach. These labs allow communities to test new sustainable mobility solutions, such as electric vehicle charging infrastructure, car-sharing programs, and mobility apps. Engaging in co-creation and co-design processes with researchers, policymakers, and other stakeholders is essential as it ensures that local knowledge, needs, and contexts are integrated into the design and implementation of emission reduction strategies, maximizing their effectiveness and relevance to the community. Energy cooperatives offer a way for communities to collectively implement and benefit from experimental research on renewable energy generation, storage, and demand-side management (Clusa et al. 2022). These cooperatives empower community members to take ownership of energy projects and drive emission reduction efforts. Local communities can collaborate with researchers to develop behavior change campaigns to reduce carbon-intensive activities (McKenzie-Mohr et al. 2011). By leveraging research insights on effective messaging, social norms, and incentives, these campaigns can encourage community members to adopt alternative modes of transportation and make sustainable choices.

6 Summary

Throughout the chapter, we explored effective strategies for translating scientific research into action within the context of emission reduction efforts. We emphasized the significance of collaboration, data sharing, and informed decision-making as key factors in driving meaningful change. We also highlighted the need to make scientific research practical and applicable in real-world situations. This entails bridging the gap between theoretical findings and their implementation in everyday scenarios, ensuring that research outcomes can be effectively utilized to address emission-related challenges. By highlighting the successes, we sought to create a focus area where efforts and attention could be directed toward implementing effective emission reduction strategies. To improve the actionableness of emission control research, we provided specific recommendations for scientists, policymakers, industries, and local communities. For scientists, we suggested leveraging advanced technologies, development of real-time emissions monitoring sensors for accurate and timely data collection. Policymakers were urged to play an active role by initiating innovation challenges that incentivize creative solutions to emission reduction. We emphasized the importance of engaging with scientists and researchers to foster collaboration and ensure evidence-informed policymaking. Involving stakeholders is the key to garnering diverse perspectives and fostering broad support for emission control measures. We also advised industries to seek partnerships with academic and research institutions to tap into the expertise and resources available. Investing in technology transfer programs and conducting pilot studies can facilitate the practical implementation of emission reduction strategies. In collaboration with scientists in monitoring and analyzing emissions, local communities can gain valuable insights for targeted interventions. By participating in field trials and pilot projects, community members can actively contribute to the development and evaluation of emission control measures.

References

Ahn, Caroline. 2023. *Indigo Ag's second crop of soil carbon credits grows 5X, validating agriculture as a meaningful solution to climate change*. Indigoag.com (blog). Indigo Ag. February 23, 2023.

Allen, David T., Felipe J. Cardoso-Saldaña, Yosuke Kimura, Qining Chen, Zhanhong Xiang, Daniel Zimmerle, Clay Bell, Chris Lute, Jerry Duggan, and Matthew Harrison. 2022. A Methane Emission Estimation Tool (MEET) for predictions of emissions from upstream oil and gas well sites with fine scale temporal and spatial resolution: Model structure and applications. *Science of The Total Environment* 829: 154277.

Amann, Markus, Jiang Kejun, H.A.O. Jiming, Shuxiao Wang, Zhuang Xing, Deng Yi Xiang, L. Hong, et al. 2008. *GAINS Asia. Scenarios for cost-effective control of air pollution and greenhouse gases in China*. International Institute for Applied Systems Analysis.

American Association for the Advancement of Science. 2023. Combining theory and experiment in electrocatalysis: Insights into materials design. *Science*. May 30, 2023.

Anju, Agrawal, S. Pandey Ravi, and Sharma Bechan. 2010. Water pollution with special reference to pesticide contamination in India. *Journal of Water Resource and Protection* 2: 432.

Baranzini, Andrea, Jeroen C.J.M. Van den Bergh, Stefano Carattini, Richard B. Howarth, Emilio Padilla, and Jordi Roca. 2017. Carbon pricing in climate policy: Seven reasons, complementary instruments, and political economy considerations. *Wiley Interdisciplinary Reviews: Climate Change* 8 (4): e462.

Bbdo, Amv. 2023. Smart Energy GB Quietly Starts a Revolution… Under the Stairs. *BBOnline*, May 30, 2023.

Benjaafar, Saif, Yanzhi Li, and Mark Daskin. 2012. Carbon footprint and the management of supply chains: Insights from simple models. *IEEE transactions on automation science and engineering* 10 (1): 99–116.

Bibra, Ekta Meena, Elizabeth Connelly, Marine Gorner, Christopher Lowans, Leonardo Paoli, Jacopo Tattini, and Jacob Teter. 2021. *Global EV Outlook 2021: Accelerating ambitions despite the pandemic*. International Energy Agency.

Birol, Fatih, and David Malpass. 2021. *It's critical to tackle coal emissions*. International Energy Agency.

Bogdanov, Dmitrii, Ashish Gulagi, Mahdi Fasihi, and Christian Breyer. 2021. Full energy sector transition towards 100% renewable energy supply: Integrating power, heat, transport and industry sectors including desalination. *Applied Energy* 283: 116273.

Bruno, Cynthia. 2022. *How to make cover crops work for you*. Indigoag.com (blog). Indigo Ag. February 16, 2022.

Bui, Mai, Claire S. Adjiman, André Bardow, Edward J. Anthony, Andy Boston, Solomon Brown, Paul S. Fennell, et al. 2018. Carbon capture and storage (CCS): The way forward. *Energy & Environmental Science* 11 (5): 1062–1176.

Cardoso-Saldaña, Felipe J. 2023. Tiered leak detection and repair programs at simulated oil and gas production facilities: Increasing emission reduction by targeting high-emitting sources. *Environmental Science & Technology*.

Center for Climate and Energy Solutions. 2020. *What We Can Do – Center for Climate and Energy Solutions*. March 26, 2020.

Clusa, Ncba. 2022. *The role of energy cooperatives in advancing clean energy*. NCBA CLUSA. August.

Daikin Global. 2023. *Daikin conducts life-cycle assessment of refrigerant reclamation*. Accessed May 29, 2023.

Degot, Charlotte, Sylvain Duranton, Michel Frédeau, and Rich Hutchinson. 2022. *Reduce carbon and costs with the power of AI*. BCG Global. October.

Dilling, Lisa, and Maria Carmen Lemos. 2011. Creating usable science: Opportunities and constraints for climate knowledge use and their implications for science policy. *Global Environmental Change* 21 (2): 680–689.

Driga, Anna Maria, and Athanasios S. Drigas. 2019. Climate change 101: How everyday activities contribute to the ever-growing issue. *International Journal of Recent Contributions from Engineering, Science & IT* 7 (1): 22–31.

Frate, Guido Francesco, Lorenzo Ferrari, and Umberto Desideri. 2021. Energy storage for grid-scale applications: Technology review and economic feasibility analysis. *Renewable Energy* 163: 1754–1772.

Gabrielli, Paolo, Matteo Gazzani, and Marco Mazzotti. 2020. The role of carbon capture and utilization, carbon capture and storage, and biomass to enable a net-zero-CO_2 emissions chemical industry. *Industrial & Engineering Chemistry Research* 59 (15): 7033–7045.

González-Arias, Judith, Francisco M. Baena-Moreno, Laura Pastor-Pérez, Daniel Sebastia-Saez, Luz M. Gallego, Fernández, and T.R. Reina. 2022. Biogas upgrading to biomethane as a local source of renewable energy to power light marine transport: Profitability analysis for the county of Cornwall. *Waste Management* 137: 81–88.

Govaerts, Bram, Nele Verhulst, A. Castellanos-Navarrete, Ken D. Sayre, John Dixon, and L. Dendooven. 2009. Conservation agriculture and soil carbon sequestration: between myth and farmer reality. *Critical Reviews in Plant Science* 28 (3): 97–122.

Greene, David L., Judith M. Greenwald, and Rebecca E. Ciez. 2020. US fuel economy and greenhouse gas standards: What have they achieved and what have we learned? *Energy Policy* 146: 111783.

Head, Brian W. 2016. Toward more "evidence-informed" policy making? *Public Administration Review* 76 (3): 472–484.

Hossain, Sazzad, and Mohammad Maksudur Rahman. 2021. Solar energy prospects in Bangladesh: Target and current status. *Engineering* 13 (8): 322–332.

Houghton, Richard A. (1995). Land-use change and the carbon cycle. Global Change Biology 1 (4): 275–287.

Ice, Javier Colato and Lindsey. 2023. Charging into the Future: The Transition to Electric Vehicles: Beyond the Numbers: U.S. Bureau of Labor Statistics. February 15, 2023.

Irlam, Lawrence. 2017. Global costs of carbon capture and storage. *Global CCS institute* 16.

Isomura, Keisuke. 2017. Development of chemical process for recovering high-quality rare-earth oxides from HV motor magnets. *SAE International Journal of Materials and Manufacturing* 10 (2): 218–225.

John, and John. 2023. *Multiple tips in order to cut electric bill by 75 percent*. ElectricRate – Compare Electric Companies and Find the Lowest Rate. April.

Kemp, Chandler E., and Arvind P. Ravikumar. 2021. New technologies can cost effectively reduce oil and gas methane emissions, but policies will require careful design to establish mitigation equivalence. *Environmental Science & Technology* 55 (13): 9140–9149.

Khan, M. Rezwan, and Sufi Iqbal. 2014. Solar PV-diesel hybrid mini cold storage for rural Bangladesh. In *In 2014 3rd International Conference on the Developments in Renewable Energy Technology (ICDRET)*, 1–3. IEEE.

Kok, Marcel, Bert Metz, Jan Verhagen, and Sascha Van Rooijen. 2008. Integrating development and climate policies: National and international benefits. *Climate Policy* 8 (2): 103–118.

Kuyper, Jonathan, Heike Schroeder, and Björn-Ola Linnér. 2018. The Evolution of the UNFCCC. *Annual Review of Environment and Resources* 43: 343–368.

Kweku, Darkwah Williams, Odum Bismark, Addae Maxwell, Koomson Ato Desmond, Kwakye Benjamin Danso, Ewurabena Asante Oti-Mensah, Asenso Theophilus Quachie, and Buanya Beryl Adormaa. 2018. Greenhouse effect: Greenhouse gases and their impact on global warming. *Journal of Scientific research and reports* 17 (6): 1–9.

Lal, Rattan, and J.M. Kimble. 1997. Conservation tillage for carbon sequestration. *Nutrient cycling in agroecosystems* 49: 243–253.

Liu, Guorui, Jiayu Zhan, Minghui Zheng, Li Li, Chunping Li, Xiaoxu Jiang, Mei Wang, Yuyang Zhao, and Rong Jin. 2015. Field pilot study on emissions, formations and distributions of PCDD/Fs from cement kiln co-processing fly ash from municipal solid waste incinerations. *Journal of Hazardous Materials* 299: 471–478.

Martin-Roberts, Emma, Vivian Scott, Stephanie Flude, R. Gareth Johnson, Stuart Haszeldine, and Stuart Gilfillan. 2021. Carbon capture and storage at the end of a lost decade. *One Earth* 4 (11): 1569–1584.

McDowell, Julie. 2019. Cover crops and carbon sequestration: Benefits to the producer and the planet. *CropWatch. March* 11: 2019. (accessed at 2023, Sep 8). https://cropwatch.unl.edu/2019/cover-crops-and-carbon-sequestration-benefits-producer-and-planet

McKenzie-Mohr, Doug. 2011. *Fostering sustainable behavior: An introduction to community-based social marketing*. New Society Publishers.

Mngomezulu, Nomi. 2023. *Advantages of route optimization software – Dovetail*. Dovetail. February 1, 2023.

Nalini. 2023. *How chemical manufacturers can reduce their carbon footprint?* Deskera Blog. February 28, 2023.

Nikitas, Alexandros, Ioannis Kougias, Elena Alyavina, and Eric Njoya Tchouamou. 2017. How can autonomous and connected vehicles, electromobility, BRT, hyperloop, shared use mobility and mobility-as-a-service shape transport futures for the context of smart cities? *UrbanScience* 1 (4): 36.

Ningthoujam, Rina, Yengkhom Disco Singh, Punuri Jayasekhar Babu, Akriti Tirkey, Srimay Pradhan, and Mrinal Kumar Sarma. 2022. Nanocatalyst in remediating environmental pollutants. *Chemical Physics Impact* 4 (May): 100064.

O'Connor, Fiona M., O. Boucher, N. Gedney, C.D. Jones, G.A. Folberth, R. Coppell, P. Friedlingstein, et al. 2010. Possible role of wetlands, permafrost, and methane hydrates in the methane cycle under future climate change: A review. *Reviews of Geophysics* 48 (4): 4.

Østergaard, Poul Alberg, Neven Duic, Younes Noorollahi, Hrvoje Mikulcic, and Soteris Kalogirou. 2020. Sustainable development using renewable energy technology. *Renewable Energy* 146: 2430–2437.

Penney, Veronica. 2021. Electric Cars Are Better for the Planet – and Often Your Budget, Too. *The New York Times*, January 23, 2021.

Poullikkas, Andreas. 2015. Sustainable options for electric vehicle technologies. *Renewable and Sustainable Energy Reviews* 41: 1277–1287.

Pratt, Gill. 2021. Toyota Europe outlines path to 100% CO_2 reduction by 2035. *Toyota EU. February* 12: 2021.

Preston, Carolyn, Corwyn Bruce, and Michael Monea. 2018. *An update on the integrated CCS project at SaskPower's boundary dam power station*, 21–26. 14th Greenhouse Gas Control Technologies Conference Melbourne.

Qazi, Atika, Fayaz Hussain, Nasrudin A.B.D. Rahim, Glenn Hardaker, Daniyal Alghazzawi, Khaled Shaban, and Khalid Haruna. 2019. Towards sustainable energy: A systematic review of renewable energy sources, technologies, and public opinions. *IEEE access* 7: 63837–63851.

Rahman, MdMomtazur, Imran Khan, David Luke Field, Kuaanan Techato, and Kamal Alameh. 2022. Powering agriculture: Present status, future potential, and challenges of renewable energy applications. *Renewable Energy* 188: 731–749.

Raza, Arshad, Raoof Gholami, Reza Rezaee, Vamegh Rasouli, and Minou Rabiei. 2019. Significant aspects of carbon capture and storage: A review. *Petroleum* 5 (4): 335–340.

Rogelj, Joeri, Michel Den Elzen, Niklas Höhne, Taryn Fransen, Hanna Fekete, Harald Winkler, Fu Roberto Schaeffer, Keywan Riahi Sha, and Malte Meinshausen. 2016. Paris Agreement climate proposals need a boost to keep warming well below 2 C. *Nature* 534 (7609): 631–639.

Rogelj, Joeri, Drew Shindell, Kejun Jiang, Solomone Fifita, Piers Forster, Veronika Ginzburg, Collins Handa, et al. 2018. *Mitigation pathways compatible with 1.5 C in the context of sustainable development*. Intergovernmental Panel on Climate Change.

Rosa, Marika, Paula Ferreira, Martin Jakob, Corinna Klessmann, Migle Grigiene, Thobias Sach, Anna Billerbeck, Gábor Horváth, Gustav Resch, and Franziska Schöniger. 2021. *Technical assistance in realisation of the 5th report on progress of renewable energy in the EU final update report*. Task 1&2.

Runst, Petrik, and Anita Thonipara. 2020. Dosis facit effectum why the size of the carbon tax matters: Evidence from the Swedish residential sector. *Energy Economics* 91: 104898.

Shafi, Uferah, Rafia Mumtaz, José García-Nieto, Syed Ali Hassan, Syed Ali Raza Zaidi, and Naveed Iqbal. 2019. Precision agriculture techniques and practices: From considerations to applications. *Sensors* 19 (17): 3796.

Shakoor, Awais, Saba Shakoor, Abdul Rehman, Fatima Ashraf, Muhammad Abdullah, Sher Muhammad Shahzad, Taimoor Hassan Farooq, et al. 2021. Effect of animal manure, crop type, climate zone, and soil attributes on greenhouse gas emissions from agricultural soils—A global meta-analysis. *Journal of Cleaner Production* 278: 124019.

Shiau, Ching-Shin Norman, Jeremy J. Michalek, and Chris T. Hendrickson. 2009. A structural analysis of vehicle design responses to Corporate Average Fuel Economy policy. *Transportation Research Part A: Policy and Practice* 43 (9-10): 814–828.

Siksnelyte-Butkiene, Indre, Edmundas Kazimieras Zavadskas, and Dalia Streimikiene. 2020. Multi-criteria decision-making (MCDM) for the assessment of renewable energy technologies in a household: A review. *Energies* 13 (5): 1164.

Silva, Ana L., Joana C. Patrício, Armando C. Prata, Damià Barcelò Duarte, and Teresa Rocha-Santos. 2021. An urgent call to think globally and act locally on landfill disposable plastics under and after covid-19 pandemic: Pollution prevention and technological (Bio) remediation solutions. *Chemical Engineering Journal* 426: 131201.

Simionescu, Mihaela, Wadim Strielkowski, and Manuela Tvaronavičienė. 2020. Renewable energy in final energy consumption and income in the EU-28 countries. *Energies* 13 (9): 2280.

Sinha, Rakesh Kumar, and Nitin Dutt Chaturvedi. 2019. A review on carbon emission reduction in industries and planning emission limits. *Renewable and Sustainable Energy Reviews* 114: 109304.

Sintayehu, Dejene W. 2018. Impact of climate change on biodiversity and associated key ecosystem services in Africa: A systematic review. *Ecosystem Health and Sustainability* 4 (9): 225–239.

Skendžić, Sandra, Monika Zovko, Ivana Pajač Živković, Vinko Lešić, and Darija Lemić. 2021. The impact of climate change on agricultural insect pests. *Insects* 12 (5): 440.

Smit, Berend, Ah-Hyung Alissa Park, and Greeshma Gadikota. 2014. The grand challenges in carbon capture, utilization, and storage. *Frontiers in Energy Research*: 55.

Sovacool, Benjamin K., Paula Kivimaa, Sabine Hielscher, and Kirsten Jenkins. 2017. Vulnerability and resistance in the United Kingdom's smart meter transition. *Energy Policy* 109: 767–781.

Steinhilber, Simone. 2016. *Exploring options for the harmonisation of renewable energy support policies in the EU using multi-criteria decision analysis*. Fraunhofer Verlag.

Stéphenne, Karl. 2014. Start-up of world's first commercial post-combustion coal fired CCS project: contribution of Shell Cansolv to SaskPower Boundary Dam ICCS project. *Energy Procedia* 63: 6106–6110.

Tabak, Rachel G., Elaine C. Khoong, David W. Chambers, and Ross C. Brownson. 2012. Bridging research and practice. *American Journal of Preventive Medicine* 43 (3): 337–350.

Tcvetkov, Pavel, Alexey Cherepovitsyn, and Sergey Fedoseev. 2019. Public perception of carbon capture and storage: A state-of-the-art overview. *Heliyon* 5 (12): e02845.

Intergovernmental Panel on Climate Change. 2022. *The Evidence Is Clear: The Time for Action Is Now. We Can Halve Emissions by 2030.*

Thomas, Gary E., and Knut Stamnes. 2002. *Radiative transfer in the atmosphere and ocean*. Cambridge University Press.

Townsend, A.L.E.X., N.A.B.E.E.L.A. Raji, and A.L.E.X. Zapantis. 2020. *The value of carbon capture and storage (CCS)*. Docklands: Global CCS Institute.

Tschörtner, Jenny, Bin Lai, and Jens O. Krömer. 2019. Biophotovoltaics: Green power generation from sunlight and water. *Frontiers in Microbiology* 10 (April).

Vanlandingham, Gary, Torey Silloway, Valerie Chang, Meredith Klein, Darcy White, Amanda Hoey, Elizabeth Davies, et al. 2014. *Evidence-based policymaking*. Pewtrusts.org. 2014.

White, Emily. 2023. *GPS route planning software: How telematics elevates your fleet*. Verizon Connect. May 10, 2023.

Wildbolz, Caroline. 2007. *Life cycle assessment of selected technologies for CO_2 transport and sequestration. Diplom Thesis. Zurich: Swiss Federal Institute of Technology.*

Williamson, Katie, Aven Satre-Meloy, Katie Velasco, and Kevin Green. 2018. *Climate change needs behavior change: Making the case for behavioral solutions to reduce global warming*. Arlington: Rare.

Yan, Yongliang, Tohid N. Borhani, Sai Gokul Subraveti, Kasturi Nagesh Pai, Vinay Prasad, Arvind Rajendran, Paula Nkulikiyinka, et al. 2021. Harnessing the power of machine learning for carbon capture, utilisation, and storage (CCUS) – A state-of-the-art review. *Energy & Environmental Science* 14 (12): 6122–6157.

Yankovitz, David, Robert Kumpf, and Aijaz Hussain. 2022. *Reducing carbon, fueling growth: Lowering emissions in the chemical industry*. Deloitte: Deloitte Insights. June 2, 2022.

Yasaka, Yoshihito, Selim Karkour, Koichi Shobatake, Norihiro Itsubo, and Fumiaki Yakushiji. 2022. Life-cycle assessment of refrigerants for air conditioners considering reclamation and destruction. *Sustainability* 15 (1): 473.

Chapter 5
Actionable Science for Hurricane

Ziheng Sun and Qian Huang

Contents

Z. Sun (✉)
Center for Spatial Information Science and Systems, Department of Geography and Geoinformation Science, George Mason University, Fairfax, VA, USA
e-mail: zsun@gmu.edu

Q. Huang
Eastern Tennessee State University, Johnson City, TN, USA

Z. Sun (ed.), *Actionable Science of Global Environment Change*,
https://doi.org/10.1007/978-3-031-41758-0_5

1 Introduction

Hurricanes are another type of billion-dollar natural hazard and hit our society almost every year. The frequency varies regionally, and climate change is one of the biggest driving factors. For example, Hurricane Dorian, a powerful Category 5 hurricane, struck the Bahamas and the Southeastern United States in August 2019. It was one of the strongest hurricanes ever recorded in the Atlantic, with sustained winds reaching 185 mph (298 km/h) (Mercy Corps 2020). The storm unleashed extreme winds, storm surges, and torrential rainfall, resulting in extensive infrastructure damage, widespread flooding, and loss of life. The islands of Abaco and Grand Bahama were particularly hard hit, with entire communities reduced to rubble. The powerful winds tore off the roof, shattered windows, and toppled walls (Fig. 5.1). The storm surge inundated the entire area, leaving a landscape of debris and devastation. Many families lost not only their homes but also cherished belongings and a sense of security. Power outages were widespread, leaving millions of residents without electricity for days, and disrupting critical services such as communication, transportation, and healthcare. The agricultural sector suffered substantial losses as well. High winds destroyed crops, including citrus groves, and damaged agricultural infrastructure. Regarding real estate, many coastal properties and beachfront homes were severely damaged or destroyed by the storm surge and powerful winds. Structures along the coastline faced flooding, erosion, and structural damage, rendering them uninhabitable or requiring extensive repairs (Shultz et al. 2020). This profoundly impacted property values and the housing market in affected areas.

Fig. 5.1 Widespread destruction in the Bahamas after Hurricane Dorian. (Source: BBC https://www.bbc.com/news/world-latin-america-49553770)

In the state of Florida, the tourism industry, a vital part of local economy, also experienced significant disruptions. The threat of the hurricane and subsequent damage caused cancellations of vacations and trips, leading to a decline in tourist arrivals. The storm's aftermath impacted popular tourist destinations, such as beach resorts, hotels, and amusement parks, including flooding, property damage, and temporary closures. The loss of revenue from reduced tourism activity had economic implications for businesses and individuals employed in the hospitality and tourism sectors. The devastation caused by Hurricane Dorian in Florida resulted in billions of dollars in damages (Marchante 2019). It took months, even years, for affected communities to recover and rebuild, with many residents and businesses facing significant challenges and hardships.

As Hurricane Dorian approached the Bahamas, meteorologists utilized state-of-the-art satellite imagery, hurricane hunter aircraft, and advanced computer models to monitor and predict its path and intensity (Fig. 5.2). The forecasts provided by the National Weather Service (NWS) and National Oceanic and Atmospheric Administration (NOAA) allowed emergency management agencies and residents to take necessary precautions in advance. Advanced weather models accurately predicted Dorian's trajectory, giving authorities time to issue evacuation orders and implement emergency plans. Real-time data from weather satellites and radar systems provided critical information on the storm's movement and intensity, aiding decision-making at every stage.

Scientists collaborated with emergency management agencies to assess the damage and prioritize relief efforts. Aerial imagery and remote sensing technologies were employed to survey the affected areas, enabling scientists to create detailed damage assessment maps. This information guided the distribution of resources and facilitated targeted assistance to the most severely impacted regions. Scientists also analyzed the factors that contributed to Hurricane Dorian's intensity and slow movement. Their research helped improve the understanding of hurricane behavior and the potential influence of climate change on storm characteristics, which can inform future mitigation strategies. Scientists with expertise in disaster response and search and rescue techniques joined forces with emergency response teams to locate and rescue individuals affected by the hurricane. They utilized advanced technologies, such as drones and satellite imagery, to identify the areas with stranded or trapped individuals. They also guided safety protocols and efficient search strategies to maximize the effectiveness of rescue operations.

Researchers specializing in medical and public health fields played a critical role in providing medical support and expertise during the response phase (Powell et al. 2012). They coordinated with healthcare professionals and emergency medical teams to establish temporary medical facilities, triage centers, and field hospitals to treat injured individuals and provide necessary healthcare services. They helped address the health risks and challenges associated with post-hurricane conditions, such as waterborne diseases or exposure-related illnesses. The knowledge in engineering, geospatial analysis, and telecommunications supported the response efforts by providing technical assistance and deploying specialized equipment. They helped set up temporary communication networks, restore critical infrastructure,

Fig. 5.2 Dorian's progression from the Bahamas toward the United States. (Source: NOAA 2019)

and ensure the functionality of emergency response systems. For example, people with knowledge of geographic information systems (GIS) provided mapping and spatial analysis support, helping responders identify accessible routes, establish evacuation centers, and plan efficient resource distribution (Sharma et al. 2016). Moreover, experts in the field of psychology and mental health worked alongside mental health professionals to address the psychological impacts of the disaster, such as trauma, stress, and anxiety (Vernberg et al. 2008). They usually offered counseling services, developed coping strategies, and conducted research to better understand the psychological effects of hurricanes on affected communities, helping to ensure adequate mental health support during the response phase. Regarding aftermath, scientists conducted extensive studies to evaluate the storm's impacts and identify lessons for future resilience. They analyzed data collected from weather stations, ocean buoys, and storm surge gauges to refine forecast models and improve predictions of storm surge heights. The findings from these analyses contribute to ongoing efforts to enhance hurricane forecasting and emergency preparedness.

We introduced during Hurricane Dorian how science played a crucial role in all stages of a major hurricane event. The utilization of advanced technologies, data analysis, and scientific expertise contributed to more accurate forecasts, timely warnings, and informed decision-making. By integrating scientific knowledge into emergency response and recovery efforts, communities can better prepare, mitigate, and recover from devastating storms like Dorian. However, not all scientific research contributes to actions during hurricanes. On the contrary, the majority of research is not utilized or engaged by policymakers, emergency responders, or the public before, during, or after hurricanes. Here are some reasons. Many scientists dedicate their careers to understand the formation, behavior, and track of hurricanes and use sophisticated models and data analysis techniques to predict the path and intensity. While these predictions are vital for preparedness and evacuation efforts, they do not provide direct interventions or solutions to mitigate the impact of hurricanes. Hurricane impacts often manifest after their formation is already known, rendering the information about their formation less critical. Moreover, despite advancements in scientific understanding, scientists have limited control over the occurrence and intensity of hurricanes. The ability of scientists to take direct action to prevent or modify hurricanes is highly restricted. Even though scientists propose several strategies to intervene or modify hurricanes, such as cloud seeding or dispersing chemicals, these activities raise serious ethical concerns and have substantial practical limitations. Altering a hurricane's trajectory or intensity can have unintended consequences and potentially lead to even more significant damage or disrupt natural ecosystems. Then come the uncertainties of scientific models. The inherent complexity of hurricanes, including temperature, humidity, wind patterns, and ocean conditions limits the ability to provide actionable guidance with high certainty. Predicting their behavior with absolute certainty is very hard, and uncertainties in modeling and data inputs can seriously affect the accuracy of predictions and may contribute to wrong conclusions.

Although most scientific research is not in action, their results are important for mitigation and preparedness, like helping decision-makers create building codes, land planning, early warning systems, and evacuation plans. However, these codes need a lot of community decisions and involve not only the science of hurricanes, but also social sciences and the local economy, culture, demography, and religious factors that have even more weight in the decision-making than science, especially in the coastal areas of developing countries. That is another important reason why science is nonactionable in many developing countries. They may have limited financial resources to invest in advanced scientific research, sophisticated monitoring systems, or early warning technologies. This lack of resources can limit the ability to gather accurate and timely data on hurricanes and hinder the development of actionable scientific guidance. They also lack resilient infrastructure, such as hurricane-resistant buildings, robust evacuation plans, or efficient communication networks. Without the necessary infrastructure in place, even if scientific knowledge is available, its application may be hindered, leading to less effective response and recovery efforts. In some communities, traditional knowledge or religious beliefs hold greater weight than scientific information, leading to challenges in disseminating and acting upon scientific recommendations. Demographic factors like high population density, informal settlements, or poverty can pose significant challenges in evacuating and assisting vulnerable populations during hurricanes. They make it difficult to implement scientific recommendations effectively and efficiently. Meanwhile, people with limited access to quality education and information about hurricanes also shadow their understanding and acceptance of scientific knowledge. Without a well-informed population, it can be challenging to communicate and promote actionable measures based on scientific research.

Overall, science has always been important in hurricane preparedness and response. However, not all sciences are equally emphasized and actionable. This chapter will dive into the reasons and come up with some suggestions that could improve the situation and make our efforts in scientific research more acceptable and can easily reach and help the communities being impacted. On the other hand, this chapter will list suggestions for the local community, decision-makers, and first responders on adopting science better and minimizing costs while maximizing the outputs. Besides, we will touch on the potential opportunities lying ahead in the future science research pathways to make hurricanes beneficial for us and convert them from threads to resources for humankind.

2 Emergency Management Phases and Hurricane Operation Network

This section will introduce emergency management phases and the current hurricane action networks in the world and explain current hurricane research purposes, final application goals, realistic limitations, potentials, and relevant successful (or failure) experiences.

2.1 *Emergency Management Phases for Hurricanes*

There are four phases of emergency management in the context of hurricanes: mitigation, preparedness, response, and recovery. These steps are interconnected and often overlap, creating a cycle that is continuously in motion (Fig. 5.3).

Mitigations involve efforts to minimize the impact of hurricanes. Mitigation strategies include developing and enforcing stringent building codes to withstand hurricane force winds, implementing land-use planning measures to avoid building in high-risk areas such as flood zones, and creating natural barriers like mangrove plantations along the coast to protect against storm surges (Godschalk et al. 1989).

Preparedness starts before the hurricane strikes and involves activities designed to build capacity and capability to respond to a hurricane (Wolshon et al. 2005). These activities can include developing, testing, and refining emergency plans, creating and maintaining emergency communication systems, and stockpiling necessary supplies and resources. Preparedness also involves training emergency personnel, establishing evacuation routes and shelters, and educating the public on how to prepare for a hurricane. This phase needs efforts from multiple departments

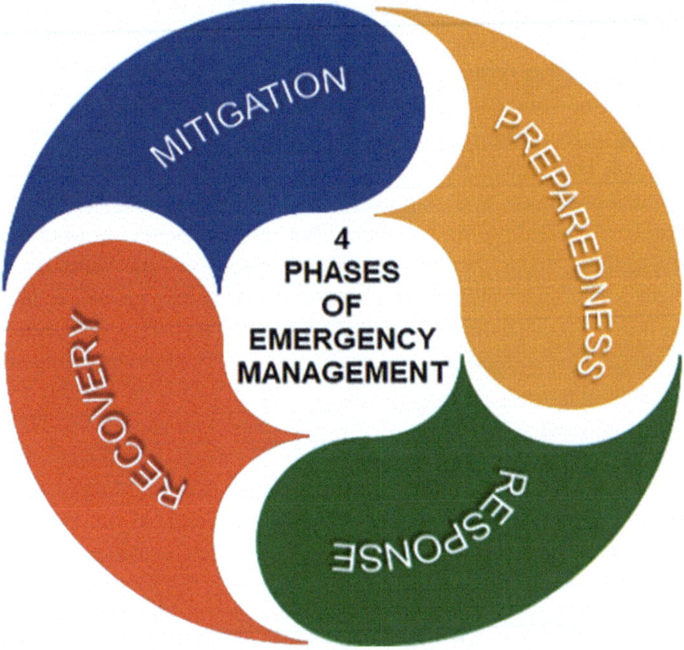

Fig. 5.3 FEMA's four phases of emergency management. (Adams et al. 2022)

at various levels of government such as the Federal Emergency Management Agency (FEMA) and local departments of emergency management, nongovernmental organizations such as the American Red Cross, community, and individuals.

When a hurricane is imminent or has already struck, the response phase starts. This includes actions taken to save lives, protect public health, and prevent further property damage. Activities in this phase include search and rescue, providing emergency medical services, disseminating public information and warnings, and meeting immediate basic needs such as food, water, and shelter (Veenema 2018).

Recovery aims to restore the affected areas to their normal or improved status (Smith and Wenger 2007). This involves repairing or reconstructing damaged infrastructure, providing medical care and counseling services, and rebuilding homes and businesses. This phase can last months to years depending on the extent of the damage and the resources available (Melemis 2015).

2.2 *The Hurricane Management in the United States*

The hurricane network in the United States includes multiple institutions and agencies working collaboratively to ensure effective management and response to hurricanes. They are the crucial operational venue for science to become actionable. Here we list some main organizations.

National Hurricane Center (NHC)

The NHC, a subdivision of the National Weather Service (NWS) and also a part of the National Oceanic and Atmospheric Administration (NOAA), is responsible for monitoring and forecasting hurricanes in the Atlantic and Eastern Pacific Ocean basins (National Hurricane Center 2023). The NHC is located in Miami, Florida. Staffed by meteorologists and scientists working year-round, the NHC provides accurate forecasts, warnings, and analyses that inform emergency planning. Using data from satellites, radar, aircraft, and weather stations, they employ computer models to track cyclones' movement (cyclone is the synonym for hurricane in the Pacific and Indian Ocean), intensity, and potential impacts. The NHC offers various forecast products, including advisories and graphical forecasts detailing storm specifics and potential hazards like storm surge, rainfall, and wind impacts.

State, local, and federal emergency management agencies utilize NHC's forecasts and warnings to develop emergency response plans, coordinate evacuations, and allocate resources for hurricane-impacted areas (Sabbaghtorkan et al. 2020). The NWS forecast offices across the United States receive the NHC's tropical cyclone forecasts and use them to issue local forecasts and warnings specific to their regions. This helps them provide accurate and localized information to the public and emergency responders. Television and radio stations, newspapers, and online news platforms rely on the NHC's updates and forecasts to inform the public about the potential impacts of hurricanes. Airlines, airports, and aviation authorities depend on the NHC's forecasts and advisories to assess the impact of hurricanes on air travel. It helps them make decisions regarding flight cancellations, diversions, and other operational adjustments to ensure passenger safety. Shipping companies, ports, and maritime authorities utilize the NHC's forecasts and advisories to plan vessel movements, reroute ships away from dangerous areas, and ensure the safety of maritime operations during hurricanes. Insurance providers and reinsurers rely on the NHC's forecasts and historical data to assess risks and determine insurance rates and coverage in hurricane-prone regions. Last but not least, the NHC's forecasts, advisories, and educational resources are accessible to the general public. Individuals living in or visiting hurricane-prone areas use the NHC's information to make decisions regarding preparedness, evacuation, and personal safety measures.

Federal Emergency Management Agency (FEMA)

The Federal Emergency Management Agency (FEMA) is a federal agency of the United States Department of Homeland Security (DHS). FEMA is led by an administrator who is appointed by the president of the United States and confirmed by the senate. The administrator is responsible for overseeing the agency's operations and coordinating disaster response and recovery efforts. FEMA's headquarters is located in Washington, D.C., but the agency has regional offices across the United States that are responsible for coordinating disaster response efforts within their respective regions. FEMA employs a workforce including emergency managers, logistics specialists, public affairs officers, analysts, and other professionals with expertise in disaster response and recovery. The exact number of employees varies based on the agency's needs and budget. It receives funding from various sources, including appropriations from the Congress, disaster relief funding, grants, and partnerships with other federal agencies and organizations. The agency's budget is subject to congressional approval and is allocated based on the level of disaster activity and projected needs. FEMA's primary mission is to coordinate and support the nation's response to and recovery from disasters, including hurricanes. The agency's duties include preparing and implementing emergency management plans and policies, coordinating federal assistance to state, local, tribal, and territorial governments, providing financial assistance and resources to affected communities for disaster response, recovery, and mitigation (FEMA 2023). It also conducts disaster response and recovery operations, including logistics, resource coordination, and technical support. They collaborate with other federal agencies, state and local governments, nonprofit organizations, and the private sector to ensure a coordinated and effective response to disasters. They are also actively promoting community preparedness and resilience through public education and outreach programs. FEMA's responsibilities extend beyond hurricanes to include other natural disasters, such as floods, earthquakes, wildfires, and severe storms.

Unlike the NHC, the role of FEMA is more on coordination and resource allocation to provide actual assistance to the impacted areas. When a hurricane is forecasted or imminent, FEMA activates its National Response Coordination Center (NRCC) and Regional Response Coordination Centers (RRCCs) to facilitate coordination among agencies at various levels (FEMA 2020). These centers serve as command and control hubs for managing resources and coordinating response efforts. Once the impact of a hurricane is expected to exceed the capabilities of state and local resources, the president can issue an emergency declaration, which unlocks federal resources and funding to support response and recovery efforts. Ahead of an approaching hurricane, FEMA prepositions resources and assets to expedite the response. This includes deploying personnel, equipment, supplies, and specialized teams to staging areas near the expected impact zone. FEMA also deploys Incident Management Assistance Teams (IMAT) to affected areas to support coordination, resource management, and incident command. IMATs work directly with state and local officials to assess needs, prioritize response efforts, and coordinate the deployment of federal resources. They also provide technical expertise and guidance on various aspects of disaster response. When Hurricane Katrina approached, 11 members of FEMA's Hurricane Liason Team were at the NHC to monitor the storm and storm advisories by August 27, 2005 (Skinner and Hodges 2006). FEMA Emergency Response Teams coordinated with the states to preposition commodities and personnel in the area, activated federal operational staging areas and mobilization centers to accept delivery of commodities and distribute them to affected areas.

FEMA provides individual assistance to homeowners, renters, and businesses affected by hurricanes. This assistance may include grants for temporary housing, home repairs, and other disaster-related expenses not covered by insurance

(Gecowets and Marquis 2008). FEMA's Individuals and Households Program (IHP) helps individuals and families recover and rebuild their lives after a hurricane. FEMA's Public Assistance program provides funding to local governments, as well as certain private nonprofit organizations, to repair, rebuild, or replace damaged public infrastructure and facilities. This includes critical infrastructure such as roads, bridges, schools, hospitals, and utilities. FEMA also participates in long-term recovery and mitigation efforts. The agency provides funding and technical assistance to help communities rebuild in a more resilient manner, incorporating measures to reduce future risks and vulnerabilities. This includes initiatives like hazard mitigation grants, floodplain management, and building code enforcement. For Hurricane Katrina, FEMA had received 1,557,937 registrations for IHP assistance from residents of Alabama, Mississippi, and Louisiana, and it made 1,380,564 applicant referrals for assistance under the housing assistance component and awarded $2,401,735,486 (Skinner and Hodges 2006, p. 13).

United States Geological Survey (USGS)

The USGS is another federal scientific agency with a broad range of capabilities and is responsible for studying and providing reliable information about the nation's natural resources, including hazards like hurricanes (Holmes et al. 2013). The USGS conducts research, monitoring, and assessments related to geology, ecosystems, natural hazards, and climate change. It operates numerous field offices, laboratories, and research centers across the United States. The agency operates a network of streamgages, tide gauges, and other monitoring stations called the National Water Information System (NWIS) that collect data on water levels, streamflow, and coastal changes (USGS 2021). The USGS operates a network of seismometers that detect and monitor earthquake activity. While not directly related to hurricanes, this network helps track seismic events that could be triggered by or have an impact on hurricanes. The USGS collects and maintains satellite imagery (Landsat) that can be used to monitor changes in land cover, coastal erosion, and other impacts of hurricanes (Kriner et al. 2021). The USGS employs scientists, geologists, hydrologists, biologists, and other professionals with expertise in various disciplines relevant to hurricane management. During hurricanes, USGS personnel are involved in data collection, field surveys, and research activities. They assess storm impacts on coastal areas, water resources, ecosystems, and geological features. USGS scientists and researchers conduct ongoing studies and monitoring activities to understand the dynamics of coastal environments, floodplains, and water systems. This helps in assessing vulnerabilities and developing mitigation strategies. They analyze data from various sources to generate flood forecasts, model storm surge scenarios, and assess potential hazards.

USGS data is crucial for understanding the impacts of hurricanes on coastal areas and for issuing flood warnings. The USGS conducts coastal vulnerability assessments to evaluate the susceptibility of coastal areas to erosion, inundation, and other hazards during hurricanes. By analyzing topography, sediment dynamics, and other factors, the agency helps identify areas at risk and provides valuable information for land-use planning and hazard mitigation (USGS 2023). After a hurricane, the USGS conducts rapid assessments to document the extent of coastal changes, including beach erosion, dune erosion, and barrier island movement. The agency uses aerial and satellite imagery, as well as ground-based surveys, to gather data on the impacts of the storm. This information is crucial for understanding the vulnerability of coastal environments and guiding recovery efforts. The USGS conducts research on the hydrological and geological aspects of hurricanes, including rainfall patterns, storm surge dynamics, and the impacts on inland waterways. This research helps improve flood forecasting models, storm surge predictions, and hazard mapping.

National Aeronautics and Space Administration (NASA)

NASA (the National Aeronautics and Space Administration) is a US government agency responsible for the nation's civilian space program and for aeronautics and aerospace research (Dembling 1959; Adams 2018). NASA studies Earth's systems, including the atmosphere, land, oceans, and ice, to better understand our planet's dynamics, climate change, and natural hazards. They collect data using satellites and aircraft, monitor environmental changes, and develop models to improve our understanding of the Earth's processes. It also pioneers new technologies and innovations that have applications both in space exploration and on Earth. They invest in research and development projects to create advanced materials, propulsion systems, robotics, and communication technologies that benefit various industries and scientific disciplines. NASA's duty is to push the boundaries of human knowledge, foster technological advancements, and explore the universe to benefit humanity (Wiles 2013).

NASA's satellites, such as the Aqua, Terra, and Suomi NPP, provide essential data for monitoring hurricanes. These satellites collect information on temperature, humidity, wind patterns, and sea surface conditions, enabling scientists and meteorologists to track the formation, intensification, and movement of hurricanes (Earth Science Data Systems 2021). Scientists at NASA use advanced remote sensing techniques to study hurricanes from space. Instruments like the Advanced Microwave Scanning Radiometer (AMSR) and the CloudSat mission help gather detailed data on precipitation, cloud properties, and the overall structure of hurricanes (Jenner 2009). This information assists in improving hurricane forecasting models and understanding the underlying processes. There are specialized research missions, such as the Hurricane and Severe Storm Sentinel (HS3) and the Cyclone Global Navigation Satellite System (CYGNSS), to study hurricanes up close (Allen 2015). These missions involve flying aircraft equipped with advanced instruments into the vicinity of hurricanes to collect valuable data on their structure, intensity, and environmental conditions. The data gathered helps improve hurricane modeling and forecasting accuracy. NASA collaborates with other agencies and research institutions to develop sophisticated computer models that simulate hurricane behavior. These models incorporate data from NASA's satellite observations and research missions, as well as other sources, to enhance hurricane forecasting and prediction capabilities. Improved models aid in early warning systems, evacuation planning, and emergency response preparedness. In addition to scientific research, NASA also provides support to disaster response efforts by supplying high-resolution satellite imagery and mapping data after a hurricane strikes and helps identify affected areas, assess damages, and support emergency management operations on the ground (Gray 2020). It assists in coordinating relief efforts, identifying critical infrastructure, and aiding in the deployment of resources to affected communities.

Department of Defense (DOD)

As one of the largest government agencies, employing hundreds of thousands of military personnel and civilians, and primarily focused on defense and military operations, the DoD also participates in supporting civilian usage and responding to emergencies (U.S. Department of Defense 2023). It operates across various domains, including land, sea, air, space, and cyberspace. The DoD carries out operations to defend the country, deter potential threats, and support allied nations. In certain situations, the DoD can support civilian agencies and organizations during emergencies, disasters, and other crises. This support is typically coordinated through established frameworks, such as the National Response Framework, where the DoD can contribute resources, expertise, and capabilities to assist in logistics, transportation, communications, and infrastructure support. It operates globally with a vast array of military bases, installations, and facilities. The DoD's funding primarily comes from the US federal budget, allocated by the Congress to cover military operations, personnel, equipment, research, and development (Austin III 2023). The DoD invests in research and development (R&D) to advance military capabilities and technologies. The R&D effort often contributes to scientific and technological advancements with potential civilian applications. It collaborates with academia, industry, and other government agencies to drive innovation in aerospace, materials science, cybersecurity, and more.

In most cases, the DoD supports civilian authorities in responding to hurricanes. They coordinate and share information with other government agencies, including the National Hurricane Center (NHC) and Federal Emergency Management Agency (FEMA). This collaboration helps enhance situational awareness, enables timely decision-making, and facilitates allocating resources where they are most needed (U.S. Department of Transportation 2006). The DoD's U.S. Northern Command (USNORTHCOM) works closely with the NHC and FEMA during hurricane events to coordinate military support, share intelligence, and provide situational updates to support the overall response effort. The DoD's assistance is coordinated through established frameworks to ensure effective collaboration and seamless integration with civilian efforts.

State and Local Emergency Management Agencies

Each state with hurricane history has its own Emergency Management Agency responsible for coordinating emergency preparedness, response, and recovery efforts (Habets et al. 2023). These agencies work closely with local governments, nonprofit organizations, and other stakeholders to ensure a coordinated and effective response to hurricanes. Florida Division of Emergency Management (FDEM) is responsible for coordinating and overseeing emergency management activities in the state of Florida (Florida Division of Emergency Management 2023). They work closely with county emergency management agencies, local governments, and nonprofit organizations to develop hurricane response plans, coordinate resources, and provide guidance during emergencies.

At the county level, emergency management offices are responsible for developing and implementing emergency response plans tailored to their specific region. They work closely with local government departments, nonprofit organizations, and community stakeholders to coordinate preparedness, response, and recovery efforts. For example, the Miami-Dade County Office of Emergency Management is established for coordinating emergency management activities within the county. They work to develop evacuation plans, establish emergency shelters, and coordinate resources during hurricane events (Hristidis et al. 2010).

Several nonprofit organizations play a significant role in hurricane management by providing support and resources during and after hurricanes. These organizations often focus on specific aspects such as disaster response, sheltering, humanitarian aid, and long-term recovery efforts. One typical example is the American Red Cross, a primary nonprofit organization involved in hurricane management. They provide emergency sheltering, distribute essential supplies, offer medical assistance, and support long-term recovery efforts (Red Cross 2022). They work in collaboration with local governments, emergency management agencies, and community partners to deliver their services effectively.

More importantly, dedicated to each community there are local community-based organizations, such as faith-based groups, community centers, and volunteer organizations, which also act as an important part of hurricane management. They provide assistance in sheltering, food distribution, debris cleanup, and volunteer coordination to support the affected communities. One example is the All Hands and Hearts, a nonprofit organization that mobilizes volunteers to assist in disaster response and recovery efforts (All Hands and Hearts 2023). They collaborate with local communities, emergency management agencies, and other organizations to provide immediate relief and help rebuild homes and infrastructure after hurricanes.

When a hurricane hits, local organizations are crucial during hurricanes (Fothergill and Peek 2004). They understand their community's unique needs due to familiarity with the geographical, social, and cultural aspects of the region (Dynes 2006). Their local knowledge aids effective communication, identification of vulnerable populations, and community safety strategies (Tierney 2007). In crises, they are often the first to respond, providing essential services like search and rescue, emergency sheltering, and relief supplies (Perry and Lindell 2003). Their closeness to the community allows quick resource mobilization. Trust and established relationships within the community enable them to raise preparedness awareness and encourage community participation (Dynes 2006). These organizations adapt swiftly to changing situations, tailoring services as needed, such as setting up shelters or assisting vulnerable populations (Quarantelli 1999). Post hurricane, they are vital for long-term recovery, assisting in damage assessment, infrastructure rebuilding, and offering support services (Perry and Lindell 2003). They complement federal agencies by offering extra resources and expertise (Fothergill and Peek 2004). Their deep understanding of their communities improves the effectiveness of recovery efforts (Tierney 2007). They are also key in sharing science-based information and guiding the response, evacuation, and restoration (Quarantelli 1999).

2.3 The Hurricane Management Operation in Other Countries

Hurricanes are a regional phenomenon, and hurricane-prone areas are commonly located in the Atlantic basin, eastern and western Pacific basin, and northern and southern Indian Ocean basins (Landsea et al. 2010). The Atlantic basin includes the Gulf of Mexico, the Caribbean Sea, and the western Atlantic Ocean. Countries such as the United States, Mexico, Central American countries, and Caribbean islands are located in this area. The eastern Pacific basin includes the coastal areas of Mexico, Central America, and the western coast of South America. Western Pacific Ocean, including the Philippines, Japan, China, Taiwan, and other countries in Southeast Asia. Also countries like India, Bangladesh, Sri Lanka, and Myanmar are also prone to tropical cyclones. Southern hemisphere countries such as Australia, Madagascar, Fiji, and New Zealand are also impacted by hurricanes (Knutson et al. 2010).

Similar to FEMA in the United States, the other countries with coastlines and hit seriously have many agencies and organizations specialized to respond to hurricanes. For instance, the National Center for Disaster Prevention (CENAPRED) is responsible for coordinating hurricane emergency response efforts in Mexico. The Mexican Red Cross provides humanitarian aid, medical assistance, and relief services during emergencies (Godwyn-Paulson et al. 2022). The Caribbean Disaster Emergency Management Agency (CDEMA) is responsible for navigating similar efforts among the Caribbean islands (CDEMA 2023). The Japan Meteorological Agency (JMA) monitors and issues warnings for tropical cyclones in Japan. IMD is the equivalent agency in India focusing on the Bay of Bengal and Arabian Sea area. The China Meteorological Administration (CMA) is specialized in monitoring and forecasting tropical cyclones in the Western North Pacific and the South China Sea (Lu et al. 2021). The ASEAN Coordinating Centre for Humanitarian Assistance on Disaster Management (AHA Centre) facilitates cooperation and coordination among western Pacific member states in disaster management. Meanwhile, each Southeast Asian country has its own National Disaster Management Offices (NDMO), responsible for coordinating cyclone response efforts at the national level. In Australia, the Bureau of Meteorology (BOM) conducts weather forecasts, warnings, and cyclone tracking mainly for the Australian region of the Southern Hemisphere (Bureau of Meteorology 2023). New Zealand MetService is responsible for weather forecasting and severe weather warnings for the island country (King et al. 2008).

Generally, the invested resources in these agencies vary depending on the intensity of the storm, the population density in affected areas, infrastructure resilience, preparedness measures, and the ability to respond effectively. The United States is definitely having the most serious damages, for example, the single Hurricane Katrina in 2005 caused damages of $192.5 billion (NCEI 2023). In other countries, the damages are also very severe, mostly in fatality. For example, the Super Typhoon Haiyan in 2013 (Lum and Margesson 2014) stormed over the Philippines and China, caused damages of $12 billion, and 6190 people lost their lives (Reid 2018). Typhoon Fitow in 2013 caused an estimated $10.4 billion in damages in the southern provinces of China (Guilford 2013). Bangladesh is another major country prone to cyclones and suffered substantial economic losses. Cyclone Bhola in 1970 resulted in one of the deadliest tropical cyclones on record, with a death toll estimated between 300,000 and 500,000 (Cerveny et al. 2017).

2.4 *The Opinion of the Impacted Society on Hurricane Research*

Public attitudes toward hurricane research and investment vary globally, influenced by factors such as frequency of hurricanes, economic capacity, and government priorities (Lazo et al. 2009). Countries with a higher frequency of hurricanes and more significant impacts, such as those in hurricane-prone regions, often allocate resources toward hurricane research and infrastructure improvements. For example, the United States has generally widespread support for hurricane research and investment in scientific advancements. Its public recognizes the importance of understanding hurricanes, improving forecasting capabilities, and enhancing response and preparedness measures (Lazo et al. 2015). The US government allocates significant funding toward hurricane research through agencies like the National Oceanic and Atmospheric Administration (NOAA), National Science Foundation (NSF), and NASA (Pielke et al. 2008). Public–private partnerships in the United States also play a role in advancing hurricane research. Despite the outputs of hurricane research, such as improved forecasting models and early warning systems, being generally well received by the public, challenges remain in effectively communicating and translating scientific findings into actionable information for the general population (Morrow et al. 2015).

The level of investment can vary depending on economic capacity and governmental priorities. In developing countries, limited access to quality education and awareness programs on the importance of science-based strategies can contribute to skepticism (Gaillard and Mercer 2013). Lack of scientific literacy and understanding can hinder the acceptance of scientific evidence and recommendations. When a science research wants to roll out to reality, there will always be a mix of reactions. Some individuals may trust and value hurricane science research, recognizing its importance in understanding and predicting the behavior of hurricanes. They may appreciate the efforts of scientists and experts in providing early warnings and guidance to mitigate risks. Others may be skeptical or doubtful of hurricane science research, influenced by misinformation or conspiracy theories. This skepticism can stem from a lack of scientific literacy, conflicting sources of information, or historical incidents where predictions were inaccurate (Lindell and Perry 2012). Some individuals may find it challenging to translate scientific research into actionable measures due to practical considerations. Factors such as limited resources, inadequate infrastructure, and competing priorities may hinder the implementation of recommended strategies. In certain communities, traditional knowledge and practices may hold a significant impact on hurricane responses. People may rely on local wisdom passed down through generations, which can sometimes conflict with or overshadow scientific recommendations (Mercer et al. 2012).

2.5 *Are Scientists and the Impacted Communities on the Same Page?*

The simple answer is no. Scientists and impacted communities are not on the same page or have the same goal most of the time, and their focus and priorities are completely different. One reason is that public outreach and communication with the public are very hard. Effectively communicating complex scientific concepts to the public, policymakers, and other stakeholders can be a challenge. Bridging the gap between scientific research and public understanding requires clear and accessible communication channels, including effective science communication and outreach programs. Implementing science-based strategies requires cooperation and collaboration among multiple stakeholders, including government agencies, policymakers, local communities, and scientific institutions. Challenges such as political agendas, conflicting interests, and bureaucratic processes can hinder the adoption and implementation of scientific recommendations. Adequate resources, including funding, infrastructure, and trained personnel, are essential to translate scientific research into actionable strategies. Being restricted on resources in some hurricane-impacted areas can pose barriers to implementing comprehensive resilience measures.

Another important factor contributing to the disconnection and making science nonactionable is the knowledge gap. Hurricane scientists possess specialized knowledge and expertise in studying and understanding hurricanes. They are trained to analyze data, develop models, and make predictions based on scientific principles. Hurricane scientists often use technical jargon, complex models, and statistical data, which may not be easily understood by nonexperts. This can lead to a lack of clarity and misinterpretation of scientific findings, making it difficult for the public to fully grasp the implications and importance of the research. This knowledge gap between scientists and the public can create a barrier in effectively communicating the complexities of hurricane science.

In addition, individuals living in hurricane-prone areas may have different perceptions of risk and prioritize immediate concerns over long-term preparedness. While scientists focus on understanding and predicting hurricanes to mitigate risks and reduce damages from their labs (most of them never visited those regions and only have perception from the collected data), the public may have more immediate concerns such as evacuation, shelter, and recovery efforts after a hurricane hits. This difference in time perspective and priorities can lead to a disconnect between scientists and the public. Other factors like cultural beliefs, traditions, and historical experiences shape public attitudes and responses to hurricanes. Traditional knowledge and practices are deeply ingrained in the community, leading to skepticism or resistance toward scientific approaches. Additionally, previous experiences with hurricanes, including instances where predictions were inaccurate or insufficient, can contribute to public mistrust or skepticism toward scientific research. Socioeconomic factors such as income disparities, access to resources, and social vulnerability can influence the perceptions and responses of individuals as well. Communities with limited resources and high levels of vulnerability may face challenges in implementing or adhering to scientific recommendations, leading to a divergence in priorities between scientists and the public.

Actually, the difference is quite obvious and sometimes even in the opposite direction on their goals between science and the public. The public, local government, and organizations often focus on immediate needs during and after a hurricane, such as ensuring public safety, providing emergency response, and facilitating recovery efforts. Their primary concern is the well-being and immediate needs of the affected communities. On the other hand, the goal of hurricane scientists is to understand the complex dynamics of hurricanes, improve prediction accuracy, and develop long-term mitigation strategies. This long-term planning perspective may not align with the immediate needs and priorities of the public and local stakeholders. For the sake of better records of hurricanes and better understand the formation and underlying mechanism, the huge storms could be a very valuable opportunity for scientists. Scientists are mainly to develop future strategies to mitigate risks, improve forecasting accuracy, and enhance preparedness, while the primary concern for the affected community during a hurricane is their immediate safety and well-being. Their focus is on evacuation, shelter, and basic needs.

As mentioned previously, a lot of hurricane research is neither actionable nor "action oriented." The community expects immediate practical solutions to protect lives, property, and infrastructure during a hurricane. They value tangible actions and strategies that directly address their immediate needs. Scientists are primarily engaged in scientific research and analysis to enhance understanding of hurricanes. Their focus is on studying the underlying processes, improving models and predictions, and contributing to the body of scientific knowledge. The community's concerns are localized and centered on their immediate environment, including their homes, neighborhoods, and local infrastructure. Their priorities revolve around protecting their community and recovering from the impact of the hurricane, while scientists take a broader perspective, considering the global impact of hurricanes and how they fit within the larger climate system. Scientists emphasize long-term planning and preparedness to minimize the impact of future hurricanes, while the community's focus is often on the immediate aftermath of a hurricane, including cleanup, rebuilding, and recovery. These differences have made the community not understand the priority and importance of science, while the scientists cannot fit their result outputs into a more make-sense position to get used and help in the real world.

2.6 Consequences of Nonactionable Science

Nonactionable science might not be a serious matter from the perspective of the science community as that is not the usual goal that they are desperately pursuing compared to publishing findings, discovering and accumulating knowledge. Making science into operation is not the designated job for scientists. However, from the perspective of the public, it is at least one part of the obligation of scientists, and nonactionable science could hurt the image of science as a whole. Meantime, nonactionable science could have extra consequences, which will further the damages to both science and the impacted communities.

Nonactionable hurricane science can result in insufficient preparedness measures, such as evacuation plans, emergency shelters, and early warning systems. This can leave communities ill-equipped to respond effectively to hurricane threats, leading to delays in evacuation, lack of essential supplies, and increased risks to life and property. If a region lacks accurate and timely hurricane forecasting information, residents may not receive adequate warnings and instructions, which can hamper their ability to take appropriate actions to protect themselves. Also, it can contribute to a higher loss of life and injuries due to inadequate understanding of hurricane dynamics, storm surge risks, or the potential for rapid intensification that can leave people exposed to dangerous conditions. They may underestimate the severity of the storm and fail to take necessary precautions. Without reliable and operable science-based guidelines, communities may lack the necessary information to implement effective mitigation strategies and building codes to withstand hurricane impacts. This can result in extensive property damage and economic losses. It also hampers the ability to implement effective risk reduction measures, like land-use planning, infrastructure improvements, and community resilience programs.

3 State-of-the-Art Hurricane Research

This section will introduce several ongoing research topics and their current progress. It will provide a general overview of the future research outputs regarding hurricanes, and we will discuss how they will be used in action and the benefits and associated risks. The research topics are categorized based on their application stage of hurricanes.

3.1 Hurricane Track Forecasting

The modern technique used for Hurricane Track Forecasting is based on numerical weather prediction (NWP) models that solve the governing equations of fluid dynamics, thermodynamics, and other physical processes to simulate the behavior of the atmosphere. These models utilize complex mathematical equations to simulate the atmosphere and predict the future behavior of hurricanes. The collected data is assimilated into the NWP models to create an initial state of the atmosphere. This initial state serves as the starting point for the model's calculations. To account for uncertainties in the initial conditions and model representation of the atmosphere, ensemble forecasting is often employed. Multiple simulations with slight variations in the initial conditions or model parameters are performed to create an ensemble of forecasts. Based on the simulated atmospheric conditions, the NWP models generate forecasts of the hurricane's track that predict the future positions of the hurricane center at regular time intervals, typically up to several days in advance. These predictions are displayed as tracks on forecast maps (Fig. 5.4).

Willoughby et al. (2007) investigated advancements and recent research progress in improving the accuracy of hurricane track predictions. They find that forecasting in the late twentieth century prevented 66–90% of the hurricane-related deaths in the United States that would have resulted from techniques used in the 1950s. The researchers highlight the use of various modeling techniques and ensemble forecasting approaches to enhance the reliability of track forecasts. They also discuss three main types of models employed in hurricane track forecasting: global numerical weather prediction models, regional atmospheric models, and ensemble forecasting systems. The current accuracy of hurricane track forecasts has significantly improved, with the National Hurricane Center (NHC) producing consistently accurate forecasts. The integration of machine learning also holds promise for improving data analysis and forecast model initialization. There are numerous studies on improving the hurricane track forecasting, especially about the landing spots and the path after hurricanes touchdown. Zhang et al. (2019) used the GOES-R series satellites for improving hurricane forecasting of Hurricane Harvey and demonstrated the value of assimilating all-sky radiances from these satellites into numerical models, leading to enhanced storm analysis and track prediction accuracy. Boussioux et al. (2022) introduce a novel multimodal machine learning framework

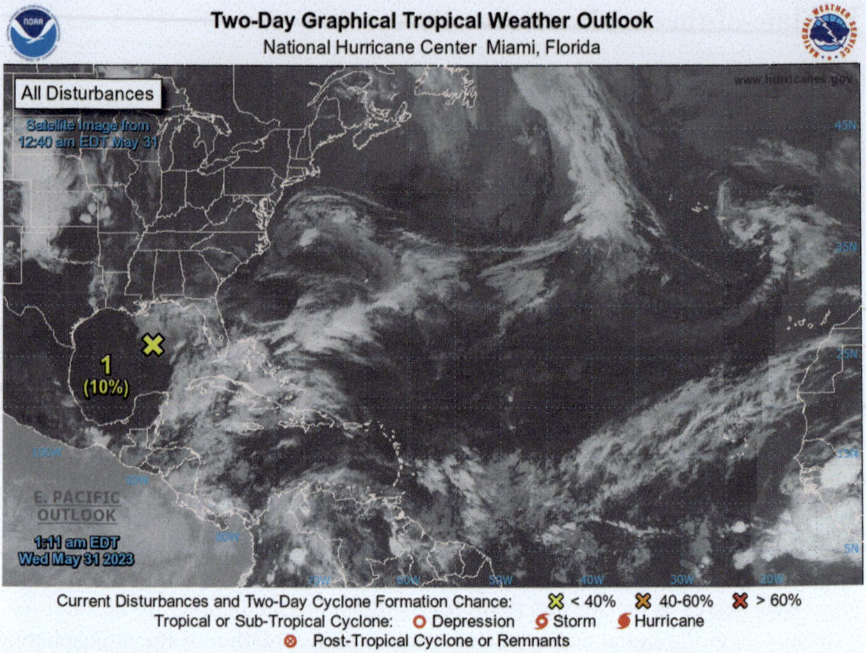

Fig. 5.4 The NHC real-time hurricane forecasting map. (Screenshot taken on May 31 from the NHC website: https://www.nhc.noaa.gov)

that combines various data sources, including meteorological and satellite data, to improve prediction accuracy. Giffard-Roisin et al. (2020) present a deep learning approach for tropical cyclone track forecasting using fused data from aligned reanalysis sources. By leveraging deep neural networks and ensemble methods, the proposed method effectively captures the complex dynamics of cyclone tracks and improves prediction accuracy. Cangialosi et al. (2020) pointed out that over the years the track forecast errors by NHC have significantly decreased, with current average errors reduced by about half since the 1990s and two-thirds since the 1970s. The improvements in track prediction are mainly attributed to advancements in numerical weather prediction (NWP) models, which provide guidance to NHC forecasters. Chen et al. (2019) evaluated the potential of the upcoming United States Next-Generation Global Prediction System (NGGPS) for hurricane prediction using the fvGFS model developed at the Geophysical Fluid Dynamics Laboratory. Retrospective forecasts initialized with the European Centre for Medium-Range Weather Forecasts (ECMWF) data show improved track forecasts compared to the ECMWF operational model for the 2017 Atlantic hurricane season. The fvGFS demonstrates better track forecast accuracy for Hurricane Maria and even lower 5-day track forecast errors for Hurricane Irma, as well as improved intensity prediction compared to both the United States and ECMWF operational models. Kim et al. (2019) proposed a Convolutional LSTM (ConvLSTM)-based spatio-temporal model to track and predict hurricane trajectories using large-scale climate data. They utilize pixel-level spatio-temporal information of tropical cyclones to model time-sequential density maps of hurricane trajectories, capturing both temporal dynamics and spatial distribution. Experimental results using a 20-year record demonstrate that their ConvLSTM-based tracking model outperforms existing methods.

As for the perception of the public about the trajectory forecasting results, Senkbeil et al. (2020) conducted surveys during the evacuations of Hurricanes Isaac (2012), Harvey (2017), and Irma (2017) to examine the perception of track forecasts by evacuees. The findings revealed that evacuees tended to perceive the hurricane tracks as being closer to their homes than what was forecasted and what actually occurred. This perception was consistent across the different hurricanes, and participants with more hurricane experience and those residing in nonmandatory evacuation zones were more likely to perceive the tracks as closer to their homes. The study suggests that evacuees in the United States generally overestimate the proximity of storms to their homes and the associated danger.

3.2 Potential Human and Climate Change Impacts on the Formation and Genesis of Hurricanes

This is an active investigation that attracts a lot of attention. The study by Knutson et al. (2010) examined the influence of climate change on hurricane activity. The researchers used climate models to simulate hurricane behavior under different warming scenarios. The results suggested that as global temperatures rise there is likely to be an increase in the frequency and intensity of hurricanes. Kossin et al. (2014) focused on the relationship between rising sea surface temperatures (SSTs) and hurricane intensification. By analyzing historical hurricane data along with SST records, the researchers found a positive correlation between warmer SSTs and increased hurricane intensity. This study highlighted the role of climate change-induced SSTs in driving stronger hurricanes. In terms of land-use changes, Emanuel (2005) investigated the potential impact of urbanization on hurricane formation. The research demonstrated that the presence of urban areas can disrupt the natural flow of air, altering local wind patterns and potentially affecting the initiation and development of hurricanes. Patricola and Wehner (2018) used climate models to project that future hurricanes are likely to become more intense and have increased rainfall rates due to warming ocean temperatures and changes in atmospheric conditions. This research emphasized the role of anthropogenic climate change in shaping the characteristics of hurricanes. Murakami et al. (2012) explored the impact of global warming on tropical cyclone tracks. The researchers found a poleward shift in the average latitude of cyclone formation and an increase in the frequency of intense cyclones in certain regions. These changes were attributed to the expansion of the tropical belt and changes in large-scale atmospheric circulation patterns. Aerosol pollution and its influence on hurricanes have been studied by Rosenfeld et al. (2008). Their research showed that aerosols can act as cloud condensation nuclei, affecting cloud properties and precipitation processes within hurricanes. This study suggested that human-generated aerosols, such as those from industrial emissions, may contribute to changes in hurricane behavior. Yang et al. (2018) found that high concentrations of aerosols can enhance cloud development and precipitation within hurricanes, potentially influencing their intensity and rainfall patterns. Pant and Cha (2019) examined the climate-dependent hurricane risks in different regions of the United States, considering the potential impact of future climate scenarios across eight locations along the US south and east coast. Three different metrics, including wind speed, annual individual building loss ratio, and regional loss, are used to assess the hurricane risk, and their findings indicate that future hurricane risk is generally higher than present for all locations. Reed et al. (2022) attempted to attribute the hurricane to human-induced climate change, characterized by increasing greenhouse gas emissions, as they contributed to a rise in global average surface temperature and higher sea surface temperatures in the North Atlantic basin during the 2020 hurricane season, particularly for storms of tropical storm and hurricane strength, showcasing the link between climate change and hurricane hazards such as intensity and rainfall. Strauss et al. (2021) tried to quantify the effect of anthropogenic sea level rise on the damages caused by Hurricane Sandy by simulating water levels and damage under different estimates of lower sea levels. The results show that approximately $8.1 billion of Sandy's damages are attributable to climate-mediated anthropogenic sea level rise, along with an extension of the flood area affecting additional people. This approach can be applied to assess the impacts of sea level rise on other past and future coastal storms.

These studies collectively contribute to our understanding of the potential human impacts on the formation and genesis of hurricanes. They highlight the influence of climate change, land-use changes, and aerosol pollution on hurricane dynamics. By integrating findings from these studies, scientists aim to improve hurricane forecasting, risk assessment, and disaster management strategies.

3.3 *Preparing Energy System for Hurricane Damages*

Eskandarpour and Khodaei (2016) present a machine learning-based prediction method for assessing the potential outage of power grid components during an approaching hurricane. The method utilizes logistic regression with a second-order function and parameter fitting to determine a decision boundary that categorizes the components into damaged or operational states. Bennett et al. (2021) developed an energy system optimization model that incorporates hurricane risks, combining storm probabilities with infrastructure fragility curves. They applied the model to Puerto Rico and found that accounting for hurricane trends increased projected electricity costs for 2040 by 32% based on historical frequencies and 82% for increased frequencies. However, transitioning to renewables and natural gas reduced costs and emissions, offering benefits regardless of climate mitigation policies.

Alemazkoor et al. (2020) proposed a transparent and efficient framework to link physics-based projections of hurricane activity under climate change with electric power distribution infrastructure risk models. By integrating these projections into an outage forecast model, they found that long-term power outage risk is primarily driven by the uncertainty in future hurricane frequency. The study highlights the need for robust approaches in managing power systems and setting reliability standards to effectively address climate change impacts. Muhs and Parvania (2019) developed a spatio-temporal hurricane impact analysis (STHIA) tool to quantify hurricane damage to the power grid. The model generates stochastic hurricane scenarios based on historical data and maps outage scenarios to distribution components using fragility curves. The STHIA model helps identify potential risks and can be used for contingency planning and grid-hardening measures.

Each research contribution has varying levels of actionableness, but they all provide valuable insights and tools that can be utilized by stakeholders in the power industry. These findings offer opportunities to improve emergency response, optimize energy systems, plan infrastructure investments, and enhance resilience against hurricane events. For example, Eskandarpour and Khodaei's (2016) machine learning-based prediction method provides actionable insights by categorizing power grid components into damaged or operational states, helping utilities make informed decisions during approaching hurricanes. This approach can guide emergency response and resource allocation. Bennett et al.'s energy system optimization model incorporates hurricane risks and provides valuable information for policymakers. By considering storm probabilities and infrastructure fragility curves, the model offers actionable recommendations such as transitioning to renewables and natural gas, which can reduce costs and emissions, which directly informs energy planning and policy decisions. Alemazkoor's framework linking physics-based projections of hurricane activity with power distribution infrastructure risk models has practical implications for long-term power outage risk management. By quantifying the uncertainty in future hurricane frequency, the study can guide infrastructure investments and resilience strategies. The STHIA tool enables power industry stakeholders to quantify hurricane damage to the power grid and identify potential risks. This tool supports proactive decision-making in preparing for and mitigating hurricane impacts.

However, while the research has significant value, it is important to consider the factors that may hinder their immediate implementation or use in practical applications. For Eskandarpour and Khodaei (2016), there are several challenges in applying this approach in real-time operations, like the availability and quality of data required during an approaching hurricane. Real-time data collection and integration from multiple sources may be complex and resource-intensive, making it difficult to implement this method in operational settings. The prediction model needs to be validated and tested extensively in diverse scenarios to ensure its reliability and generalizability. As for Bennett et al.'s optimization model, its implementation may face practical challenges like the coordination and collaboration among various stakeholders involved in the energy transition process. Transitioning to renewables and natural gas requires significant infrastructure investments and policy support.

Implementation barriers such as regulatory constraints, funding limitations, and public acceptance can impede the adoption of the model's recommendations. The complexity of the energy system and the need for long-term planning also make it challenging to incorporate all relevant factors accurately. Regarding the Alemazkoor framework, one key challenge lies in the availability of high-quality climate projections at the regional or local scale, which may have limitations and uncertainties. Translating climate projections into actionable insights for infrastructure planning requires interdisciplinary collaboration and expertise in both climate science and power system engineering. Implementing such a framework would demand significant data collection, model development, and validation efforts.

3.4 Advancements in Surge and Flood Modeling and Prediction

Surge and flood modeling can estimate and forecast the extent and severity of coastal storm surges and inland flooding associated with weather events such as hurricanes, tropical cyclones, or heavy rainfall. These models combine meteorological data, hydraulic principles, and topographic information to simulate the behavior of water during extreme weather events. These models use numerical simulations to calculate the water elevation and flow patterns along the coast during a storm event and take into account the rainfall patterns, watershed characteristics (such as slope, soil type, and land cover), and the hydrological processes governing the movement of water within the drainage basin. These models can estimate the extent and depth of flooding in different areas, helping to identify flood-prone regions and evaluate potential flood impacts. The model outputs need to be compared with observed data or historical events to assess the accuracy and reliability of the model predictions. Using the validated model allows us to make real-time or future predictions of storm surge or flood hazards based on the current or projected meteorological conditions.

Mulia et al. (2023) proposed a deep learning approach based on generative adversarial networks (GAN) to enhance the accuracy of storm surge modeling during typhoon events. The method translates the outputs of parametric models into more realistic atmospheric forcing fields resembling those produced by numerical weather prediction (NWP) models. The GAN model is trained using historical typhoon events, and its performance is evaluated by simulating storm surges for recent events. The results show that the storm surge model accuracy with GAN-generated forcing is comparable to that of the NWP models and outperforms parametric models. The proposed approach offers a rapid and efficient method for storm forecasting and has the potential to incorporate diverse data sources for further improvements in forecasting accuracy. Thomas et al. (2019) examined the effects of Hurricane Matthew's timing and forward speed on flooding using a simulation model. The interaction between storm surge and tides is found to contribute to spatial variability in water levels, and changes in the storm's forward speed affect the extent and volume of inundation. Yang et al. (2020) presented the rapid forecasting and mapping system (RFMS) to quickly generate accurate coastal inundation maps within a minute based on storm advisories. The system utilizes a storm surge database consisting of high-resolution simulations of 490 optimal storms and employs an efficient interpolation scheme to predict surge responses based on various landfall parameters. The RFMS was applied to southwest Florida and the Florida Panhandle, demonstrating good agreement with observed data and high-resolution simulations. It has the potential for real-time forecasting during hurricanes, mitigation planning, and future flood mapping for coastal resilience. Ayyad et al. (2022) evaluated the Stevens Flood Advisory System (SFAS), an ensemble prediction system, for its performance in forecasting storm tide and resurgence during Tropical Cyclone Isaias in 2020. The system accurately captured the uncertainties associated with the storm surge event, and the super-ensemble spread provided a better estimate of uncertainties than sub-ensembles based on individual meteorological forcing systems. While the central forecast underpredicted peak water levels and the resurgence peak, some ensemble members predicted these events, highlighting the advantages of ensemble averaging. The SFAS demonstrated better accuracy and spread compared to the National Hurricane Center's forecast for this particular storm. Bilskie et al. (2020) optimized the northern Gulf of Mexico (NGOM3) research-grade mesh to create a computationally efficient real-time mesh for a hurricane storm surge early warning system. The real-time mesh, called NGOM-RT, is generated using a mesh decimation scheme that reduces mesh nodes and elements while preserving the representation of the bare-earth topography. Comparisons between NGOM-RT and NGOM3 show virtually no difference in simulated times series and peak water levels for several hurricanes, while the NGOM-RT mesh allows for faster model simulations, being 1.5–2.0 times faster than using NGOM3 on the same number of compute cores. Wing et al. (2019) coupled a continental-scale hydraulic model with forecasts of streamflow, rainfall, and coastal surge height to provide medium-term flood inundation forecasts for Hurricane Harvey. The results show that the hydraulic model successfully captures the benchmark flood extent, with a high level of skill in replicating observations, demonstrating the practical use of fully hydrodynamic approaches in large-scale forecast frameworks for accurate projections of inundation extent. Yang et al. (2020) introduced the Integrated Scenario-based Evacuation (ISE) framework, which integrates coastal flooding, inland flooding, and wind in a comprehensive manner to support evacuation decision-making. Considering inland areas for evacuation can substantially reduce risk without significantly increasing travel times for evacuees, highlighting the importance of treating different hazards together as a system. Also, there are many studies on the aftermath assessment, for example, Lagomasino et al. (2021) investigated the impact of Hurricane Irma on mangrove forests in southwest Florida. Their results find the importance of tidal restoration and hydrological management in vulnerable coastal areas to reduce mangrove mortality and enhance resilience against future cyclones.

However, although the mentioned research are published findings, they are still relatively nonactionable. Mulia's approach may be limited due to the requirement for historical typhoon events for training the GAN model. This reliance on historical data may restrict its effectiveness in predicting storm surges for future events or regions with limited historical data. Thomas's finding could be immediately actionable to inform flood risk assessment and mitigation strategies. Wing study's continental-scale nature may limit its practical applicability at smaller spatial scales or for localized flood events. The resources and data requirements for implementing such a large-scale hydraulic model may also be challenging for operational use. As for the Ayyad study, while the SFAS accurately captures uncertainties associated with storm surges, its immediate applicability may be limited to the specific event studied. The study highlights the advantages of ensemble prediction systems, but further development and validation are necessary before broader implementation.

4 Successful Use Cases

To highlight the actionable science, we investigated several recent hurricanes and attempted to find the example of successful application of science achievements. With this section, we aim to summarize the common places and the key points that are required to make science more applicable in reality.

4.1 Hurricane Sandy (2012)

Prior to Hurricane Sandy's landfall, scientists from the National Hurricane Center, the US Geological Survey, and academic institutions collaborated to develop improved storm surge models. The collaboration focused on enhancing storm surge models, which are sophisticated numerical tools that simulate the effects of hurricanes on coastal areas. By incorporating more accurate data, refining the models' algorithms, and utilizing advanced computational techniques, the scientists were able to improve the accuracy and reliability of the storm surge predictions. These improved models provided valuable insights into the potential magnitude and extent of the storm surge, which was crucial for emergency management agencies and coastal communities to understand the risks and take appropriate actions. The results provided by the improved storm surge models enabled emergency management agencies to issue timely evacuation orders. These models accurately predicted the unprecedented storm surge that would impact the northeast coast of the United States. As a result, emergency management agencies were able to issue timely evacuation orders and coastal communities were better prepared for the storm's impacts.

With a clear understanding of the potential impacts of the storm surge, authorities could make informed decisions about which areas were at the highest risk and needed to evacuate. This helped ensure the safety of residents in vulnerable coastal communities by allowing them to evacuate before the storm made landfall. The work also contributed to better coastal preparedness. The accurate storm surge predictions provided by the improved models allowed coastal communities to prepare for the storm's impacts. They could implement necessary measures such as reinforcing infrastructure, securing property, and establishing emergency response plans. By having a clear understanding of the potential severity of the storm surge, communities were better equipped to mitigate risks and protect lives and property.

4.2 Hurricane Harvey (2017)

During Hurricane Harvey, scientists and engineers from the National Weather Service, the US Army Corps of Engineers, and local authorities successfully utilized advanced rainfall prediction models and real-time monitoring systems to issue accurate flood warnings and manage reservoirs. These efforts helped guide evacuation orders, prioritize rescue operations, and prevent catastrophic flooding in certain areas.

The Texas Water Development Board's Flood Operations Analysis team utilized high-resolution satellite imagery to map and monitor flood extents, which provided critical information for various operations during the response. The accurate mapping of flood extents helped identify areas where people were stranded or in need of assistance. This information aided search and rescue teams in prioritizing their efforts and reaching affected individuals more efficiently. By having a clear understanding of the flood extent, emergency responders and resource management teams were able to allocate personnel, equipment, and supplies more effectively. This allowed for a more targeted and efficient response to the disaster.

4.3 Hurricane Maria (2017)

Following the devastation caused by Hurricane Maria in Puerto Rico, scientists and engineers collaborated to develop advanced techniques for improving the resilience of the power grid. They studied the impacts of Hurricane Maria and analyzed power system failures. Based on the research findings, scientists and engineers worked on designing and implementing the identified measures, designing the integration of renewable energy sources, developing smart grid technologies, and engineering microgrid systems. The design phase involved considering factors such as system compatibility, scalability, cost-effectiveness, and local conditions. Before deployment, the developed techniques underwent rigorous testing and evaluation to ensure their effectiveness and reliability. This involved simulating hurricane scenarios, assessing performance under different conditions, and making necessary refinements. Finally, they help carry out the plan to complete the integration of renewable energy sources, the implementation of smart grid technologies, and the development of microgrid systems.

These efforts aimed to enhance the reliability and rapid restoration of electricity in the event of future hurricanes. This involved installing solar panels and wind turbines in strategic locations to generate clean energy. By diversifying the energy sources, the power grid becomes less reliant on traditional fossil fuel-based generation, which is susceptible to disruptions during hurricanes. Renewable energy sources provide a more sustainable and resilient energy supply. Smart grid technologies were deployed to enhance the monitoring, control, and automation of the power grid. These technologies include advanced sensors, communication systems, and data analytics. By leveraging real-time data and intelligent algorithms, the smart grid enables better situational awareness of the power system, early detection of faults, and efficient management of electricity distribution. This helps in identifying and resolving issues promptly, minimizing power outages, and enabling faster restoration after a hurricane.

Microgrids are localized power systems that can operate independently from the main grid. In the context of improving power grid resilience, microgrid systems were developed to provide electricity to critical facilities and communities during emergencies. These systems integrate renewable energy sources, energy storage technologies (such as batteries), and advanced control systems. By establishing microgrids in key locations, essential services such as hospitals, emergency response centers, and critical infrastructure can continue to operate even if the main grid experiences disruptions. Throughout the process, collaboration between scientists, engineers, utility companies, government agencies, and community stakeholders was crucial. It ensured that the solutions aligned with the specific needs and requirements of Puerto Rico and addressed the challenges faced by the power grid. Engaging stakeholders facilitated knowledge sharing, decision-making, and implementation support. The implementation of advanced techniques to improve the resilience of the power grid in Puerto Rico brought several benefits, making it more resilient in the face of subsequent hurricanes. The new grid system with renewable energy integration and smart grid technologies offers increased flexibility and adaptability. During hurricanes, when conventional power plants might be offline or damaged, renewable energy sources can continue to generate electricity. Smart grid technologies enable better management and optimization of energy resources, allowing for dynamic adjustments in response to changing conditions. This adaptability helps to mitigate the impact of power disruptions and supports the rapid recovery of the grid system.

5 How to Make Hurricane Research More Actionable?

5.1 *Suggestions for Researchers*

After observing the usage of science results in hurricane operations, we conclude several key steps that will increase the actionableness of your research. First, scientists and engineers should conduct a detailed analysis of the specific impacts of the hurricane, such as power system failures, to identify the areas for improvement. Understand stakeholder needs and priorities by conducting interviews, surveys, and workshops to gather input on the challenges they face during hurricanes and the type of information and tools they require. Identify key decision-makers and stakeholders who can influence policy, planning, and implementation. Then foster collaborations with interdisciplinary teams, including meteorologists, hydrologists, engineers, social scientists, and policymakers, to bring diverse expertise and perspectives to the research.

To improve hurricane preparedness, response, and recovery, it is important to establish partnerships with government agencies, research institutions, and nonprofit organizations. These collaborations enable the exchange of expertise and knowledge on a global scale. Open data policies and sharing of relevant datasets, such as historical hurricane data, meteorological observations, remote sensing data, and infrastructure inventories, are crucial for informed decision-making. Standardized data formats and metadata protocols can enhance data sharing and interoperability among stakeholders. The adoption of emerging technologies, including crowd-sourced data, citizen science initiatives, and low-cost sensor networks, can augment existing data collection efforts. By integrating real-time observations, remote sensing data, and high-resolution topographic information into models, the accuracy and resolution of hurricane forecasting, flood modeling, and infrastructure resilience assessments can be improved. Validating and calibrating models using historical hurricane events and field data collected during and after hurricanes, while conducting sensitivity analyses, help understand the uncertainties associated with different model parameters and input data. These steps contribute to more reliable findings and better-informed decision-making processes. By establishing partnerships, promoting open data policies, embracing emerging technologies, and enhancing model accuracy through validation and calibration, the field of hurricane research can be advanced, leading to improved preparedness and response strategies.

At this stage, scientists should continue to translate research findings into practical decision support tools that can be easily understood and utilized by stakeholders, and develop user-friendly interfaces, interactive maps, and visualization tools to communicate complex scientific information to decision-makers and the general public. It is also important to integrate predictive models with real-time monitoring systems to provide timely information on hurricane track, intensity, storm surge, and rainfall, and further customize decision support tools to address specific needs of different user groups, such as emergency managers, infrastructure operators, and community leaders. Once tools and systems are developed and ready for testing, conduct scenario-based analysis to assess the potential impacts of different hurricane scenarios on infrastructure, communities, and ecosystems. Collaborate with stakeholders to identify critical assets, vulnerable populations, and areas prone to flooding, and conduct detailed risk assessments. Quantify the potential economic, social, and environmental impacts of hurricanes to support informed decision-making. Develop probabilistic risk assessment frameworks to capture uncertainties and inform risk mitigation strategies. Meanwhile, engage with policymakers and participate in policy and planning processes related to hurricane preparedness, response, and recovery. Provide expert input and scientific evidence to inform the development of regulations, building codes, and land-use policies that enhance hurricane resilience. Advocate for the integration of scientific knowledge into disaster management frameworks and the allocation of resources for research and implementation. Participate in advisory committees and expert panels to provide technical guidance and support evidence-based decision-making. Eventually, go ahead to communicate research findings through peer-reviewed publications, technical reports, policy briefs, and public presentations. Engage with the media to disseminate accurate and understandable information about hurricanes, their impacts, and preparedness measures. Develop educational materials and conduct workshops to raise awareness among the general public, schools, community organizations, and policymakers. Collaborate with local communities to co-produce knowledge, respecting local knowledge systems and incorporating community perspectives into research and action plans.

5.2 *Suggestions for Local Government and Communities*

To make it smooth to communicate with scientists, the stakeholders also need to do some work. The first thing is to identify local vulnerabilities and needs. They need to conduct a comprehensive assessment of local vulnerabilities to hurricanes, including geographical, social, and infrastructure vulnerabilities, talk to community leaders, local government officials, and relevant stakeholders to understand their specific needs and priorities in relation to hurricane preparedness, response, and recovery. Identify critical infrastructure, high-risk areas, and vulnerable populations to inform targeted actions.

To enhance hurricane preparedness and response at the local level, collaboration between various stakeholders is vital. This includes local government agencies, community organizations, nonprofit groups, academic institutions, and relevant stakeholders. Establishing dedicated local working groups or committees with representatives from different sectors and areas of expertise can effectively address hurricane-related issues. Partnerships with universities, research institutions, and meteorological agencies provide access to scientific expertise and resources. Advocating for local data collection efforts, such as meteorological observations, rainfall measurements, flood gauges, and infrastructure inventories, is essential.

Encouraging transparency and accessibility, stakeholders should share local data among themselves.

The utilization of emerging technologies like community-based monitoring systems and citizen science initiatives can play a significant role in gathering local data. Collaboration with research institutions and meteorological agencies enables access to resources such as hurricane modeling tools, flood maps, and risk assessment frameworks. Working closely with scientists and researchers, it is important to translate scientific knowledge into practical guidance for local decision-making. Leveraging existing resources developed by national or regional agencies, such as storm surge models, flood forecasting systems, and early warning tools, is crucial. Developing and regularly updating localized emergency response plans tailored to hurricanes is necessary. These plans should consider local vulnerabilities, available resources, and community needs. Integrating scientific data, predictive models, and risk assessments into emergency response plans can inform decision-making during hurricane events. Conducting drills and exercises helps test the effectiveness of response plans and identifies areas for improvement. Incorporating scientific knowledge into infrastructure planning and design is essential to enhance resilience against hurricanes. This includes integrating risk assessments and modeling results into land-use planning processes to identify suitable development areas and restrict construction in high-risk zones. Promoting the use of resilient building codes, flood-resistant construction techniques, and nature-based solutions further reduces vulnerability to hurricanes.

Meanwhile, conduct public awareness campaigns to educate the community about hurricane risks, preparedness measures, evacuation routes, and emergency contacts, and organize community workshops, training sessions, and informational meetings to disseminate scientific knowledge about hurricanes and their impacts. Foster partnerships with schools, community organizations, and local media to amplify outreach efforts and engage a wide range of audiences. Establish monitoring systems to track the effectiveness of actions taken based on scientific knowledge. Regularly evaluate the impact of implemented strategies and initiatives, gathering feedback from the community and stakeholders. Adjust approaches as needed based on lessons learned and changing local conditions.

6 Summary

The effective utilization of scientific knowledge is crucial in enhancing preparedness, response, and recovery efforts for hurricanes. By making science actionable, local governments, communities, and stakeholders can improve their ability to mitigate risks, protect lives, and reduce damages caused by these powerful storms. Throughout this chapter, we have explored strategies and key points on how to make science more actionable for hurricanes. The key points to make science more actionable for hurricanes include fostering collaborative partnerships between local government agencies, community organizations, research institutions, and stakeholders to leverage scientific expertise and resources. Enhance local data collection efforts and promote the sharing of data among stakeholders to facilitate informed decision-making. Work with research institutions and meteorological agencies to access scientific tools, models, and resources tailored to local contexts. Develop localized emergency response plans that incorporate scientific data, predictive models, and risk assessments to inform decision-making during hurricane events, and integrate scientific knowledge into infrastructure planning and design, incorporating resilient building codes and nature-based solutions to reduce vulnerability. Also, the science community should conduct public awareness campaigns and community engagement activities to raise awareness about hurricane risks and disseminate knowledge.

In the future, the community should focus on enhancing hurricane track forecasting accuracy and the lead time for warnings to provide more precise and timely information to communities at risk, and invest in further research to improve flood modeling and prediction capabilities, incorporating real-time data and high-resolution modeling techniques. Explore community-driven science initiatives, citizen science, and participatory approaches to engage communities in data collection, monitoring, and decision-making processes. Deepen our understanding of the impacts of climate change on hurricanes and develop strategies to adapt to changing hurricane characteristics and associated risks. By continuing to advance actionable science for hurricanes and addressing these future areas of focus, we can further strengthen our resilience to hurricanes and improve our ability to protect lives, infrastructure, and ecosystems in the face of these powerful natural disasters.

References

Adams, Jennifer. 2018. *What is NASA?* NASA. Brian Dunbar. http://www.nasa.gov/audience/forstudents/5-8/features/nasa-knows/what-is-nasa-58.html.

Adams, J. Hunter, Mark Southard, Steven Forbes, Butch Leinweber, and Daniel Nix. 2022. Effective training improves disaster response. *Opflow* 48 (January): 10–14. https://doi.org/10.1002/opfl.1632.

Alemazkoor, Negin, Benjamin Rachunok, Daniel R. Chavas, Andrea Staid, Arghavan Louhghalam, Roshanak Nateghi, and Mazdak Tootkaboni. 2020. Hurricane-induced power outage risk under climate change is primarily driven by the uncertainty in projections of future hurricane frequency. *Scientific Reports* 10 (1): 15270.

All Hands and Hearts. 2023. *All hands and hearts | Volunteer-powered disaster relief*. https://www.allhandsandhearts.org/.

Allen, Bob. 2015. *Cyclone global navigation satellite system (CYGNSS)*. Text. NASA. May 26, 2015. http://www.nasa.gov/cygnss.

Austin, Lloyd J., III. 2023. *Department of defense releases the President's fiscal year 2024 defense budget*. U.S. Department of Defense. https://www.defense.gov/News/Releases/Release/Article/3326875/department-of-defense-releases-the-presidents-fiscal-year-2024-defense-budget/https%3A%2F%2Fwww.defense.gov%2FNews%2FReleases%2FRelease%2FArticle%2F3326875%2Fdepartment-of-defense-releases-the-presidents-fiscal-year-2024-defense-budget%2F.

Ayyad, Mahmoud, Philip M. Orton, Hoda El Safty, Ziyu Chen, and Muhammad R. Hajj. 2022. Ensemble forecast for storm tide and resurgence from Tropical Cyclone Isaias. *Weather and Climate Extremes* 38: 100504.

Bennett, Jeffrey A., Claire N. Trevisan, Joseph F. DeCarolis, Cecilio Ortiz-García, Marla Pérez-Lugo, Bevin T. Etienne, and Andres F. Clarens. 2021. Extending energy system modelling to include extreme weather risks and application to hurricane events in Puerto Rico. *Nature Energy* 6 (3): 240–249.

Bilskie, Matthew V., Scott C. Hagen, and Stephen C. Medeiros. 2020. Unstructured finite element mesh decimation for real-time Hurricane storm surge forecasting. *Coastal Engineering* 156: 103622.

Boussioux, Léonard, Cynthia Zeng, Théo Guénais, and Dimitris Bertsimas. 2022. Hurricane forecasting: A novel multimodal machine learning framework. *Weather and Forecasting* 37 (6): 817–831.

Bureau of Meteorology. 2023. *Tropical cyclones*. corporateName=Bureau of Meteorology. http://www.bom.gov.au/cyclone/.

Cangialosi, John P., Eric Blake, Mark DeMaria, Andrew Penny, Andrew Latto, Edward Rappaport, and Vijay Tallapragada. 2020. Recent progress in tropical cyclone intensity forecasting at the National Hurricane Center. *Weather and Forecasting* 35 (5): 1913–1922.

CDEMA. 2023. *What is CDEMA? – CDEMA*. https://www.cdema.org/about-us/what-is-cdema.

Cerveny, Randall S., Pierre Bessemoulin, Christopher C. Burt, Mary Ann Cooper, Zhang Cunjie, Ashraf Dewan, Jonathan Finch, et al. 2017. WMO assessment of weather and climate mortality extremes: Lightning, tropical cyclones, tornadoes, and hail. *Weather, Climate, and Society* 9 (3): 487–497. https://doi.org/10.1175/WCAS-D-16-0120.1.

Chen, Jan-Huey, Shian-Jiann Lin, Linus Magnusson, Morris Bender, Xi Chen, Linjiong Zhou, Baoqiang Xiang, Shannon Rees, Matthew Morin, and Lucas Harris. 2019. Advancements in hurricane prediction with NOAA's next-generation forecast system. *Geophysical Research Letters* 46 (8): 4495–4501.

Dembling, Paul G. 1959. National coordination for space exploration-the National Aeronautics and Space Act of 1958. *JAG Journal* 16. https://heinonline.org/HOL/LandingPage?handle=hein.journals/naval1959&div=11&id=&page=

Dynes, Russell. 2006. Social capital: Dealing with community emergencies. *Homeland Security Affairs* 2 (2): 1–26.

Earth Science Data Systems, NASA. 2021. Tropical cyclones. Resource Spotlight. Earthdata. Earth Science Data Systems, NASA. April 23, 2021. https://www.earthdata.nasa.gov/resource-spotlight/tropical-cyclones.

Emanuel, Kerry. 2005. Increasing destructiveness of tropical cyclones over the past 30 years. *Nature* 436 (7051): 686–688.

Eskandarpour, Rozhin, and Amin Khodaei. 2016. Machine learning based power grid outage prediction in response to extreme events. *IEEE Transactions on Power Systems* 32 (4): 3315–3316.

FEMA. 2020. Mission areas and core capabilities | FEMA.Gov. https://www.fema.gov/emergency-managers/national-preparedness/mission-core-capabilities.

———. 2023. Office of Response and Recovery | FEMA.Gov. https://www.fema.gov/about/offices/response-recovery.

Florida Division of Emergency Management. 2023. *Florida division of emergency management | About the division*. https://www.floridadisaster.org/dem/.

Fothergill, Alice, and Lori A. Peek. 2004. Poverty and disasters in the United States: A review of recent sociological findings. *Natural Hazards* 32: 89–110.

Gaillard, Jean-Christophe, and Jessica Mercer. 2013. From knowledge to action: Bridging gaps in disaster risk reduction. *Progress in Human Geography* 37 (1): 93–114.

Gecowets, Gregory A., and Jefferson P. Marquis. 2008. *Applying lessons of hurricane Katrina*. Washington, DC: National Defense University/Institute for National Strategic Studies.

Giffard-Roisin, Sophie, Mo Yang, Guillaume Charpiat, Christina Kumler Bonfanti, Balázs Kégl, and Claire Monteleoni. 2020. Tropical cyclone track forecasting using fused deep learning from aligned reanalysis data. *Frontiers in Big Data* 3: 1.

Godschalk, David R., David J. Brower, and Timothy Beatley. 1989. *Catastrophic coastal storms: Hazard mitigation and development management*. Duke University Press.

Godwyn-Paulson, P., M.P. Jonathan, P.F. Rodríguez-Espinosa, S. Abdul Rahaman, P.D. Roy, G. Muthusankar, and C. Lakshumanan. 2022. Multi-hazard risk assessment of coastal municipalities of Oaxaca, Southwestern Mexico: An index based remote sensing and geospatial technique. *International Journal of Disaster Risk Reduction* 77: 103041.

Gray, Ellen. 2020. *NASA disaster response – NASA prepares for hurricane zeta*. October 28, 2020. https://blogs.nasa.gov/disaster-response/.

Guilford, Gwynn. 2013. The scary reasons why China should spend $11 billion a year on 'climate-proofing.' *Quartz*. October 22, 2013. https://qz.com/137612/the-scary-reasons-why-china-should-spend-11-billion-a-year-on-climate-proofing.

Habets, Margot, Sarah Jackson, Savannah Baker, Qian Huang, Leah Blackwood, Maggie Kemp, and Susan Cutter. 2023. Assessing the quality of state hazard mitigation plans based on hazard, risk, and vulnerability assessments. *Journal of Homeland Security and Emergency Management*. in press.

Holmes, Robert R., Lucile M. Jones, Jeffery C. Eidenshink, Jonathan W. Godt, Stephen H. Kirby, Jeffrey J. Love, Christina A. Neal, et al. 2013. *US geological survey natural hazards science strategy–promoting the safety, security, and economic well-being of the nation*. Reston: US Department of the Interior, US Geological Survey.

Hristidis, Vagelis, Shu-Ching Chen, Tao Li, Steven Luis, and Yi Deng. 2010. Survey of data management and analysis in disaster situations. *Journal of Systems and Software* 83 (10): 1701–1714. https://doi.org/10.1016/j.jss.2010.04.065.

Jenner, Lynn. 2009. NASA – NASA's A-train of satellites 'on track' with hurricane research. *Feature*. Brian Dunbar. https://www.nasa.gov/mission_pages/hurricanes/features/atrain.html.

Kim, Sookyung, Hyojin Kim, Joonseok Lee, Sangwoong Yoon, Samira Ebrahimi Kahou, Karthik Kashinath, and Mr Prabhat. 2019. Deep-hurricane-tracker: Tracking and forecasting extreme climate events. In *2019 IEEE winter conference on applications of computer vision (WACV)*, 1761–1769. IEEE.

King, D.N.T., Arthur Skipper, and Wiremu Bill Tawhai. 2008. Māori environmental knowledge of local weather and climate change in Aotearoa–New Zealand. *Climatic Change* 90 (4): 385–409.

Knutson, Thomas R., John L. McBride, Johnny Chan, Kerry Emanuel, Greg Holland, Chris Landsea, Isaac Held, James P. Kossin, A.K. Srivastava, and Masato Sugi. 2010. Tropical cyclones and climate change. *Nature Geoscience* 3 (3): 157–163.

Kossin, James P., Kerry A. Emanuel, and Gabriel A. Vecchi. 2014. The poleward migration of the location of tropical cyclone maximum intensity. *Nature* 509 (7500): 349–352.

Kriner, Erica, Paige Aldenberg, James Byrne, Brianne Kendall, and Nicholas Roberge. 2021. St. Joseph Peninsula disasters: Using NASA earth observations to investigate land cover, shoreline change, and sediment transport in St. Joseph Peninsula after Hurricane Michael.

Lagomasino, David, Temilola Fatoyinbo, Edward Castañeda-Moya, Bruce D. Cook, Paul M. Montesano, Christopher S.R. Neigh, Lawrence A. Corp, Lesley E. Ott, Selena Chavez, and Douglas C. Morton. 2021. Storm surge and ponding explain mangrove dieback in southwest Florida following Hurricane Irma. *Nature Communications* 12 (1): 4003.

Landsea, Christopher W., Gabriel A. Vecchi, Lennart Bengtsson, and Thomas R. Knutson. 2010. Impact of duration thresholds on Atlantic tropical cyclone counts. *Journal of Climate* 23 (10): 2508–2519.

Lazo, Jeffrey K., Rebecca E. Morss, and Julie L. Demuth. 2009. 300 billion served: Sources, perceptions, uses, and values of weather forecasts. *Bulletin of the American Meteorological Society* 90 (6): 785–798.

Lazo, Jeffrey K., Ann Bostrom, Rebecca E. Morss, Julie L. Demuth, and Heather Lazrus. 2015. Factors affecting hurricane evacuation intentions. *Risk Analysis* 35 (10): 1837–1857. https://doi.org/10.1111/risa.12407.

Lindell, Michael K., and Ronald W. Perry. 2012. The protective action decision model: Theoretical modifications and additional evidence. *Risk Analysis: An Official Publication of the Society for Risk Analysis* 32 (4): 616–632. https://doi.org/10.1111/j.1539-6924.2011.01647.x.

Lu, Xiaoqin, Hui Yu, Ming Ying, Bingke Zhao, Shuai Zhang, Limin Lin, Lina Bai, and Rijin Wan. 2021. Western North Pacific tropical cyclone database created by the China Meteorological Administration. *Advances in Atmospheric Sciences* 38 (4): 690–699. https://doi.org/10.1007/s00376-020-0211-7.

Lum, Thomas, and Rhoda Margesson. 2014. *Typhoon Haiyan (Yolanda): US and international response to Philippines disaster*. Library of Congress, Washington, DC.

Marchante, Michelle. 2019. Florida didn't take a direct hit – But Here's the damage Dorian caused on the coast. *Miami Herald*. September 4, 2019. https://www.miamiherald.com/news/weather/hurricane/article234691682.html.

Melemis, Steven M. 2015. Relapse prevention and the five rules of recovery. *The Yale Journal of Biology and Medicine* 88 (3): 325–332.

Mercer, Jessica, Ilan Kelman, Björn Alfthan, and Tiina Kurvits. 2012. Ecosystem-based adaptation to climate change in Caribbean small Island developing states: Integrating local and external knowledge. *Sustainability* 4 (8): 1908–1932. https://doi.org/10.3390/su4081908.

Mercy Corps. 2020. The facts: Hurricane Dorian's devastating effect on The Bahamas – Bahamas | ReliefWeb. August 15, 2020. https://reliefweb.int/report/bahamas/facts-hurricane-dorian-s-devastating-effect-bahamas.

Morrow, Betty H., Jeffrey K. Lazo, Jamie Rhome, and Jesse Feyen. 2015. Improving storm surge risk communication: Stakeholder perspectives. *Bulletin of the American Meteorological Society* 96 (1): 35–48.

Muhs, John W., and Masood Parvania. 2019. Stochastic spatio-temporal hurricane impact analysis for power grid resilience studies. In *2019 IEEE power & energy society Innovative Smart Grid Technologies Conference (ISGT)*, 1–5. IEEE.

Mulia, Iyan E., Naonori Ueda, Takemasa Miyoshi, Takumu Iwamoto, and Mohammad Heidarzadeh. 2023. A novel deep learning approach for typhoon-induced storm surge modeling through efficient emulation of wind and pressure fields. *Scientific Reports* 13 (1): 7918.

Murakami, Hiroyuki, Yuqing Wang, Hiromasa Yoshimura, Ryo Mizuta, Masato Sugi, Eiki Shindo, Yukimasa Adachi, et al. 2012. Future changes in tropical cyclone activity projected by the new high-resolution MRI-AGCM. *Journal of Climate* 25 (9): 3237–3260.

National Hurricane Center. 2023. *National hurricane center*. Accessed 27 May 2023. https://www.nhc.noaa.gov/.

NCEI. 2023. NOAA National Centers for environmental information (NCEI) U.S. Billion-Dollar weather and climate disasters. https://www.ncei.noaa.gov/access/billions/, https://doi.org/10.25921/stkw-7w73.

Pant, Sami, and Eun Jeong Cha. 2019. Potential changes in hurricane risk profile across the United States coastal regions under climate change scenarios. *Structural Safety* 80: 56–65.

Patricola, Christina M., and Michael F. Wehner. 2018. Anthropogenic influences on major tropical cyclone events. *Nature* 563 (7731): 339–346.

Perry, Ronald W., and Michael K. Lindell. 2003. Preparedness for emergency response: Guidelines for the emergency planning process. *Disasters* 27 (4): 336–350.

Pielke, Roger A., Joel Gratz, Christopher W. Landsea, Douglas Collins, Mark A. Saunders, and Rade Musulin. 2008. Normalized hurricane damage in the United States: 1900–2005. *Natural Hazards Review* 9 (1): 29–42. https://doi.org/10.1061/(ASCE)1527-6988(2008)9:1(29).

Powell, Tia, Dan Hanfling, and Lawrence O. Gostin. 2012. Emergency preparedness and public health: The lessons of hurricane Sandy. *The Journal of the American Medical Association* 308 (24): 2569–2570.

Quarantelli, Enrico Louis. 1999. Disaster related social behavior: Summary of 50 years of research findings.

Red Cross. 2022. *What services does the red cross provide during a disaster?* https://www.redcross.org/about-us/news-and-events/news/2022/what-services-does-the-red-cross-provide-during-a-disaster.html.

Reed, Kevin A., Michael F. Wehner, and Colin M. Zarzycki. 2022. Attribution of 2020 hurricane season extreme rainfall to human-induced climate change. *Nature Communications* 13 (1): 1905.

Reid, Kathryn. 2018. *2013 typhoon Haiyan: Facts, FAQs, and how to help*. World Vision (blog). November 9, 2018. https://www.worldvision.org/disaster-relief-news-stories/2013-typhoon-haiyan-facts.

Rosenfeld, Daniel, Ulrike Lohmann, Graciela B. Raga, Colin D. O'Dowd, Markku Kulmala, Sandro Fuzzi, Anni Reissell, and Meinrat O. Andreae. 2008. Flood or drought: How do aerosols affect precipitation? *Science* 321 (5894): 1309–1313.

Sabbaghtorkan, Monir, Rajan Batta, and Qing He. 2020. Prepositioning of assets and supplies in disaster operations management: Review and research gap identification. *European Journal of Operational Research* 284 (1): 1–19.

Senkbeil, Jason, Jacob Reed, Jennifer Collins, Kimberly Brothers, Michelle Saunders, Walker Skeeter, Emily Cerrito, Saurav Chakraborty, and Amy Polen. 2020. Perceptions of hurricane-track forecasts in the United States. *Weather, Climate, and Society* 12 (1): 15–29.

Sharma, Ashok Kumar, Surya Parkash, and Varun Joshi. 2016. *Geographical information systems for disaster response and management*, 6. National Institute of Disaster Management.

Shultz, James M., Duane E. Sands, James P. Kossin, and Sandro Galea. 2020. Double environmental injustice – Climate change, hurricane Dorian, and The Bahamas. *New England Journal of Medicine* 382 (1): 1–3. https://doi.org/10.1056/NEJMp1912965.

Skinner, Richard L., and Marcia Moxey Hodges. 2006. *A performance review of FEMA's disaster management activities in response to Hurricane Katrina*. Department of Homeland Security, Office of Inspections and Special Reviews.

Smith, Gavin P., and Dennis Wenger. 2007. Sustainable disaster recovery: Operationalizing an existing agenda. In *Handbook of disaster research*, Handbooks of sociology and social research, ed. Havidán Rodríguez, Enrico L. Quarantelli, and Russell R. Dynes, 234–257. New York: Springer. https://doi.org/10.1007/978-0-387-32353-4_14.

Strauss, Benjamin H., Philip M. Orton, Klaus Bittermann, Maya K. Buchanan, Daniel M. Gilford, Robert E. Kopp, Scott Kulp, Chris Massey, Hans de Moel, and Sergey Vinogradov. 2021. Economic damages from Hurricane Sandy attributable to sea level rise caused by anthropogenic climate change. *Nature Communications* 12 (1): 2720.

Thomas, Ajimon, J.C. Dietrich, T.G. Asher, M. Bell, B.O. Blanton, J.H. Copeland, A.T. Cox, C.N. Dawson, J.G. Fleming, and R.A. Luettich. 2019. Influence of storm timing and forward speed on tides and storm surge during Hurricane Matthew. *Ocean Modelling* 137: 1–19.

Tierney, Kathleen J. 2007. From the margins to the mainstream? Disaster research at the crossroads. *Annual Review of Sociology* 33: 503–525.

U.S. Department of Defense. 2023. *United States Department of defense fiscal year 2022 agency financial report*. Home. Accessed 30 May 2023. https://comptroller.defense.gov/.

U.S. Department of Transportation. 2006. *Catastrophic hurricane evacuation plan evaluation: A report to congress*. https://www.fhwa.dot.gov/reports/hurricanevacuation/index.htm.

USGS. 2021. *USGS streamgaging network | U.S. Geological Survey*. https://www.usgs.gov/mission-areas/water-resources/science/usgs-streamgaging-network.

———. 2023. *About | U.S. Geological Survey*. Accessed 29 May 2023. https://www.usgs.gov/special-topics/hurricanes/about.

Veenema, Tener Goodwin. 2018. *Disaster nursing and emergency preparedness*. Springer Publishing Company.

Vernberg, Eric M., Alan M. Steinberg, Anne K. Jacobs, Melissa J. Brymer, Patricia J. Watson, Joy D. Osofsky, Christopher M. Layne, Robert S. Pynoos, and Josef I. Ruzek. 2008. Innovations in disaster mental health: Psychological first aid. *Professional Psychology: Research and Practice* 39 (4): 381.

Wiles, Jennifer. 2013. *Why we explore*. Text. NASA. June 13, 2013. http://www.nasa.gov/exploration/whyweexplore/why_we_explore_main.html.
Willoughby, Hugh E., E.N. Rappaport, and F.D. Marks. 2007. Hurricane forecasting: The state of the art. *Natural Hazards Review* 8 (3): 45–49.
Wing, Oliver E.J., Christopher C. Sampson, Paul D. Bates, Niall Quinn, Andrew M. Smith, and Jeffrey C. Neal. 2019. A flood inundation forecast of Hurricane Harvey using a continental-scale 2D hydrodynamic model. *Journal of Hydrology* X (4): 100039.
Wolshon, Brian, Elba Urbina, Chester Wilmot, and Marc Levitan. 2005. Review of policies and practices for hurricane evacuation. I: Transportation planning, preparedness, and response. *Natural Hazards Review* 6 (3): 129–142. https://doi.org/10.1061/(ASCE)1527-6988(2005)6:3(129).
Yang, Xin, Lijing Zhou, Chuanfeng Zhao, and Jing Yang. 2018. Impact of aerosols on tropical cyclone-induced precipitation over the mainland of China. *Climatic Change* 148: 173–185.
Yang, Kun, Vladimir A. Paramygin, and Y. Peter Sheng. 2020. A rapid forecasting and mapping system of storm surge and coastal flooding. *Weather and Forecasting* 35 (4): 1663–1681.
Zhang, Fuqing, Masashi Minamide, Robert G. Nystrom, Xingchao Chen, Shian-Jian Lin, and Lucas M. Harris. 2019. Improving Harvey forecasts with next-generation weather satellites: Advanced hurricane analysis and prediction with assimilation of GOES-R all-sky radiances. *Bulletin of the American Meteorological Society* 100 (7): 1217–1222.

Chapter 6
Actionable Science for Wildfire

Ziheng Sun

Contents

Z. Sun (✉)
Center for Spatial Information Science and Systems, Department of Geography and Geoinformation Science, George Mason University, Fairfax, VA, USA
e-mail: zsun@gmu.edu

Z. Sun (ed.), *Actionable Science of Global Environment Change*,
https://doi.org/10.1007/978-3-031-41758-0_6

1 Introduction

Wildfires pose significant dangers and have wide-ranging impacts on human life, public safety, and economy. The destruction of forests, residential areas, and infrastructure leads to the serious loss of property and assets (Fearnside 2005). It affects industries such as agriculture, forestry, tourism, and recreation. Smoke pollution can also affect air quality, leading to economic losses in sectors like transportation and tourism (Fig. 6.1). Firefighting efforts and post-fire rehabilitation entail substantial costs for governments and communities (Schumann et al. 2020). The 2018 Camp Fire in California resulted in 85 fatalities and an estimated $16.5 billion in direct economic losses (Iglesias et al. 2022), including the destruction of thousands of homes, businesses, and infrastructure. Direct exposure to flames, heat, and smoke can lead to injuries and fatalities. The inhalation of smoke particles and pollutants can cause respiratory problems and exacerbate preexisting health conditions (Ling and van Eeden 2009). Evacuations and displacement of communities further impact people's well-being and mental health. For example, the 2020 Australian bushfires resulted in the loss of over 30 human lives, widespread injuries, and the displacement of thousands of residents (Filkov et al. 2020). Wildfires can destroy natural habitats, leading to the loss of biodiversity. The combustion of vegetation releases large amounts of carbon dioxide into the atmosphere, contributing to climate change. The loss of vegetation also increases the risk of soil erosion and impacts water quality. The 2019 Amazon rainforest fires caused significant damage to one of the world's most important ecosystems (Arruda et al. 2019), leading to habitat loss

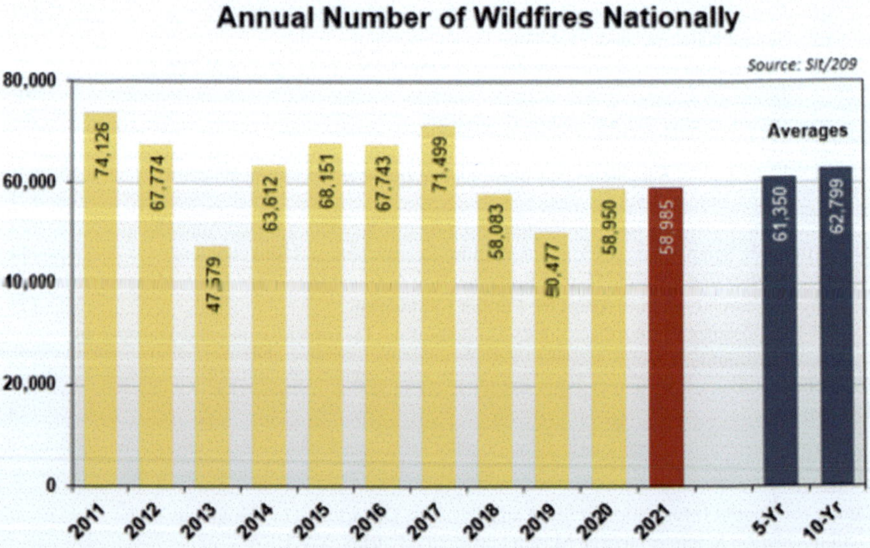

Fig. 6.1 Wildfire trends in the United States. (Image courtesy: https://www.earthdata.nasa.gov/resource-spotlight/wildfires)

for numerous plant and animal species. Wildfires can also damage critical infrastructure and utility systems, such as power lines, communication networks, and water supply facilities (Jahn et al. 2022). Disruptions in these services can have far-reaching consequences, affecting daily life, emergency response capabilities, and the functionality of essential services, such as the 2017 Thomas Fire in California destroying power lines, and resulting in power outages for thousands of residents and impacting communication systems (Kolden and Henson 2019).

Science helps us comprehend the complex behavior of wildfires, including how they ignite, spread, and interact with the environment. This knowledge is essential for developing strategies to predict fire behavior and make informed decisions regarding firefighting tactics, resource allocation, and evacuation measures. Science-based meteorological models and weather monitoring systems also help forecast fire weather conditions, such as high temperatures, low humidity, and strong winds (Rummukainen 2012). Accurate predictions enable early warning systems and provide critical information to fire management agencies, allowing them to prepare and allocate resources effectively (Grasso and Singh 2011). Understanding the ecological role of fire in different ecosystems aids in developing strategies for prescribed burning and ecosystem management. Science-based research contributes to identifying fire-adaptive species, managing invasive plants, and restoring fire-dependent ecosystems (Dennis-Parks 2004). It helps strike a balance between suppressing wildfires for public safety and allowing natural fire regimes to maintain ecosystem health (Moritz et al. 2014). Science also allows for the assessment of fire risk by considering various factors such as vegetation type, fuel moisture content, topography, and historical fire data (Yebra et al. 2013). The findings of scientists can aid in post-fire assessments to evaluate the impact on ecosystems, water quality, and soil erosion and develop strategies for post-fire rehabilitation and restoration of affected areas, including reforestation efforts, erosion control measures, and habitat restoration (Bento-Gonçalves et al. 2012). However, all these need to rely on the collaboration between continued scientific research and fire management agencies and communities to advance our knowledge and improve wildfire control strategies.

Without actionable science, the communities will face increased risks and vulnerabilities, leading to significant negative impacts on their lives. Nonactionable research will result in inadequate understanding of fire behavior and ineffective evacuation planning, which can lead to delays or failures in issuing timely evacuation orders. For example, in the 2018 Camp Fire in California, the lack of actionable research for the power grid companies like PG&E and poor evacuation planning contributed to the loss of 85 lives (Conway 2021). Actionable science can greatly help in developing evidence-based evacuation plans, identifying evacuation zones, determining evacuation timelines, and improving public communication during wildfire events (Seeger et al. 2018). It can enable the identification of high-risk areas and helps prioritize resource allocation for prevention measures, such as fuel management, prescribed burns, and public education campaigns.

However, a lot of wildfire research is not practical for real-world application. They often involve complex scientific models, laboratory experiments, or simulations that may not directly translate into operational strategies for fire management agencies (Sayad et al. 2019). The challenge lies in bridging this gap and effectively communicating research findings to practitioners in a format that is applicable and useful (Enquist et al. 2017). In addition, conducting research takes time, and the timeline for scientific studies may not align with the urgent needs of wildfire management. The dynamic and rapidly evolving nature of wildfires requires immediate decision-making and response, which may not allow for the integration of recent scientific findings into operational practices. On the other hand, wildfires present numerous practical challenges, such as unpredictable weather conditions, rugged terrain, limited resources, and the need for quick decision-making. These constraints can make it challenging to implement certain research findings that require extensive resources, specialized equipment, or ideal conditions that may not be feasible during a wildfire event. Also, the struggle of science to be realistic is always there as wildfire behavior is influenced by a multitude of factors, including weather, topography, fuel conditions, and human factors (Christianson 2014). The complexity and uncertainty associated with wildfire dynamics make it difficult to develop universally applicable and actionable research findings that can be applied across diverse landscapes and fire situations.

This chapter will first examine the current practice in wildfire prevention, responding, and recovery, and find the success and failures of application of scientific results, like the use of fire weather forecasts, fuel management strategies based on ecological research, and the development of fire behavior models that aid in fire suppression efforts (Finney and Cohen 1998). The failures will involve situations where scientific knowledge was not effectively integrated into operational practices, resulting in inadequate fire suppression strategies, evacuation planning, or post-fire recovery efforts. These examples demonstrate how actionable scientific findings have contributed to effective wildfire prevention and response. It will touch on issues like the research-practice gap, time constraints, practical limitations, complexity and uncertainty in wildfire dynamics, and policy or institutional barriers. Eventually give out a list of suggestions for scientists and stakeholders to go forward to work together and improve the actionableness of wildfire research. Emphasizing the importance of conducting applied research in real-world settings can enhance the practicality and relevance of wildfire scientific findings. This could involve conducting experiments, field studies, and simulations that mimic operational conditions and directly address the challenges faced during wildfire prevention, response, and recovery.

2 Current Practice in Wildfire Management

Effective wildfire management requires a comprehensive understanding of the current practices employed. This section looks into the existing approaches and strategies employed by fire management agencies, aiming to evaluate their success and failures in applying scientific research to wildfire management. By examining real-world examples, both successful and failed ones, we can gain insights into the practical application of scientific findings and identify areas for improvement (Thompson et al. 2019). From the utilization of fire weather forecasts to fuel management strategies and fire behavior modeling, this section provides an overview of the current state of wildfire management and sets the foundation for understanding the challenges and opportunities in enhancing the actionableness of wildfire research.

2.1 Emergency Response and Firefighting Tactics

Here we overview the current responding code from initial detection and rapid mobilization to on-the-ground firefighting techniques, such as fire suppression, containment, and perimeter control. The initial step is the timely detection and reporting of wildfires. This can be achieved through various methods, such as lookout towers, aerial surveillance, remote sensing technologies, and public reports. For instance, advanced satellite systems like NASA's Moderate Resolution Imaging Spectroradiometer (MODIS) can detect active fire hotspots, alerting authorities to the presence of wildfires (Giglio et al. 2006). In addition, community members who spot smoke or flames can report them to local fire departments, initiating the emergency response process.

Once a wildfire is reported, incident management teams assess the situation by gathering critical information about the fire's location, size, behavior, and potential threats. This assessment guides the development of an incident action plan, including objectives, strategies, and tactics for managing the fire. For example, fire behavior analysts analyze factors such as fuel conditions, weather patterns, and topography to understand how the fire might spread and develop appropriate response strategies.

With the incident action plan in place, firefighting resources are mobilized to the affected area. This includes firefighters, fire engines, bulldozers, aircraft, and support personnel. For instance, during the devastating 2020 wildfires in California (Keeley and Syphard 2021), resources from local, state, and federal agencies were deployed, including CAL FIRE crews, National Guard units, and specialized firefighting aircraft like air tankers and helicopters (Gagnon 2021). Firefighters employ various tactics to suppress and contain the fire. These tactics involve both direct and indirect approaches. In a direct attack, firefighters engage the fire head-on using hand tools, hoses, and fire retardants. In contrast, indirect attack tactics focus on creating control lines to halt the fire's advance. This can involve constructing firebreaks, removing vegetation, and conducting tactical firing operations to remove fuel. During the Australian bushfire crisis in 2019–2020, firefighters used these tactics to combat the rapidly spreading fires (Ward et al. 2020).

Perimeter control is vital to prevent the fire from spreading beyond predetermined boundaries (Tymstra et al. 2010). Firefighters work to establish and reinforce control lines around the fire's perimeter. This can include clearing vegetation, creating wider firebreaks, and implementing strategic backburning operations. These actions aim to limit the fire's spread and protect communities and valuable assets.

For instance, during the 2021 Oregon Bootleg Fire (Marsavin et al. 2023), firefighters used bulldozers and hand crews to construct control lines and protect nearby communities.

Once the fire is contained, mop-up operations begin. Firefighters carefully extinguish hotspots and smoldering embers along the fire's edge to prevent reignition. Rehabilitation efforts focus on restoring the impacted area by rehabilitating damaged ecosystems, stabilizing soil, and implementing erosion control measures. This ensures that the fire is fully extinguished and minimizes long-term environmental impacts. For instance, after the devastating wildfires in Australia, rehabilitation efforts involved reseeding burnt areas, restoring habitats, and supporting the recovery of affected wildlife.

Throughout the entire firefighting, incident command structures, such as the Incident Command System (ICS) (Chang 2017), facilitate coordination, communication, and decision-making among firefighting agencies, emergency management personnel, and other stakeholders. These structures provide a unified framework for managing the incident and ensure efficient resource allocation and strategic planning. It is important to acknowledge that the effectiveness of emergency response and firefighting tactics depends on various factors, including fire behavior, weather conditions, terrain, available resources, and community preparedness.

2.2 *Wildfire Prevention and Preparedness*

Wildfire prevention and preparedness usually involves assessing factors such as fuel load, vegetation type, weather patterns, and proximity to communities and critical infrastructure (Tymstra et al. 2020). Based on the assessment, wildfire risk zones can be identified, guiding the development of prevention and preparedness strategies. For instance, in high-risk areas, regulations may be implemented to restrict activities that could spark wildfires, such as campfire bans or restrictions on outdoor burning (Barlow and Carlos 2004). Fuel management plays a crucial role in reducing the availability and continuity of combustible materials, including fuel reduction techniques like prescribed burning, mechanical clearing, and vegetation management around structures. Prescribed burning involves controlled fires set under specific conditions to remove accumulated fuels and promote healthier ecosystems (Prichard et al. 2021). Mechanical clearing may involve the use of equipment like mowers, chippers, and chainsaws to create firebreaks and reduce fuel loads. These measures aim to create defensible space and limit the potential for wildfires to spread rapidly.

Deploying early warning systems is comprised of the installation of wildfire detection technologies like cameras, satellite monitoring, and automated weather stations. These systems provide real-time information on fire activity, enabling prompt response and evacuation when necessary. For instance, in Australia, the Victorian Bushfire Information Line and the Country Fire Authority (CFA) use a combination of technologies to monitor fire behavior and issue warnings to affected communities (Teague et al. 2010). A good example system is the Canada Wildland Fire Information System (CWFIS) (Anderson 2005), which is an online platform that provides comprehensive information on wildfires in Canada (Fig. 6.2). It is a collaborative effort between federal, provincial, and territorial fire management agencies to enhance fire management, public safety, and awareness. The system offers various tools and resources for monitoring and reporting wildfires across the country. The FWI System is a key component of CWFIS and assesses the potential behavior of wildfires based on weather conditions. It includes several indices, such as the Fine Fuel Moisture Code (FFMC), Duff Moisture Code (DMC), and Buildup Index (BUI), which help assess fire danger and predict fire behavior. The system provides daily reports on fire weather conditions, fire behavior predictions, and fire danger ratings. These reports help fire management agencies and other stakeholders make informed decisions regarding fire suppression, resource allocation, and public safety. The CWFIS features an interactive fire danger map that displays real-time fire danger ratings across Canada. This map helps users visualize areas of high fire risk and assists in allocating firefighting resources and implementing fire restrictions. The system offers real-time active fire mapping, which provides the location, extent, and intensity of ongoing wildfires across the country. This information is crucial for situational awareness, incident response, and public safety. It has fire perimeter mapping, which provides the boundary of the burned areas for large

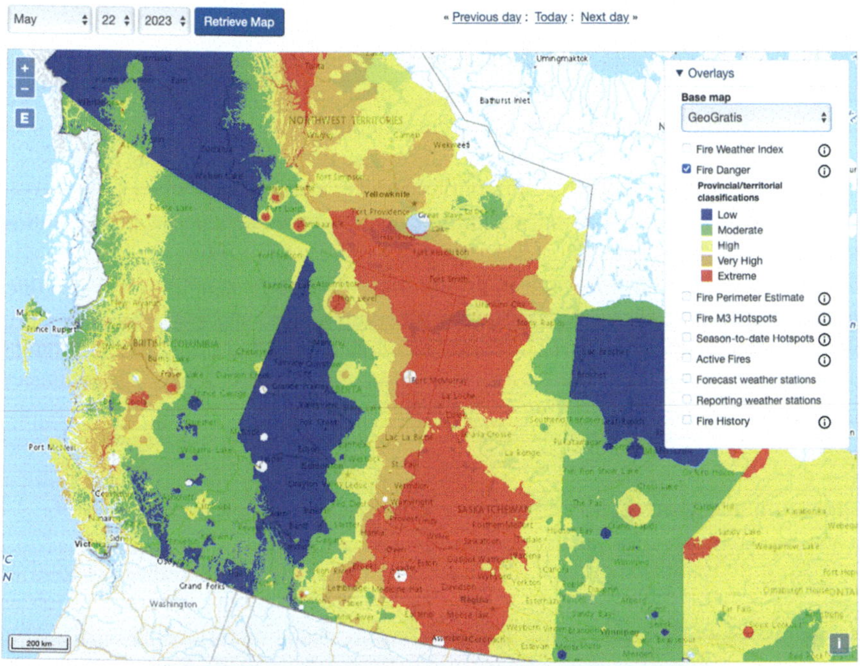

Fig. 6.2 Canada Wildland Fire Information System (CWFIS) fire danger map interface

wildfires. This mapping helps assess the impact of fires, monitor fire progression, and support post-fire analysis and recovery efforts.

Wildfire prevention and preparedness involve fostering collaboration and resource sharing among firefighting agencies and neighboring communities. Mutual aid agreements and partnerships enable the sharing of personnel, equipment, and resources during wildfire incidents. This cooperative approach ensures a coordinated response and enhances the capacity to address large-scale wildfires. An example of mutual aid is the Pacific Northwest Wildfire Compact in the United States, where states collaborate to provide assistance during wildfire emergencies.

Developing robust emergency plans and evacuation procedures is crucial for protecting lives and property. This involves working closely with emergency management agencies, local governments, and communities to establish evacuation routes, assembly points, and communication protocols. Public awareness campaigns and drills help prepare residents to respond effectively during evacuation orders. For instance, during the 2018 Tubbs Fire in California, coordinated evacuation efforts saved lives and facilitated efficient movement of residents to safe locations (Kramer et al. 2019).

2.3 Community Preparedness and Education

Importance of Community Engagement and Involvement

Engaging communities in wildfire prevention and preparedness efforts ensures that residents understand the risks, are equipped with necessary knowledge, and actively participate in mitigation strategies. Recent examples highlight the significance of community involvement. For instance, in the aftermath of the devastating 2019–2020 Australian bushfires, affected communities actively engaged in recovery initiatives, including tree planting, habitat restoration, and community-led fire preparedness workshops. In the United States, community-based organizations like Fire Safe Councils have been instrumental in promoting fire-adaptive communities by organizing educational programs, community clean-up events, and fuel reduction projects (Everett and Fuller 2011). These examples demonstrate how community engagement strengthens resilience, fosters collective responsibility, and enhances the overall effectiveness of wildfire management efforts.

Education and Outreach Programs on Wildfire Prevention and Response

Educating the public about wildfire risks and promoting preparedness is vital as well. Public education campaigns raise awareness about safe practices, such as proper disposal of cigarette butts, the use of fire-resistant materials in construction, and creating defensible space around homes (Weber et al. 2019). Community workshops, informational materials, and interactive websites can provide valuable resources for individuals to understand and mitigate wildfire risks. For example, in fire-prone regions like California, organizations like CAL FIRE conduct outreach programs to educate residents on wildfire prevention and preparedness measures.

Collaborative Approaches Between Scientists, Communities, and Agencies

Collaboration across various parties is essential for leveraging diverse expertise, local knowledge, and shared resources. Real-world examples highlight the success of such collaborations. In California, the UC Berkeley Fire Center partnered with local communities and fire agencies to develop the Firewise Communities program, which engages residents in creating defensible spaces around their properties and implementing fire-resistant landscaping. This collaborative effort has resulted in increased community preparedness and reduced wildfire risk (Smith et al. 2016). Similarly, in Australia, the Bushfire and Natural Hazards Cooperative Research Centre (Sharples et al. 2016) works closely with firefighters, emergency services, and local communities to co-produce research and develop practical solutions. This collaboration has led to improved fire behavior predictions, enhanced early warning systems, and community-led initiatives like the Fireballs in the Sky citizen science project. These examples demonstrate how collaborative approaches foster innovation, build trust, and enhance the resilience of communities in the face of wildfires.

2.4 Post-Fire Recovery and Resilience

This section focuses on the phases following wildfires, where communities and ecosystems work toward recovery and building resilience. We examine the key factors that contribute to successful post-fire recovery, including ecological restoration, community support, and long-term planning. The section emphasizes the importance of integrating science-based approaches into recovery efforts, such as assessing soil health, replanting native vegetation, and implementing erosion control measures. Additionally, it highlights the significance of engaging local communities in decision-making processes to ensure their needs and perspectives are considered. By adopting a holistic and collaborative approach, post-fire recovery and resilience efforts can mitigate the long-lasting impacts of wildfires and foster the restoration of both natural and human systems.

Assessment of Post-Fire Impacts on Ecosystems and Communities

Immediately after a wildfire, emergency response teams conduct an initial assessment to determine the safety of the affected area and identify any immediate threats to households. Once it is safe to enter the impacted area, damage assessment teams, including experts from various fields, conduct detailed surveys to evaluate the extent of damage to individual households. This includes assessing structural damage, loss of personal belongings, and potential hazards such as fallen trees or unstable structures. Concurrently, teams work closely with impacted households to understand their immediate and long-term needs. This involves conducting interviews and surveys to assess the requirements for temporary shelter, food, water, medical assistance, and other essential services. Ecologists and environmental scientists evaluate the impacts of the wildfire on the surrounding ecosystems. This includes studying the loss of vegetation, changes in soil quality, and potential threats to wildlife habitats. Field surveys, remote sensing techniques, and data analysis help in assessing the ecological impacts. Social scientists and community organizations collaborate to assess the social and psychological impacts on affected households. This involves understanding the emotional trauma, displacement, and community disruptions caused by the fire. Surveys, interviews, and focus groups are conducted to gather information and provide support to those affected. The collected data from the assessments are analyzed to generate comprehensive reports. These reports highlight the findings, including the extent of damage, immediate needs of households, and the ecological and social impacts. The reports are crucial in informing policymakers, agencies, and organizations involved in the recovery and rebuilding process.

Real wildfire examples, such as the 2018 California Camp Fire, illustrate how this assessment process plays out. Teams on the ground assessed the damage to individual households, identified immediate needs like shelter and medical assistance, evaluated the ecological impacts on nearby forests and wildlife habitats, and worked closely with communities to understand the social and psychological impacts (Knapp et al. 2021). The gathered information helped in providing targeted support and guiding the recovery efforts to ensure the resilience and well-being of the affected households.

Restoration and Rehabilitation Strategies for the Affected Areas

The post-fire assessment helps in understanding the specific needs and challenges of the affected areas. Based on the assessment, a restoration plan is developed. This plan outlines the goals, objectives, and strategies for restoring the natural ecosystem (Steelman and Burke 2007). It includes actions such as reseeding native plants, implementing erosion control measures, and enhancing wildlife habitats. The plan also considers the resilience and adaptability of the ecosystem in the face of future fire events. Restoring vegetation is a critical aspect of the rehabilitation process. This involves planting native species, including trees, shrubs, and grasses, to stabilize the soil, prevent erosion, and provide habitat for wildlife. Seed collection, nursery propagation, and strategic planting techniques are utilized to ensure successful establishment. Burned areas are prone to erosion, which can further degrade the ecosystem. Soil stabilization techniques, such as mulching, terracing, and erosion control blankets, are implemented to reduce erosion risks and promote soil health (Ahmad et al. 2020). These measures help prevent sedimentation in nearby water bodies and support the recovery of native plant species. Also, efforts are made to restore and enhance wildlife habitats in the fire-affected areas, including creating nesting sites, installing bird boxes, and constructing structures like snag trees to provide shelter and breeding areas for wildlife, which aims to support the recovery of diverse species and promote ecological balance. Throughout the restoration process, ongoing monitoring is conducted to assess the success of implemented strategies and make necessary adjustments. Monitoring includes tracking vegetation regrowth, evaluating soil stability, and assessing wildlife presence. Adaptive management allows for modifications to the restoration plan based on scientific observations and emerging knowledge. Meanwhile, restoration efforts often involve active engagement with local communities, landowners, and stakeholders. This collaboration promotes a sense of ownership and encourages community participation in the restoration process such as volunteer programs, educational initiatives, and partnerships with community organizations to foster long-term stewardship of the restored areas.

Long-Term Planning for Wildfire Resilience

Effective long-term planning requires appropriate land use practices and zoning regulations to reduce wildfire vulnerability such as limiting development in high-risk areas, implementing setbacks and defensible space requirements around structures, and encouraging fire-resistant building materials and designs (Schumann et al. 2020). Zoning ordinances and building codes are updated and enforced to ensure adherence to wildfire resilience guidelines. To reduce the risk of wildfire ignition and spread, the planning needs to emphasize fuel management strategies like implementing controlled burns, mechanical treatments, and vegetation thinning programs to reduce the accumulation of flammable materials, such as dead trees, shrubs, and brush. Strategic fuel breaks are created to interrupt the path of wildfires and provide opportunities for firefighting operations. Critical infrastructure, including power lines, transportation networks, and communication systems, are assessed and modified to enhance their resilience to wildfires. Planning also addresses water availability and accessibility for firefighting purposes. In addition, planning incorporates the establishment and enhancement of early warning systems to provide timely information and alerts to communities and deploy weather monitoring stations, remote sensing technologies, and community notification systems. Monitoring includes tracking changes in fuel loads, evaluating the success of fuel management projects, and assessing the resilience of ecosystems. This information helps guide adaptive management approaches to continuously improve wildfire resilience strategies. Other emergency preparedness efforts like evacuation planning, community drills, and the development of evacuation routes and shelters will also be included. The planning should recognize the dynamic nature of wildfires and the need for ongoing assessment and adaptation. Regular evaluations of planning strategies and practices are conducted to identify successes, challenges, and areas for improvement. This iterative process also allows for the integration of new scientific findings to improve the planning.

2.5 *Challenges and Limitations of Current Practice*

Climate change has contributed to longer and more intense wildfire seasons, making it challenging to manage wildfires effectively (Hessburg et al. 2021). The increased frequency of wildfires puts a strain on resources and makes it difficult to allocate them appropriately. Limited resources can lead to delays in response time, inadequate suppression efforts, and difficulty in implementing prevention and mitigation measures. Many communities in wildfire-prone areas are located in the wildland–urban interface, where homes and structures intermingle with natural vegetation. This poses a significant challenge as it increases the risk to both human lives and property during wildfire events. Challenges such as limited evacuation routes, lack of preparedness among residents, and inadequate communication systems can impede evacuation efforts and put lives at risk. There are also limitations to conducting prescribed burns, including air quality concerns, regulatory barriers, and public acceptance issues. Meanwhile, invasive species and forest health issues can exacerbate wildfire risks. The spread of invasive plants and pests can increase fuel loads and make ecosystems more susceptible to fire. Managing these factors requires long-term strategies and collaboration between various stakeholders. Many individuals residing in wildfire-prone areas lack awareness and understanding of wildfire risks, prevention measures, and evacuation procedures. Insufficient education and outreach efforts can hinder effective preparedness and response during wildfire events (Keim 2008). Besides, fragmented communication systems, jurisdictional challenges, and differing priorities can impede seamless cooperation and hinder response efforts. Liability concerns and legal complexities associated with wildfire management can impact prescribed burning, land management decisions, and insurance coverage, making it challenging to implement effective strategies. Activities like assessing post-fire impacts, securing funding for restoration efforts, and addressing social and economic impacts on affected communities are always complex and resource-intensive processes.

2.6 *Shift Toward a More Proactive and Science-Based Approach*

Advanced technologies and monitoring systems can improve early detection and warning capabilities. This includes the use of satellite imagery, remote sensing, and weather monitoring tools to detect and predict fire behavior accurately (Yuan et al. 2015). Timely and reliable information allows for proactive decision-making and efficient resource allocation. By integrating scientific data and community input, decision-makers can develop targeted strategies for fuel management, land use planning, and infrastructure protection. Implementing proactive fuel management practices, including prescribed burning, can reduce fuel loads and create defensible spaces around communities. By strategically conducting controlled burns during favorable conditions, the risk of uncontrolled wildfires can be minimized. Collaboration with stakeholders, including landowners, agencies, and communities, is essential to address concerns, increase acceptance, and expand prescribed burning efforts. Continual investment in research and innovation is vital for advancing wildfire management practices. This includes studying fire behavior, climate change impacts, ecosystem resilience, and technological advancements. By integrating the latest scientific findings into management strategies, decision-makers can adapt and improve their approaches over time. Updating policies and governance frameworks to align with a proactive and science-based approach is crucial. This involves incentivizing and supporting proactive measures, such as prescribed burning and fuel management, through regulatory reforms, funding mechanisms, and insurance incentives. It also requires integrating climate change considerations and long-term planning into land and resource management policies. However, adequate and sustained funding is essential to support research, infrastructure development, community programs, and firefighting resources. Securing long-term funding commitments from government agencies and exploring innovative funding models can ensure the continuity of proactive wildfire management efforts.

3 Advanced Research for Wildfire

Through innovative technologies, such as satellite imagery, remote sensing, and computer modeling, scientists can accurately monitor fire behavior, predict fire spread, and assess fire risk. Recent research efforts focus on studying the impacts of climate change on wildfire frequency and severity, exploring new firefighting techniques, and developing proactive strategies for prevention and mitigation. This section will introduce some cutting-edge topics in this type of research.

3.1 Remote Sensing Research for Early Detection and Monitoring

Barmpoutis et al. (2020a) provide an overview of optical remote sensing technologies used in early fire warning systems, focusing on flame and smoke detection algorithms. It categorizes the systems into terrestrial, airborne, and spaceborne-based, and discusses the strengths and weaknesses of optical remote sensing for fire detection. The findings aim to contribute to future research projects and the development of improved early warning fire systems. Xu and Xu (2017) explore the use of the geostationary Himawari-8 satellite to generate real-time information about ongoing wildfires in Australia. The satellite's high-temporal-resolution multispectral imagery allows for large-scale monitoring and detection of wildfires. The case study of the 2015 Esperance wildfire demonstrates the satellite's effectiveness in detecting wildfires, even in the presence of smoke and moderate cloud cover. It also enables the real-time monitoring of fire spread rates and directions, offering potential for automated detection of abnormal fire behavior. Yuan et al. (2017) present a novel forest fire detection method using unmanned aerial vehicles (UAVs) equipped with vision-based systems. The method combines color and motion features to identify fire candidate regions in images captured by the UAV's camera. A color-based fire detection algorithm extracts fire-colored pixels, while two types of optical flow algorithms compute motion vectors of the fire candidate regions. Experimental results demonstrate the effectiveness of the proposed method in accurately extracting and tracking fire pixels in aerial video sequences, improving forest fire detection accuracy while minimizing false alarms. Hua and Shao (2017) provide an overview of forest fire monitoring (FFM) using satellite- and drone-mounted infrared remote sensing (IRRS). The review encompasses different IRRS algorithms, with a focus on spatial contextual methods that can be applied using commonly available satellite data. Medium-resolution IRRS data and specific algorithms are identified as effective tools for landscape-scale monitoring and early warning of forest fires. Sherstjuk et al. (2018) present a fire monitoring and detection system for tactical forest firefighting operations utilizing unmanned aerial vehicles (UAVs), remote sensing, and image processing. Çolak and Sunar (2020) analyzed fire risk in the Menderes region, İzmir, Turkey, using remote sensing technology by integrating pre-fire remote sensing data with ancillary data in GIS, with which the spatial and temporal patterns of forest fire risk were evaluated. Land surface temperature (LST) changes and in situ meteorological measurements were used to assess the rapid fire risk, and a linear model incorporating six fire risk variables was applied to generate a fire risk map. The model was validated by overlaying historical forest fire data on the fire risk map, demonstrating its effectiveness in identifying high- and moderate-high-risk areas. Lee et al. (2017) revealed that traditional methods of wildfire monitoring, such as manned airplanes and satellite images, have limitations in terms of cost, temporal resolution, and spatial resolution. To address these challenges, a wildfire detection system utilizing unmanned aerial vehicles (UAVs) and deep convolutional neural networks (CNNs) was developed, providing cost-effective, high-resolution images for early wildfire detection. The system demonstrated high accuracy across a wide range of aerial photographs, enabling more effective wildfire monitoring and response efforts.

The Fire Information for Resource Management System (FIRMS) (Fig. 6.3) is a comprehensive online tool developed by NASA that provides valuable information and real-time monitoring of wildfires worldwide (Davies et al. 2008). FIRMS utilizes satellite data to detect and track active fires, providing users with up-to-date information on fire locations, intensities, and associated data such as fire radiative power and thermal anomalies. The system integrates data from various satellite sensors, including MODIS and VIIRS (Riggs et al. 2017), to provide a comprehensive and accurate picture of wildfire activity. It offers a user-friendly interface that allows users to access fire information through an interactive map. The map displays fire hotspots and allows users to zoom in and obtain detailed information about specific fires. Additionally, FIRMS provides data on fire emissions, smoke plumes, and

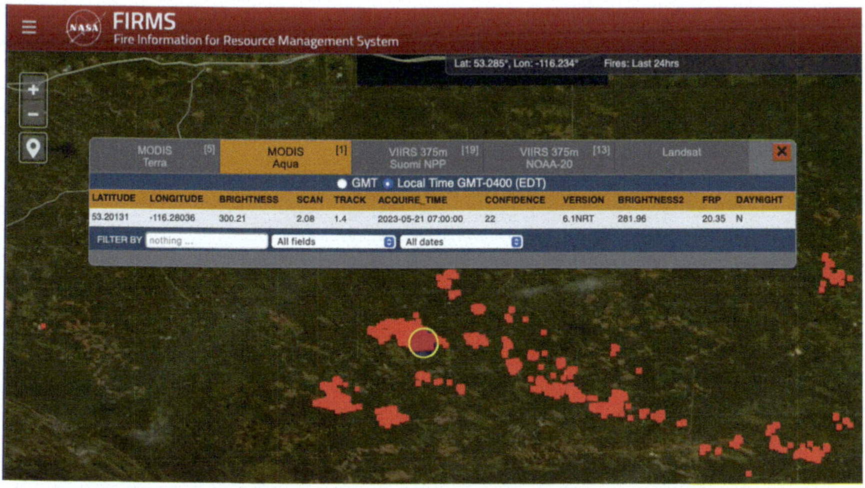

Fig. 6.3 FIRMS interface (NASA Fire Information for Resource Management System) – MODIS and VIIRS active fire/thermal anomaly data may be from fire, hot smoke, agriculture, or other sources

other fire-related parameters, enabling scientists, emergency responders, and land managers to assess the impact of wildfires on the environment, air quality, and human health. The data provided by FIRMS are essential for wildfire management and resource allocation. The system enables early detection of wildfires, facilitating rapid response and firefighting efforts. It helps authorities identify high-risk areas, monitor fire behavior, and make informed decisions regarding evacuation orders and resource deployment. Moreover, FIRMS aids in post-fire analysis and recovery efforts by providing historical fire data and assessing fire severity.

Another major fire information platform from NASA, Worldview, offers access to a wide range of satellite imagery, including fire data, for monitoring and analyzing wildfires across the globe (Fig. 6.4). It also provides a user-friendly interface that allows users to visualize and analyze fire-related information in near real time. The fire data available in Worldview is derived from the similar satellite sensors like MODIS and VIIRS, which capture thermal signatures and detect active fires. Worldview allows users to monitor the location, extent, and intensity of active fires. This information is crucial for assessing fire behavior, identifying areas at risk, and monitoring the progression of fire events over time. It enables fire managers, emergency responders, and land management agencies to make informed decisions regarding fire suppression efforts, resource allocation, and evacuation strategies. During and after wildfire events, Worldview's fire data can aid in disaster response and recovery efforts. It helps assess the extent of fire-affected areas, track fire perimeter growth, and identify areas of high severity. This information assists in evaluating the damage caused by wildfires, assessing infrastructure vulnerability, and prioritizing post-fire recovery and rehabilitation activities. It empowers decision-makers, researchers, and the general public with timely and comprehensive

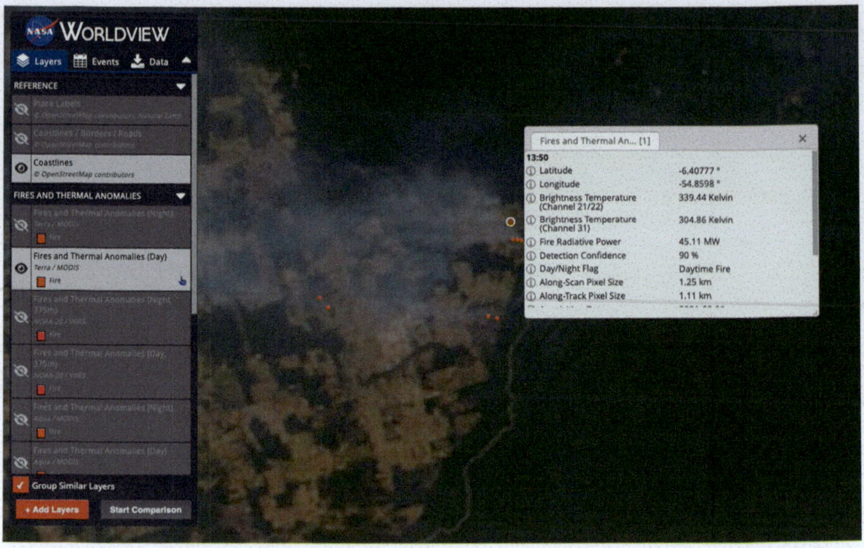

Fig. 6.4 NASA Worldview fire and thermal detection

information to mitigate the risks associated with wildfires and promote effective fire management strategies.

While remote sensing-based systems like FIRMS and Worldview are valuable assets, there are certain limitations that cannot be ignored. Satellite imagery used in these systems usually have coarse spatial resolution, making it challenging to detect and monitor small-scale fires or fires in remote areas accurately. Additionally, the temporal resolution may vary depending on the satellite sensor, resulting in delays in detecting and reporting fire events. Cloud cover and smoke can obstruct satellite imagery, reducing the effectiveness of fire detection and monitoring. Thick smoke can obscure fire signatures and make it challenging to accurately assess the extent and intensity of fires. Similarly, cloud cover can limit the availability of clear imagery, especially in regions with persistent cloud cover or during certain seasons. Remote sensing-based fire products may encounter false positives (incorrectly identifying non-fire features as fires) or false negatives (missing actual fire events). Various factors can contribute to these errors, including the presence of hotspots unrelated to fires (e.g., industrial activities) or the inability to detect fires due to limitations in sensitivity or atmospheric conditions. Additionally, remote sensing may have limitations in providing detailed information on fire behavior, such as fire spread rate, fireline intensity, or ember showers. These details are crucial for fire management and decision-making but may require ground-based observations or other specialized tools for accurate assessment. Depending on the remote sensing system and data processing workflows, there may be a delay in accessing and disseminating fire data. Real-time data availability can be critical for timely decision-making during active fire events, and any delays in data processing or accessibility can hinder effective fire management efforts. Another important factor is that remote sensing-based fire products rely on satellite observations, and ground validation for verifying the accuracy and reliability of the detected fire events. However, ground-based observations may not always be feasible due to remote or inaccessible fire locations, posing challenges in validating the remote sensing-derived fire data. These reasons are greatly limiting the actionableness of using remote sensing tools in practical wildfire firefighting.

3.2 Sensor Networks and Real-Time Data Collection

There are many ground operational networks that can support wildfire early warning and monitoring. RAWS (Remote Automated Weather Stations) is a network of weather stations (Fig. 6.5) strategically placed in wildfire-prone areas (Nauslar et al. 2018). These stations continuously monitor weather conditions such as temperature, humidity, wind speed, and precipitation, providing valuable data for assessing fire danger and supporting early warning systems. The National Weather Service also operates a network of weather stations across the United States. These stations provide real-time weather data, including temperature, humidity, wind speed, and atmospheric conditions, which are critical for monitoring and predicting fire behavior. These networks provide data on rainfall, streamflow, soil moisture, and other factors that can help assess fire risks and predict potential fire behavior. EONET (Earth Observatory Natural Event Tracker), managed by NASA, is a global system that collects and shares information on various natural hazards, including wildfires (Ward 2015). It aggregates data from multiple sources, including ground-based sensors, satellite imagery, and other remote sensing technologies, to provide real-time updates on wildfire events worldwide. FLIR (Forward-Looking Infrared) networks utilize infrared technology to detect and monitor heat signatures associated with wildfires (Khan et al. 2009). These networks consist of ground-based or aerial-based sensors that can detect hotspots and track fire progression, providing valuable information for early detection and response. Besides these officially maintained networks, citizen science initiatives involve engaging the public in data collection and monitoring efforts. Platforms such as iNaturalist and eBird allow individuals to report wildfire observations and contribute to a collective understanding of fire events. These initiatives can supplement ground-based sensing networks and provide additional data points for monitoring and early warning systems.

Barmpoutis et al. (2020b) proposed the use of 360-degree sensor cameras for early fire detection. The approach involves converting equirectangular projection format images to stereographic images and utilizing DeepLab V3+ networks (Chen et al. 2018) for flame and smoke segmentation. Experimental results demonstrate the system's effectiveness, achieving a high F-score fire detection rate of 94.6% and showcasing its potential contribution to early fire detection while reducing the number of required sensors. Ahlawat and Chauhan (2020) highlight the utilization of wireless sensor networks (WSNs) for forest fire detection and information monitoring. The authors propose an efficient real-time setup that collects information from

Fig. 6.5 RAWS station with wildfire closeby. (Image courtesy: https://www.nifc.gov/about-us/what-is-nifc/remote-automatic-weather-stations)

different locations and uploads it to a remote web server. Using Wi-Fi and NodeMCU micro-controller with built-in ESP 8266 Wi-Fi module, communication is established within the network and the proposed solution is implemented on the Arduino Integrated Development Environment (IDE) (Srivastava et al. 2018). Abdullah et al. (2017) present a compact, energy-efficient sensor network that combines various sensory inputs for continuous monitoring of forest environments and early detection of fires, and successfully tested in a real-life firefighting trial, showing promising results for coordinated firefighting scenarios. Lutakamale and Kaijage (2017) present a wildfire monitoring and detection system that utilizes a wireless sensor network that monitors temperature, humidity, and smoke to detect fires, and immediately sends a warning message with the probable location to the responsible authority via a cellular network. The system prototype, developed using Arduino microcontroller and various sensors, demonstrates the capability to detect wildfires in real time, making it an effective solution for early wildfire detection and reporting. Kadir et al. (2019) propose the development of wireless sensor networks (WSNs) for detecting forest fire hotspots in Indonesia, focusing on the high-risk region of Riau Province. WSNs are used as ground sensor systems to collect environmental data, which is then analyzed in the data center to identify fire hotspots and potential fire risks. The deployment of sensors in strategic locations, along with mathematical analysis, enhances the feasibility and effectiveness of early warning and alert systems for forest fire detection and prevention in Indonesia. Doolin and Sitar (2005) present the design and field testing of a wireless sensor system for monitoring wildfires, using environmental sensors to collect temperature, humidity, and barometric pressure data. The system performed well during prescribed burns, capturing the passage of the flame front, temperature changes, humidity decreases, and barometric pressure drops. The recorded data indicated the development of locally significant weather conditions even during relatively cool grass fires, with maximum temperature reaching 95 °C, minimum relative humidity of 9%, and a significant drop in barometric pressure. Somov (2011) conducted a survey of approaches for early wildfire detection using wireless sensor networks (WSNs), with a focus on real deployments and hardware prototypes. The methods are categorized into gas sensing, environmental parameter sensing, and video monitoring, and are analyzed based on cost, power consumption, and implementation complexity. Slavkovikj et al. (2014) discussed the current systems and methods for utilizing social media data in wildfire detection and management, highlighting their potential and examining approaches from other hazard management systems. They also proposed a general social sensor-based platform for wildfire detection and management. Barrado et al. (2010) presented a pervasive application for fighting forest fires that utilizes unmanned aircraft, personal electronic devices (PEDs), and a three-layered communication network. The system enables firefighters to obtain temperature maps of burned areas, locate hot spots, and receive commands from their manager in real time, contributing to more effective decision-making and firefighting efforts.

Although these research all demonstrated promising results, they may not be immediately actionable in real-world wildfire responding due to various reasons. First, the proposed approaches involve advanced techniques, complex algorithms, or specialized hardware that are not readily available or easily implemented in practical firefighting operations. The results were obtained in controlled experimental environments or small-scale deployments, but scalability and successful deployment in larger, real-world scenarios could pose significant barriers. The implementation of certain solutions requires significant financial resources, infrastructure, or expertise that are infeasible within the budget or operational capabilities of firefighting agencies. The proposed systems may need further validation, testing, and refinement to ensure their reliability, robustness, and resilience in challenging and dynamic wildfire environments. The compatibility and integration of the proposed solutions with existing firefighting systems, protocols, and networks may need to be addressed for seamless adoption and practical implementation.

3.3 Application of Computer Models and Simulations in Predicting Wildfire Behavior

Duff and Tolhurst (2015) examine the development of operational models that simulate fire suppression as part of decision support systems. The authors summarize the progress in modeling approaches, discuss their strengths and limitations, and offer insights into future research directions. Hanson et al. (2000) focus on recent developments in computer models of wildfires and their potential application in mitigating the threat. The article also discusses the need for an operational wildfire prediction center to harness existing capabilities and develop new tools for addressing this natural process. Monedero et al. (2019) developed the Wildfire Analyst™ Pocket Edition application (WFA Pocket), a mobile tool designed for firefighters, providing real-time, interactive 3D maps that display fire characteristics and estimated progression based on user input data. The application integrates GIS capabilities, can be used online or offline, and retrieves fuel, weather, and canopy data from online servers. Marsavin et al. (2023) used Convolutional Long Short-Term Memory (ConvLSTM) networks to model fire progression dynamics in space-time and achieved impressive effectiveness. Zhai et al. (2020) presented a learning-based wildfire spread model that combines real-time rate of spread (RoS) measurement with machine learning and a level-set method to predict short-term wildfire spread. The model is validated through comparisons with experimental measurements and applied to a real-scale shrubland fire scenario. Results demonstrate the capability of the proposed method to predict fire spread without relying on empirical RoS models, offering potential benefits for modeling real wildfires. Papadopoulos and Pavlidou (2011) investigated the use of discrete event models and simulators to study complex phenomena in ecosystems, with a specific focus on forecasting forest fire propagation. Twenty-three simulators are reviewed, and the FARSITE simulator model is identified as the most noteworthy and extensively evaluated in a test environment. Rashid et al. (2020) introduced the CompDrone framework, which combines computational wildfire modeling with social-media-driven drone sensing (SDS) for improved wildfire monitoring. By leveraging techniques from cellular automata, constrained optimization, and game theory, CompDrone addresses the challenges of limited social signals and predicting optimal drone dispatch regions. Porterie et al. (2005) developed a physical two-phase to simulate wildland fire behavior and emissions, considering the dynamics, turbulence, soot formation, and radiation. The model successfully captured the rate of spread and fuel consumption ratio of a prescribed savanna fire, demonstrating good qualitative agreement with in situ experimental data. Bakhshaii and Johnson (2019) explained the evolution of wildfire models, specifically the transition to mechanistic combustion models and large-eddy simulation (LES) coupled with computational fluid dynamics (CFD) or mesoscale weather models. These integrated models, which consider fuel, terrain, and weather conditions, represent the next generation of wildfire modeling and are designed for specific spatial and temporal scales. Lopes et al. (2002) developed FireStation for simulating fire spread over complex topography. It incorporates a semi-empirical model for fire rate of spread, wind field simulation, and a user-friendly graphical interface. The system aims to facilitate operational fire behavior prediction and has shown promising results when compared to experimental data.

However, similar to all the other numerical models, wildfire models have the same restrictions when being used to guide real-world operations. The uncertainties due to limitations in input data, parameterization, and the inherent complexity of fire behavior can affect the accuracy and reliability of the model predictions. Wildfire behavior involves a range of complex physical processes, including combustion, radiation, and turbulence. Capturing all these processes accurately in numerical models can be challenging, and simplifications or assumptions may be necessary, which can introduce uncertainties. The resolution of numerical models may not capture fine-scale variations in fire behavior, such as spot fires or localized wind patterns. This can lead to limitations in accurately predicting fire spread and behavior at smaller scales. Validating and calibrating numerical models require accurate and extensive field data, which may not always be available. Limited validation can impact the reliability and confidence in model outputs. Meanwhile, numerical models for wildfire behavior are sensitive to input parameters, such as fuel moisture, wind speed, and topography. Small errors or uncertainties in these parameters can significantly affect the model outputs, leading to inaccuracies in fire spread predictions. On the cost-wise side, running numerical models for wildfire behavior can be computationally demanding and time-consuming, especially for large-scale simulations or simulations with high spatial and temporal resolutions (Rodriguez-Aseretto et al. 2013). This can limit the practicality and real-time applicability of the models in operational firefighting scenarios. From the operational perspective, the effective use of numerical model outputs for wildfire operations relies on the expertise and interpretation of the end users. Understanding and properly interpreting the model outputs require knowledge and experience in wildfire behavior, which may not be available to all personnel involved in firefighting operations. Also, translating complex model outputs into meaningful and actionable information for decision-makers can be a challenge. Clear communication and effective visualization of the model results are crucial to ensure the usability and understanding of the information by operational personnel. Scientists have to tackle all these issues to make their research more actionable.

3.4 *Next-Generation Firefighting Techniques*

Scientists never stop finding new solutions to more effectively contain fires. There are many new potential or emerging technologies that might be the next game changer. For example, drones equipped with specialized sensors and cameras can provide real-time situational awareness, thermal imaging, and aerial surveillance of wildfire incidents (Sousa et al. 2020). UAS can assist in identifying fire hotspots, monitoring fire behavior, and guiding firefighting efforts more effectively. The recent development of new fire-resistant materials, such as fire-resistant gels, foams, and coatings, can be applied to structures, equipment, and vegetation to provide enhanced fire protection. These materials can reduce the flammability of surfaces and slow down the spread of fire. Other innovations in fire-resistant fabrics and personal protective equipment (PPE) can greatly improve the safety and effectiveness of firefighters (Song et al. 2016). Advanced materials can provide increased heat resistance, improved breathability, and better protection against radiant heat and flames. Researchers are actively looking for new fire-suppression agents, including environmentally friendly alternatives. These agents aim to improve firefighting effectiveness by increasing extinguishing capabilities, reducing environmental impact, and enhancing safety for both firefighters and the ecosystem. At a larger scale, implementing fire-resistant landscaping practices, such as strategically planting fire-resistant vegetation, creating firebreaks, and using noncombustible materials around structures, can help reduce the spread of wildfires and protect vulnerable areas. As for the inaccessible area for humans, autonomous or remotely operated robotic systems designed for firefighting can access hazardous areas and perform tasks that may be too dangerous for human firefighters. These robots can deploy fire suppressants, gather data, and assist in fire suppression efforts.

Aydin et al. (2019) explore the use of fire extinguishing balls in conjunction with drones and remote sensing technologies as a supplemental approach to traditional firefighting methods. The proposed system includes scouting unmanned aircraft systems (UAS) for detection and monitoring, communication UAS for establishing communication channels, and firefighting UAS for autonomously delivering fire extinguishing balls. The experiments conducted so far indicate that while smaller-sized fire extinguishing balls may not be effective for building fires, they show promise in extinguishing short grass fires, which has guided the authors toward focusing on wildfire fighting. The paper also discusses the development of heavy payload drones and the progress in building an apparatus to carry fire-extinguishing balls attached to drones. Bordado and Gomes (2007) overviewed that synthetic polymers and superabsorbent polymers have shown significant advancements and potential in various fields, including agriculture and fire suppression. However, it is important to consider their environmental impact, proper application, and potential limitations in specific scenarios. New aerosol-based fire extinguishing systems (Rohilla et al. 2022) have gained popularity due to their effectiveness and ease of use. These aerosols contain fine particles that can quickly suppress fires by interrupting the chemical chain reaction. They are particularly useful in enclosed spaces and electrical fires. Traditional foam agents used in firefighting contain harmful chemicals such as perfluorooctanoic acid (PFOA) and perfluorooctane sulfonate (PFOS). Innovations have led to the development of eco-friendly foam agents that are free from these toxic substances (Pierau et al. 2022). These foam agents maintain their fire suppression capabilities while reducing environmental impact. Water mist systems use fine droplets of water to suppress fires (Lazzarini et al. 2000). These systems are effective in controlling fires by cooling the flames, reducing the oxygen supply, and preventing the fire from spreading. Water mist systems are especially useful in environments where water damage needs to be minimized, such as data centers and heritage buildings. Clean agent fire suppressants, such as halocarbon-based gases, are used to extinguish fires without leaving residue or causing damage to sensitive equipment (Sebastian2022). These agents work by displacing oxygen and interrupting the combustion process. They are commonly used in areas where water or foam-based suppression systems may cause more harm than the fire itself. Another technology is powder-based fire extinguishers that have been used for a long time, but advancements have led to the development of more effective and specialized powders (Du et al. 2019). These powders, such as monoammonium phosphate (MAP) and potassium bicarbonate, are capable of suppressing various types of fires, including those involving flammable liquids, electrical equipment, and combustible metals.

Future technology like nanotechnology has opened up new possibilities for fire suppression (Mosina et al. 2020). Nanoparticles, such as graphene and nanoclay, have shown promise in enhancing the extinguishing properties of traditional fire suppressants. They improve heat transfer, increase the surface area coverage, and enhance the overall fire suppression capabilities. Other relevant technical breakthroughs like fire-resistant coatings (Gan et al. 2020) can be used to provide passive fire protection by delaying the spread of flames and reducing heat transfer. These coatings can be applied to various surfaces, including walls, ceilings, and structural elements. They help to buy critical time for evacuation and firefighting efforts. Advancements in sensor technology and artificial intelligence have led to the development of smart fire detection and suppression systems (Neumann et al. 2018). Researchers are also exploring the use of bio-based materials, such as plant extracts and biodegradable compounds, as fire suppressants (Kalali et al. 2019). These eco-friendly alternatives aim to reduce the environmental impact of fire extinguishing agents while maintaining effective fire suppression properties.

4 Case Studies and Success Stories

4.1 Examples of Successful Application of Actionable Science in Wildfire Management

Advanced technologies, such as remote sensing, weather forecasting, and satellite imagery, are widely utilized to develop early warning systems for wildfires. These systems enable authorities to detect and predict fire behavior, allowing for early evacuation and proactive firefighting strategies. The Fire Integrated Real-Time Intelligence System (FIRIS) in California (Altintas 2021) combines satellite data, weather information, and ground sensors to provide real-time situational awareness during wildfires, assisting fire managers in decision-making and resource allocation. For example, in the Thomas Fire, which is one of the largest wildfires in California's history and burned over 281,000 acres (Dahill 2019), FIRIS was utilized to monitor fire behavior, track its progression, and assess the potential threats to communities. The system integrated data from satellites, weather stations, and ground sensors to provide accurate information to incident commanders. This allowed firefighting resources to be deployed effectively and facilitated timely evacuation orders. During the Mendocino Complex Fire, comprising the Ranch Fire and the River Fire, and the largest recorded wildfire complex in California, consuming more than 459,000 acres (Scalingi 2020), FIRIS provided up-to-date information on fire behavior, hotspots, and fire spread patterns and helped incident managers make informed decisions on resource allocation, air operations, and firefighter safety. In the Bobcat Fire that burned over 115,000 acres in the Angeles National Forest

(Seeberger 2020), FIRIS was used extensively to monitor fire behavior, identify critical fire perimeters, and assess potential threats to infrastructure, communities, and sensitive ecosystems. The system's data and visualizations aided incident commanders in determining containment strategies and allocating firefighting resources effectively.

The Canadian Forest Fire Behavior Prediction (FBP) System is widely used to estimate fire behavior in Canada. It incorporates factors like fuel moisture, wind speed, and slope to predict fire spread and intensity, aiding in proactive fire management. During the devastating Fort McMurray wildfire in Alberta, Canada, the FBP System is used to predict fire behavior and aid firefighters and incident management teams to understand the fire's spread, plan evacuations, and allocate resources effectively. In British Columbia, Canada, the FBP System's ability to predict fire behavior helps in effective resource allocation and evacuation planning. By understanding how a fire is likely to spread, incident management teams can allocate firefighting resources strategically and evacuate areas at risk in a timely manner, ensuring the safety of residents and responders. In Ontario, Canada, the FBP System is employed during large-scale wildfires to assist with resource management and evacuation planning. During the Parry Sound 33 wildfire in 2018, the FBP System provided valuable information about fire behavior, which helped authorities make decisions about evacuation orders and allocate firefighting resources effectively.

The National Fire Plan in the United States emphasizes fuel reduction efforts, such as the use of controlled burns and mechanical treatments, to reduce fire risk. The implementation of these strategies has proven successful in mitigating wildfire impacts. The NFP provided support for fire suppression efforts and post-fire rehabilitation for many wildfires such as Hayman Fire (2002) (Graham 2003), Shasta-Trinity Complex Fire (2008), Wallow Fire (2011), and Rim Fire (2013). The Hayman Fire started on June 8, 2002, in Park County, Colorado. Despite an aggressive initial attack response, the fire rapidly spread due to high winds, low humidity, and dry fuel conditions. The severe drought and continuous fuel across the landscape contributed to extreme fire behavior, including torching trees and prolific spotting, resulting in the fire crossing U.S. Highway 77. The Firewise USA program, initiated by the National Fire Protection Association (NFPA), encourages communities to implement wildfire mitigation measures. Participating communities receive science-based guidance on defensible space creation and community planning to reduce wildfire vulnerability. The Firewise USA program has been implemented in various communities of California to enhance their resilience to wildfires. For instance, the Lake Almanor Peninsula Firewise Community in Plumas County has actively participated in the program, implementing measures such as vegetation management, community education, and collaboration with local fire agencies to reduce the risk of wildfires. Also, the Firewise USA program has made a significant impact in communities like Boulder County, Colorado. The Coal Creek Canyon Fire Protection District, a Firewise community, has actively worked to create defensible spaces by conducting wildfire assessments, hosting educational workshops, and coordinating fuel reduction projects. These efforts have helped safeguard homes and reduce the potential for wildfire damage.

4.2 *Real-World Stories Highlighting the Benefits of Science-Based Approaches*

First, we may take a look at the Rim Fire (2013). The Rim Fire, which started on August 17, 2013, in California's Stanislaus National Forest (Jenner 2013), became one of the largest wildfires in the state's history, burning over 257,000 acres. Fire behavior analysts utilized scientific methods to study the fire's behavior, taking into account weather patterns, topography, fuel conditions, and historical fire data. By analyzing these factors, they were able to predict the fire's potential spread and intensity. Advanced fire modeling techniques, such as the Weather Research and Forecasting model coupled with fire behavior models (WRF-SFIRE) (Mandel et al. 2014), were employed to simulate fire behavior under different weather scenarios. This allowed fire managers to anticipate fire growth patterns and strategically allocate firefighting resources. Satellite-based sensors, such as NASA MODIS, provided real-time data on the fire's perimeter, heat signatures, and smoke plumes. This information was crucial in identifying fire hotspots and prioritizing firefighting efforts. The Rim Fire saw the deployment of aerial firefighting resources, including air tankers and helicopters. Scientifically informed strategies were used to determine the most effective locations for fire retardant drops and water bucket deployments. This targeted approach helped create firebreaks and slow the fire's progression. Firefighters and fire managers used scientific knowledge to strategically construct containment lines, considering factors such as topography, fuel conditions, and predicted fire behavior. These containment lines served as physical barriers to prevent the fire's spread and protect communities and critical infrastructure. After the fire was contained, scientific approaches were employed to assess the impacts on the ecosystem and develop restoration plans. Scientists studied the fire's effects on vegetation, soil erosion, and wildlife habitat to guide post-fire rehabilitation efforts. This involved activities such as reseeding native plants, erosion control measures, and monitoring of ecosystem recovery. The success in containing and managing the Rim Fire was a result of collaborative efforts between fire managers, scientists, and various agencies. Scientists provided valuable insights and recommendations based on their expertise, which informed decision-making processes throughout the firefighting and restoration efforts.

Let us shift our attention to the wildfires in other countries. The Black Saturday Bushfires, which occurred on February 7, 2009, in the state of Victoria, Australia, were one of the most devastating wildfire events in the country's history, resulting in the loss of 173 lives and the destruction of thousands of homes (Whittaker et al. 2013). Advanced predictive models, such as the Phoenix RapidFire software, were employed to simulate fire behavior and potential ember attacks under different weather scenarios. This information was crucial in understanding the risks and aiding in decision making related to firefighting efforts, including the deployment of resources and prioritizing high-risk areas. Science-based early warning systems, such as the Victorian Fire Risk Register (VFRR) and the Country Fire Authority (CFA) FireReady app (Bowen 2020), provided real-time fire updates and warnings to communities. These systems utilized scientific data, including weather forecasts, fire behavior models, and satellite imagery, to issue timely alerts, enabling residents to evacuate early and emergency services to respond more effectively. The knowledge of fire behavior and resource effectiveness informed the deployment of aerial firefighting resources, such as water-bombing aircraft and helicopters.

5 Actionable Science Suggestions for Wildfire Researchers and Stakeholders

Based on the observation and analysis, combining with our formula from Chap. 1, we provide some suggestions for scientists to consider to improve the actionableness of wildfire research.

5.1 *Addressing the Real Gap in Implementing Actionable Science for Wildfires*

While fundamental research is valuable, scientists should also focus on applied research that directly addresses practical challenges faced in wildfire management. This includes studying specific fire behavior phenomena, developing and testing new tools and technologies, and evaluating the effectiveness of different management strategies. Researchers should prioritize the dissemination of their findings in a format that is accessible and useful for practitioners. This includes publishing research in peer-reviewed journals, but also developing concise and practical summaries, guidelines, and toolkits that can be easily understood and implemented by those working in the field. Meantime, by focusing on applied research that directly addresses practical challenges faced in wildfire management, scientists can provide solutions and insights that are immediately relevant to the field. This type of research takes into account the specific needs and constraints of practitioners, helping them make informed decisions and take effective actions. Studies that investigate the behavior of firebrands (burning embers) during wildfires can have more practical implications for firefighters and community planning (Caton et al. 2017). Understanding real demands by ground teams like how firebrands are transported by winds and ignite spot fires can lead to improved firefighting strategies, such as positioning fire engines strategically to prevent spot fire ignitions. About technology, adopting and testing new tools such as unmanned aerial systems (UAS), or drones, equipped with thermal cameras and multispectral sensors, allows researchers to provide practical solutions for enhanced fire detection, monitoring, and response. By collaborating with engineers and technologists, scientists can translate their knowledge into tangible innovations that can be readily implemented in the field. Last but not least, the world is constantly changing, and so will the firefighting strategy. Science guidance should have room for change of plans and think about as many situations as possible. Evaluating the effectiveness of different management strategies used in wildfire prevention, suppression, and mitigation provides evidence-based insights for decision-making. By assessing the outcomes of these strategies, scientists can inform policymakers and practitioners about the most effective approaches, leading to more efficient and targeted wildfire management efforts.

5.2 *Research Gaps and Areas for Further Exploration*

While significant progress has been made in wildfire research, there are still several research gaps and areas for further exploration that are the key questions from the operators but not yet fully answered by scientists. We hope researchers can prioritize these areas to get them solved to tackle the urgent demands on the field. These gaps arise due to various challenges, limitations, and complexities associated with wildfires. The first gap is we are still trying to understand the influence of climate change on wildfire behavior and dynamics (De Rigo et al. 2017). Research needs to explore the complex interactions between climate drivers, such as temperature, precipitation, and wind patterns, and their effects on fuel availability, fire frequency, and intensity. However, predicting future climate scenarios and their specific impacts on wildfires is challenging as climate models have uncertainties and localized effects can vary significantly. On the other side, while immediate fire impacts, such as direct damage to ecosystems and infrastructure, are well-studied, there is a need for research on the long-term ecological and socioeconomic effects of wildfires. This includes studying the recovery and regeneration processes of fire-affected landscapes, as well as the socioeconomic impacts on local communities and their resilience in the aftermath of wildfires. Long-term studies require sustained monitoring efforts and may take years to gather meaningful data. Third major question is about the role of humans in the whole cycle. Understanding human behavior, attitudes, and decision-making during wildfires is critical for effective fire management. Research should explore the social, cultural, and economic factors that influence wildfire preparedness, evacuation decisions, and community resilience. However, collecting data on human behavior and conducting comprehensive social science research in high-stress disaster situations can be challenging due to ethical considerations and logistical constraints. In addition, the WUI (Wildland-Urban Interface) (Radeloff et al. 2005), where human developments meet wildland areas, is particularly vulnerable to wildfires. Further research is needed to understand the dynamics of fire spread in the WUI, the effectiveness of different mitigation strategies, and the social and economic factors that influence decision-making and community engagement. However, implementing actionable solutions in the WUI requires coordination among multiple stakeholders, including homeowners, local governments, and land management agencies, which can present logistical and political challenges.On the technology challenge, advances in remote sensing technologies and predictive modeling have greatly enhanced our understanding of wildfire behavior. However, there is still a need for research to improve the accuracy, resolution, and timeliness of remote sensing data, as well as the reliability of predictive models. Additionally, translating these technological advancements into operational tools that can be effectively used by fire management agencies and practitioners requires further development and refinement. As for public health, wildfires generate significant amounts of smoke, which poses health risks and air quality concerns. Research should focus on developing effective smoke management strategies, including improved smoke forecasting, modeling, and communication systems. However, implementing smoke management practices involves coordination among multiple agencies, consideration of local air quality regulations, and public education efforts, which can present logistical and policy challenges.

As these questions are challenging but we are very urgently craving for answers, research on these topics will receive very high expectations for action conversion rate whenever there is a breakthrough. It is essential for researchers, policymakers, and practitioners to collaborate and actively communicate research findings to ensure that breakthroughs lead to actionable outcomes. By considering real-world challenges, stakeholder needs, and the practical feasibility of implementing research outcomes, scientists can increase the likelihood of their work translating into effective wildfire management strategies.

5.3 Importance of Interdisciplinary Collaboration, In-Time Sharing, and Transparent Communication

Wildfire management is a complex and multidisciplinary field, requiring expertise from various disciplines such as ecology, meteorology, social sciences, and engineering (Bonebrake et al. 2018). Scientists should actively seek out interdisciplinary collaborations to address the diverse aspects of wildfire management and integrate different perspectives into their research. Scientists should actively engage and collaborate with firefighters, land managers, and other stakeholders involved in wildfire management. By working together, they can ensure that research aligns with the needs and realities of on-the-ground fire management, making it more relevant and applicable. Scientists should actively participate in knowledge exchange activities, such as conferences, workshops, and field demonstrations, where they can share their research findings and learn from practitioners. This two-way exchange of knowledge can help researchers gain insights into real-world challenges and refine their research to be more applicable. One example of successful knowledge exchange is the Joint Fire Science Program (JFSP) in the United States (Fig. 6.6) (Maletsky et al. 2018), which supports research projects that address the needs of fire managers and practitioners. The JFSP facilitates collaboration between scientists and land managers, ensuring that research findings are directly applicable to on-the-ground fire management.

Fig. 6.6 The science exchanges supported by JFSP. (Image courtesy: https://www.firescience.gov)

Also, another very important task (no need to say) is that scientists should actively seek funding opportunities that prioritize actionable research on wildfires. Governments, research agencies, and foundations often provide grants and funding for research projects that address pressing issues and have practical applications in managing wildfires. For example, the European Commission's Horizon 2020 program (Pollex and Lenschow 2018) has funded projects like FIREFLIES and PyroLife, which focus on improving fire management practices through interdisciplinary research and innovation.

6 Conclusion

In the field of wildfire management, there is a growing recognition of the need for actionable science to bridge the gap between researchers and operational teams. Currently, wildfire management practices rely on a combination of experience, expertise, and available knowledge, but there is room for improvement. Scientists are actively addressing this gap by conducting research that directly addresses practical challenges faced in wildfire management. The focus is on studying specific fire behavior phenomena such as fire spread, ignition patterns, and fire–atmosphere interactions, developing and testing new tools and technologies, and evaluating the effectiveness of different management strategies. By studying these specific phenomena, scientists can provide valuable insights and develop predictive models that assist in making informed decisions during firefighting operations. Scientists should also prioritize interdisciplinary collaborations, engage with stakeholders, disseminate findings in accessible formats, and actively seek funding opportunities for actionable research. By adopting these approaches, researchers can enhance the application of their work, leading to more effective and informed wildfire fighting strategies. Most importantly, scientists should actively collaborate with operational teams, firefighters, and land managers to ensure that their research aligns with the needs and realities of on-the-ground fire management. This collaboration will facilitate the integration of scientific findings into operational strategies and enhance the effectiveness of wildfire management practices.

On the other hand, stakeholders and people in wildfire-impacted areas expect significant progress and advancements in the next two decades. They anticipate that scientists will develop innovative and effective strategies to mitigate the impacts of wildfires, including improved early warning systems, better fire behavior prediction models, and more accurate risk assessments. Stakeholders also hope for increased community resilience through better land management practices, enhanced public awareness, and the implementation of fire-resistant construction techniques. Additionally, they expect scientists to contribute to the development of sustainable solutions that balance fire management with ecological conservation. Overall, stakeholders and people in wildfire-impacted areas look forward to seeing science-based approaches integrated into operational practices, resulting in more efficient and effective wildfire management and a safer environment for communities at risk.

References

Abdullah, Sakib, Sandor Bertalan, Stanislav Masar, Adem Coskun, and Izzet Kale. 2017. A wireless sensor network for early forest fire detection and monitoring as a decision factor in the context of a complex integrated emergency response system. In *2017 IEEE Workshop on Environmental, Energy, and Structural Monitoring Systems (EESMS)*, 1–5. IEEE.

Ahlawat, Harsh Deep, and R.P. Chauhan. 2020. Detection and monitoring of forest fire using serial communication and Wi-Fi wireless sensor network. In *Handbook of wireless sensor networks: Issues and challenges in current scenario's*, 464–492.

Ahmad, Nur Syabeera Begum Nasir, Firuza Begham Mustafa, and Gideon Didams. 2020. A systematic review of soil erosion control practices on the agricultural land in Asia. *International Soil and Water Conservation Research* 8 (2): 103–115.

Altintas, Ilkay. 2021. Building cyberinfrastructure for translational impact: The WIFIRE example. *Journal of Computational Science* 52: 101210.

Anderson, Kerry. 2005. The Canadian Wildland Fire Information System (CWFIS): Its role within NRCan's National carbon accounting and sustainable development framework. In *Sixth symposium on fire and forest meteorology*.

Arruda, Daniel, Hugo G. Candido, and Rúbia Fonseca. 2019. Amazon fires threaten Brazil's agribusiness. *Science* 365 (6460): 1387–1387.

Aydin, Burchan, Emre Selvi, Jian Tao, and Michael J. Starek. 2019. Use of fire-extinguishing balls for a conceptual system of drone-assisted wildfire fighting. *Drones* 3 (1): 17.

Bakhshaii, Atoossa, and Edward A. Johnson. 2019. A review of a new generation of wildfire–atmosphere modeling. *Canadian Journal of Forest Research* 49 (6): 565–574.

Barmpoutis, Panagiotis, Tania Stathaki, Kosmas Dimitropoulos, and Nikos Grammalidis. 2020a. Early fire detection based on aerial 360-degree sensors, deep convolution neural networks and exploitation of fire dynamic textures. *Remote Sensing* 12 (19): 3177.

Barmpoutis, Panagiotis, Periklis Papaioannou, Kosmas Dimitropoulos, and Nikos Grammalidis. 2020b. A review on early forest fire detection systems using optical remote sensing. *Sensors* 20 (22): 6442.

Barlow, Jos, and Carlos A. Peres 2004. "Ecological responses to El Niño–induced surface fires in central Brazilian Amazonia: management implications for flammable tropical forests." Philosophical Transactions of the Royal Society of London. Series B: Biological Sciences 359 (1443): 367–380.

Barrado, Cristina, Roc Messeguer, Juan López, Enric Pastor, Eduard Santamaria, and Pablo Royo. 2010. Wildfire monitoring using a mixed air-ground mobile network. *IEEE Pervasive Computing* 9 (4): 24–32.

Bento-Gonçalves, António, António Vieira, Xavier Úbeda, and Deborah Martin. 2012. Fire and soils: Key concepts and recent advances. *Geoderma* 191: 3–13.

Bonebrake, Timothy C., Christopher J. Brown, Johann D. Bell, Julia L. Blanchard, Alienor Chauvenet, Curtis Champion, I-Ching Chen, et al. 2018. Managing consequences of climate-driven species redistribution requires integration of ecology, conservation and social science. *Biological Reviews* 93 (1): 284–305.

Bordado, João C.M., and João F.P. Gomes. 2007. New technologies for effective forest fire fighting. *International Journal of Environmental Studies* 64 (2): 243–251.

Bowen, Janne Elizabeth Trelease. 2020. *Community-based emergency planning: An autoethnographic approach to exploring the barriers and solutions*. PhD dissertation, Monash University.

Caton, Sara E., Raquel S.P. Hakes, Daniel J. Gorham, Aixi Zhou, and Michael J. Gollner. 2017. Review of pathways for building fire spread in the wildland urban interface part I: Exposure conditions. *Fire Technology* 53: 429–473.

Chang, Hsien-Ho. 2017. A literature review and analysis of the incident command system. *International Journal of Emergency Management* 13 (1): 50–67.

Chen, Liang-Chieh, Yukun Zhu, George Papandreou, Florian Schroff, and Hartwig Adam. 2018. Encoder-decoder with atrous separable convolution for semantic image segmentation. In *Proceedings of the European conference on computer vision (ECCV)*, 801–818.

Christianson, Amy. 2014. Social science research on indigenous wildfire management in the 21st century and future research needs. *International Journal of Wildland Fire* 24 (2): 190–200.

Çolak, Emre, and Filiz Sunar. 2020. Evaluation of forest fire risk in the Mediterranean Turkish forests: A case study of Menderes region, Izmir. *International Journal of Disaster Risk Reduction* 45: 101479.

Conway, Sean Calvin. 2021. *An evaluation of human-centered impacts and social vulnerability in the case of the Camp Fire in Butte County, California*. PhD dissertation.

Dahill, Lisa E. 2019. Lent, lament, and the river: interfaith ritual in the ashes of the Thomas fire. *Liturgy* 34 (4): 4–14.

Davies, Diane K., Shriram Ilavajhala, Min Minnie Wong, and Christopher O. Justice. 2008. Fire information for resource management system: Archiving and distributing MODIS active fire data. *IEEE Transactions on Geoscience and Remote Sensing* 47 (1): 72–79.

De Rigo, Daniele, GiorgioLibertà, Tracy HoustonDurrant, Tomàs ArtésVivancos, and JesúsSan-Miguel-Ayanz. 2017. *Forest fire danger extremes in Europe under climate change: variability and uncertainty*. PhD dissertation, Publications Office of the European Union.

Dennis-Parks, Reda M. 2004. Healthy Forests Restoration Act-will it really protect homes and communities. *Ecology LQ* 31: 639.

Doolin, David M., and Nicholas Sitar. 2005. Wireless sensors for wildfire monitoring. In *Smart structures and materials 2005: Sensors and smart structures technologies for civil, mechanical, and aerospace systems*, vol. 5765, 477–484. SPIE.

Du, Dexu, Xiaohui Shen, Li Feng, Min Hua, and Xuhai Pan. 2019. Efficiency characterization of fire extinguishing compound superfine powder containing Mg (OH) 2. *Journal of Loss Prevention in the Process Industries* 57: 73–80.

Duff, Thomas J., and Kevin G. Tolhurst. 2015. Operational wildfire suppression modelling: A review evaluating development, state of the art and future directions. *International Journal of Wildland Fire* 24 (6): 735–748.

Enquist, Carolyn A.F., Stephen T. Jackson, Gregg M. Garfin, Frank W. Davis, Leah R. Gerber, Jeremy A. Littell, Jennifer L. Tank, et al. 2017. Foundations of translational ecology. *Frontiers in Ecology and the Environment* 15 (10): 541–550.

Everett, Yvonne, and Michelle Fuller. 2011. Fire safe councils in the interface. *Society and Natural Resources* 24 (4): 319–333.

Fearnside, Philip M. 2005. Deforestation in Brazilian Amazonia: History, rates, and consequences. *Conservation Biology* 19 (3): 680–688.

Filkov, Alexander I., Tuan Ngo, Stuart Matthews, Simeon Telfer, and Trent D. Penman. 2020. Impact of Australia's catastrophic 2019/20 bushfire season on communities and environment. Retrospective analysis and current trends. *Journal of Safety Science and Resilience* 1 (1): 44–56.

Finney, Mark A., and Jack D. Cohen. 1998. Expectation and evaluation of fuel management objectives. In *Proceedings of the fire, fuel treatments, and ecological restoration*, 353–366. USDA Forest Service.

Gagnon, Joseph E. 2021. *Special Wildland Firefighting Tactics (SWiFT): How a conventional military approach can bring an unconventional end to destructive wildfires*. US Army School for Advanced Military Studies.

Gan, Wentao, Chaoji Chen, Zhengyang Wang, Yong Pei, Weiwei Ping, Shaoliang Xiao, Jiaqi Dai, et al. 2020. Fire-resistant structural material enabled by an anisotropic thermally conductive hexagonal boron nitride coating. *Advanced Functional Materials* 30 (10): 1909196.

Giglio, Louis, Ivan Csiszar, and Christopher O. Justice. 2006. Global distribution and seasonality of active fires as observed with the Terra and Aqua Moderate Resolution Imaging Spectroradiometer (MODIS) sensors. *Journal of Geophysical Research: Biogeosciences* 111 (G2).

Graham, Russell T. 2003. *Hayman fire case study*. Rocky Mountain Research Station: US Department of Agriculture, Forest Service.

Grasso, Veronica F., and Ashbindu Singh. 2011. Early warning systems: State-of-art analysis and future directions. *Draft Report, UNEP* 1.

Hanson, Howard P., Michael M. Bradley, James E. Bossert, Rodman R. Linn, and Leland W. Younker. 2000. The potential and promise of physics-based wildfire simulation. *Environmental Science & Policy* 3 (4): 161–172.

Hessburg, Paul F., Susan J. Prichard, R. Keala Hagmann, Nicholas A. Povak, and Frank K. Lake. 2021. Wildfire and climate change adaptation of western north American forests: A case for intentional management. *Ecological Applications* 31 (8): e02432.

Hua, Lizhong, and Guofan Shao. 2017. The progress of operational forest fire monitoring with infrared remote sensing. *Journal of Forestry Research* 28 (2): 215–229.

Iglesias, Virginia, Jennifer K. Balch, and William R. Travis. 2022. US fires became larger, more frequent, and more widespread in the 2000s. *Science Advances* 8 (11): eabc0020.

Jahn, Wolfram, James L. Urban, and Guillermo Rein. 2022. Powerlines and wildfires: Overview, perspectives, and climate change: Could there be more electricity blackouts in the future? *IEEE Power and Energy Magazine* 20 (1): 16–27.

Jenner, Lynn. "Nighttime view of California's rim fire." (2013).

Kadir, Evizal Abdul, Hitoshi Irie, and Sri Listia Rosa. 2019. Modeling of wireless sensor networks for detection land and forest fire hotspot. In *2019 international conference on electronics, information, and communication (ICEIC)*, 1–5. IEEE.

Kalali, Ehsan N., Zhang Lu, Marjan E. Shabestari, Jeremy Croyal, and De-Yi Wang. 2019. Flame-retardant wood polymer composites (WPCs) as potential fire safe bio-based materials for building products: Preparation, flammability and mechanical properties. *Fire Safety Journal* 107: 210–216.

Keeley, Jon E., and Alexandra D. Syphard. 2021. Large California wildfires: 2020 fires in historical context. *Fire Ecology* 17 (1): 1–11.

Keim, Mark E. 2008. Building human resilience: the role of public health preparedness and response as an adaptation to climate change. *American Journal of Preventive Medicine* 35 (5): 508–516.

Khan, Jesmin F., Mohammad S. Alam, and Sharif M.A. Bhuiyan. 2009. Automatic target detection in forward-looking infrared imagery via probabilistic neural networks. *Applied Optics* 48 (3): 464–476.

Knapp, Eric E., Yana S. Valachovic, Stephen L. Quarles, and Nels G. Johnson. 2021. Housing arrangement and vegetation factors associated with single-family home survival in the 2018 Camp Fire, California. *Fire Ecology* 17: 1–19.

Kolden, Crystal A., and Carol Henson. 2019. A socio-ecological approach to mitigating wildfire vulnerability in the wildland urban interface: A case study from the 2017 Thomas fire. *Fire* 2 (1): 9.

Kramer, Heather Anu, Miranda H. Mockrin, Patricia M. Alexandre, and Volker C. Radeloff. 2019. High wildfire damage in interface communities in California. *International Journal of Wildland Fire* 28 (9): 641–650.

Lazzarini, A.K., R.H. Krauss, H.K. Chelliah, and Gregory T. Linteris. 2000. Extinction conditions of non-premixed flames with fine droplets of water and water/NaOH solutions. *Proceedings of the Combustion Institute* 28 (2): 2939–2945.

Lee, Wonjae, Seonghyun Kim, Yong-Tae Lee, Hyun-Woo Lee, and Min Choi. 2017. Deep neural networks for wild fire detection with unmanned aerial vehicle. In *2017 IEEE international conference on consumer electronics (ICCE)*, 252–253. IEEE.

Ling, Sean H., and Stephan F. van Eeden. 2009. Particulate matter air pollution exposure: role in the development and exacerbation of chronic obstructive pulmonary disease. *International Journal of Chronic Obstructive Pulmonary Disease* 4: 233–243.

Lopes, A.M.G., Miguel G. Cruz, and D.X. Viegas. 2002. FireStation—an integrated software system for the numerical simulation of fire spread on complex topography. *Environmental Modelling & Software* 17 (3): 269–285.

Lutakamale, Albert S., and Shubi Kaijage. 2017. Wildfire monitoring and detection system using wireless sensor network: A case study of Tanzania. *Wireless Sensor Network* 9: 274.

Maletsky, Lisa D., William P. Evans, Loretta Singletary, and Lorie L. Sicafuse. 2018. Joint fire science program (JFSP) fire science exchange network: A national evaluation of initiative impacts. *Journal of Forestry* 116 (4): 328–335.

Mandel, J., S. Amram, J.D. Beezley, G. Kelman, A.K. Kochanski, V.Y. Kondratenko, B.H. Lynn, B. Regev, and M. Vejmelka. 2014. Recent advances and applications of WRF–SFIRE. *Natural Hazards and Earth System Sciences* 14 (10): 2829–2845.

Marsavin, Andrey, Ralph van Gageldonk, Noah Bernays, Nathaniel W. May, Daniel A. Jaffe, and Juliane L. Fry. 2023. Optical properties of biomass burning aerosol during the 2021 Oregon fire season: Comparison between wild and prescribed fires. *Environmental Science: Atmospheres* 3 (3): 608–626.

Monedero, Santiago, Joaquin Ramirez, and Adrián Cardil. 2019. Predicting fire spread and behaviour on the fireline. Wildfire analyst pocket: A mobile app for wildland fire prediction. *Ecological Modelling* 392: 103–107.

Moritz, Max A., Enric Batllori, Ross A. Bradstock, A. Malcolm Gill, John Handmer, Paul F. Hessburg, Justin Leonard, et al. 2014. Learning to coexist with wildfire. *Nature* 515 (7525): 58–66.

Mosina, Kseniia S., Elena A. Nazarova, Aleksandr V. Vinogradov, Vladimir V. Vinogradov, Elena F. Krivoshapkina, and Pavel V. Krivoshapkin. 2020. Alumina nanoparticles for firefighting and fire prevention. *ACS Applied Nano Materials* 3 (5): 4386–4393.

Nauslar, Nicholas J., John T. Abatzoglou, and Patrick T. Marsh. 2018. The 2017 North Bay and Southern California fires: A case study. *Fire* 1 (1): 18.

Neumann, Guilherme Borba, Vitor Pinheiro De Almeida, and Markus Endler. 2018. Smart Forests: fire detection service. In *2018 IEEE symposium on computers and communications (ISCC)*, 01276–01279. IEEE.

Papadopoulos, George D., and Fotini-Niovi Pavlidou. 2011. A comparative review on wildfire simulators. *IEEE Systems Journal* 5 (2): 233–243.

Pierau, Lucie, Christine Elian, Jun Akimoto, Yoshihiro Ito, Sylvain Caillol, and Davy-Louis Versace. 2022. Bio-sourced monomers and cationic Photopolymerization: The green combination towards eco-friendly and non-toxic materials. *Progress in Polymer Science* 127: 101517.

Pollex, Jan, and Andrea Lenschow. 2018. Surrendering to growth? The European Union's goals for research and technology in the Horizon 2020 framework. *Journal of Cleaner Production* 197: 1863–1871.

Porterie, B., J.L. Consalvi, Ahmed Kaiss, and J.C. Loraud. 2005. Predicting wildland fire behavior and emissions using a fine-scale physical model. *Numerical Heat Transfer, Part A: Applications* 47 (6): 571–591.

Prichard, Susan J., Paul F. Hessburg, R. Keala Hagmann, Nicholas A. Povak, Solomon Z. Dobrowski, Matthew D. Hurteau, Van R. Kane, et al. 2021. Adapting western North American forests to climate change and wildfires: 10 common questions. *Ecological Applications* 31 (8): e02433.

Radeloff, Volker C., Roger B. Hammer, Susan I. Stewart, Jeremy S. Fried, Sherry S. Holcomb, and Jason F. McKeefry. 2005. The wildland–urban interface in the United States. *Ecological Applications* 15 (3): 799–805.

Rashid, Md Tahmid, Yang Zhang, Daniel Zhang, and Dong Wang. 2020. CompDrone: Towards integrated computational model and social drone based wildfire monitoring. In *2020 16th international conference on distributed computing in sensor systems (DCOSS)*, 43–50. IEEE.

Riggs, George A., Dorothy K. Hall, and Miguel O. Román. 2017. Overview of NASA's MODIS and visible infrared imaging radiometer suite (VIIRS) snow-cover earth system data records. *Earth System Science Data* 9 (2): 765–777.

Rodriguez-Aseretto, Dario, Daniele de Rigo, Margherita Di Leo, Ana Cortés, and Jesús San-Miguel-Ayanz. 2013. A data-driven model for large wildfire behaviour prediction in Europe. *Procedia Computer Science* 18: 1861–1870.

Rohilla, Meenakshi, Amit Saxena, Yogesh Kumar Tyagi, Inderpal Singh, Rajesh Kumar Tanwar, and Rajiv Narang. 2022. Condensed aerosol based fire extinguishing system covering versatile applications: A review. *Fire Technology* 58 (1): 327–351.

Rummukainen, Markku. 2012. Changes in climate and weather extremes in the 21st century. *Wiley Interdisciplinary Reviews: Climate Change* 3 (2): 115–129.

Sayad, Younes Oulad, Hajar Mousannif, and Hassan Al Moatassime. 2019. Predictive modeling of wildfires: A new dataset and machine learning approach. *Fire Safety Journal* 104: 130–146.

Scalingi, Paula L. 2020. Creating Wildfire-Resilient communities. In *The Palgrave handbook of climate resilient societies*, 1–28. Cham: Palgrave Macmillan.

Schumann, I.I.I., L. Ronald, Miranda Mockrin, Alexandra D. Syphard, Joshua Whittaker, Owen Price, Cassandra Johnson Gaither, Christopher T. Emrich, and Van Butsic. 2020. Wildfire recovery as a "hot moment" for creating fire-adapted communities. *International Journal of Disaster Risk Reduction* 42: 101354.

Sebastian, Roshan. 2022. A review of fire mitigation methods for li-ion battery energy storage system. *Process Safety Progress* 41 (3): 426–429.

Seeberger, Rachel A. 2020. *A new simulation-optimization model for wildland fire resource prepositioning*. Naval Postgraduate School.

Seeger, Matthew W., Laura E. Pechta, Simani M. Price, Keri M. Lubell, Dale A. Rose, Saloni Sapru, Melanie C. Chansky, and Belinda J. Smith. 2018. A conceptual model for evaluating emergency risk communication in public health. *Health Security* 16 (3): 193–203.

Sharples, Jason J., Geoffrey J. Cary, Paul Fox-Hughes, Scott Mooney, Jason P. Evans, Michael-Shawn Fletcher, Mike Fromm, Pauline F. Grierson, Rick McRae, and Patrick Baker. 2016. Natural hazards in Australia: Extreme bushfire. *Climatic Change* 139: 85–99.

Sherstjuk, Vladimir, Maryna Zharikova, and Igor Sokol. 2018. Forest fire monitoring system based on UAV team, remote sensing, and image processing. In *2018 IEEE second international conference on data stream mining & processing (DSMP)*, 590–594. IEEE.

Slavkovikj, Viktor, Steven Verstockt, Sofie Van Hoecke, and Rik Van de Walle. 2014. Review of wildfire detection using social media. *Fire Safety Journal* 68: 109–118.

Smith, Alistair MS, Crystal A. Kolden, Travis B. Paveglio, Mark A. Cochrane, David MJS Bowman, Max A. Moritz, Andrew D. Kliskey. et al. 2016. "The science of firescapes: achieving fire-resilient communities." *Bioscience 66 (2): 130–146.*

Somov, Andrey. 2011. Wildfire safety with wireless sensor networks. *EAI Endorsed Transactions on Ambient Systems* 1 (1): e4.

Song, Guowen, Sumit Mandal, and René Rossi. 2016. *Thermal protective clothing for firefighters*. Woodhead Publishing.

Sousa, Maria João, Alexandra Moutinho, and Miguel Almeida. 2020. Thermal infrared sensing for near real-time data-driven fire detection and monitoring systems. *Sensors* 20 (23): 6803.

Srivastava, Prakhar, Mohit Bajaj, and Ankur Singh Rana. 2018. Overview of ESP8266 Wi-Fi module based smart irrigation system using IOT. In *2018 Fourth international conference on advances in electrical, electronics, information, communication and bio-informatics (AEEICB)*, 1–5. IEEE.

Steelman, Toddi A., and Caitlin A. Burke. 2007. Is wildfire policy in the United States sustainable? *Journal of Forestry* 105 (2): 67–72.

Teague, Bernard, SusanPascoe, and RonaldMcLeod. "The 2009 victorian bushfires royal commission final report: Summary." (2010).

Thompson, Matthew P., Wei Yu, David E. Calkin, Christopher D. O'Connor, Christopher J. Dunn, Nathaniel M. Anderson, and John S. Hogland. 2019. Risk management and analytics in wildfire response. *Current Forestry Reports* 5: 226–239.

Tymstra, Cordy, R.W. Bryce, B.M. Wotton, S.W. Taylor, and O.B. Armitage. 2010. *Development and structure of Prometheus: The Canadian wildland fire growth simulation model*. Edmonton: Natural Resources Canada, Canadian Forest Service, Northern Forestry Centre, Information Report NOR-X-417.

Tymstra, Cordy, Brian J. Stocks, Xinli Cai, and Mike D. Flannigan. 2020. Wildfire management in Canada: Review, challenges and opportunities. *Progress in Disaster Science* 5: 100045.

Ward, Kevin. 2015. The earth observatory natural event tracker (EONET): An API for matching natural events to GIBS imagery. In *AGU fall meeting abstracts*, vol. 2015, IN53A-1821.

Ward, Michelle, Ayesha I.T. Tulloch, James Q. Radford, Brooke A. Williams, April E. Reside, Stewart L. Macdonald, Helen J. Mayfield, et al. 2020. Impact of 2019–2020 mega-fires on Australian fauna habitat. *Nature Ecology & Evolution* 4 (10): 1321–1326.

Weber, D., E. Moskwa, G.M. Robinson, D.K. Bardsley, J. Arnold, and M.A. Davenport. 2019. Are we ready for bushfire? Perceptions of residents, landowners and fire authorities on lower Eyre peninsula, South Australia. *Geoforum* 107: 99–112.

Whittaker, Joshua, Katharine Haynes, John Handmer, and Jim McLennan. 2013. Community safety during the 2009 Australian 'black Saturday'bushfires: An analysis of household preparedness and response. *International Journal of Wildland Fire* 22 (6): 841–849.

Xu, Guang, and Zhong Xu. 2017. Real-time wildfire detection and tracking in Australia using geostationary satellite: Himawari-8. *Remote Sensing Letters* 8 (11): 1052–1061.

Yebra, Marta, Philip E. Dennison, Emilio Chuvieco, David Riaño, Zylstra Philip, E. Raymond Hunt Jr., F. Mark Danson, Yi Qi, and Sara Jurdao. 2013. A global review of remote sensing of live fuel moisture content for fire danger assessment: Moving towards operational products. *Remote Sensing of Environment* 136: 455–468.

Yuan, Chi, Youmin Zhang, and Zhixiang Liu. 2015. A survey on technologies for automatic forest fire monitoring, detection, and fighting using unmanned aerial vehicles and remote sensing techniques. *Canadian Journal of Forest Research* 45 (7): 783–792.

Yuan, Chi, Zhixiang Liu, and Youmin Zhang. 2017. Aerial images-based forest fire detection for firefighting using optical remote sensing techniques and unmanned aerial vehicles. *Journal of Intelligent & Robotic Systems* 88: 635–654.

Zhai, Chunjie, Siyu Zhang, Zhaolou Cao, and Xinmeng Wang. 2020. Learning-based prediction of wildfire spread with real-time rate of spread measurement. *Combustion and Flame* 215: 333–341.

Chapter 7
Actionable Science for Sea Level Rise

Ziheng Sun

Contents

Z. Sun (✉)
Center for Spatial Information Science and Systems, Department of Geography and Geoinformation Science, George Mason University, Fairfax, VA, USA
e-mail: zsun@gmu.edu

Z. Sun (ed.), *Actionable Science of Global Environment Change*,
https://doi.org/10.1007/978-3-031-41758-0_7

1 Introduction

People who do not live in coastal communities might occasionally mumble that the sea level is always changing and what the big deal is. Compared to other instant extremely impactful and noticeable events like hurricanes, droughts, wildfires, flooding, or snow storms, sea level rise sounds like a very remote thing and not realistic threats for today. On the contrary, it is a true pressing concern in the context of global climate change and the fact that 71% of the Earth's surface is covered in water. As the oceans absorb heat due to global warming (Masson-Delmotte et al. 2018, 2022), seawater expands and increases in volume. This process, known as thermal expansion, contributes to the overall rise in sea level (Mimura 2013). The expansion of seawater is driven by the increase in ocean temperatures, which is primarily attributed to the accumulation of greenhouse gases in the atmosphere. The melting of glaciers and ice sheets is another significant contributor. As global temperatures rise, ice sheets in the Arctic and Antarctica experience accelerated melting, leading to the discharge of freshwater into the oceans. This influx of freshwater contributes to the overall rise in sea level. In addition to glaciers and ice sheets, the loss of land-based ice from mountain ranges, such as the Himalayas and the Andes, also contributes to sea level rise. The melting of these land-based ice bodies, including glaciers and snowpack, leads to the runoff of water into the oceans, further adding to sea level rise. Changes in ocean circulation patterns and dynamics can also influence sea level rise (Cazenave and Nerem 2004). Factors such as changes in wind patterns, ocean currents, and the redistribution of heat can affect regional sea level variations. For instance, coastal areas can experience relative sea level rise due to changes in ocean circulation, which may exacerbate the impacts of global sea level rise in certain regions. Rising sea levels increase the vulnerability of coastal regions to extreme events and storm surges. As sea levels rise, the same storm events can lead to higher storm surges, causing more frequent and severe coastal flooding. This poses significant risks to coastal infrastructure, ecosystems, and human populations.

Sea level rise has already put low-lying island nations into vulnerability, such as the Maldives and Tuvalu (Jaschik 2014). These nations are already experiencing the effects of rising sea levels, including coastal erosion, saltwater intrusion into freshwater resources, and the loss of habitable land. These impacts have significant social, economic, and environmental consequences. Rising sea levels contribute to increased coastal erosion, where the shoreline retreats due to the encroachment of the sea. The erosive forces of waves and currents are intensified, leading to the loss of beaches, coastal vegetation, and even land. This erosion not only affects the natural environment but also threatens the infrastructure and settlements located along the coast. As sea levels rise, saltwater can infiltrate freshwater resources, such as groundwater aquifers and surface water bodies. This intrusion of saltwater contaminates the limited freshwater sources, making them unsuitable for drinking, irrigation, and agricultural purposes. The intrusion also impacts coastal ecosystems, affecting the flora and fauna that depend on freshwater habitats. With the encroachment of the sea, low-lying island nations face the loss of habitable land. As the coastline recedes, communities are forced to relocate, leading to displacement and potential loss of cultural heritage. The loss of habitable land also has economic implications as it affects tourism, agriculture, and other sectors that rely on coastal resources. The impacts of sea level rise on low-lying island nations extend beyond physical changes. These nations often have high population densities, making it challenging to accommodate the displaced communities. The loss of land, infrastructure, and livelihoods can lead to social and economic disruptions, including increased poverty, unemployment, and social inequality. Meanwhile, rising sea levels pose significant threats to coastal ecosystems, including coral reefs, mangroves, and seagrass beds (Keyzer et al. 2020). These ecosystems provide essential habitat for marine biodiversity, act as natural buffers against storms and erosion, and contribute to carbon sequestration. The loss and degradation of these ecosystems have far-reaching ecological consequences, impacting fish populations, coastal protection, and overall ecosystem health.

The major role of science now is for increasing our awareness and understanding. Accurate and comprehensive monitoring of sea level rise is essential for understanding its magnitude and trends. Scientists use various techniques such as tide gauges, satellite altimetry, and GPS to measure sea level change (Adebisi et al. 2021). This data provides a baseline for assessing the impacts of sea level rise and monitoring the effectiveness of mitigation measures. Scientists develop computer models that simulate and predict future sea level rise scenarios based on different factors such as greenhouse gas emissions, ice melt, and ocean dynamics (Nicholls et al. 2014). These models help policymakers and coastal planners make informed decisions about adaptation strategies, coastal defense systems, and land-use planning. Climate science helps understand the drivers of sea level rise, particularly the contributions from global warming and melting ice caps and glaciers. By studying climate patterns, ocean currents, and ice dynamics, scientists can improve projections of future sea level rise and inform policy decisions. Science helps identify areas at high risk of sea level rise impacts, such as coastal erosion, flooding, and saltwater intrusion. Vulnerability assessments use scientific data to prioritize adaptation efforts, protect critical infrastructure, and ensure the resilience of coastal communities. The Intergovernmental Panel on Climate Change (IPCC) reports provide policymakers with comprehensive assessments of the scientific understanding (Church et al. 2013). These reports serve as a foundation for international climate negotiations and the development of adaptation and mitigation strategies.

However, when coming to the actions, the current research is still falling short of guiding our efforts and policies in coastal communities and island nations. In other words, most research results are just projections without actionable information. Nonactionable science could lead to inadequate or delayed implementation of adaptation measures, leaving coastal communities exposed to the risks of sea level rise and can result in increased vulnerability to coastal flooding, erosion, and saltwater intrusion. Without actionable science, coastal regions may experience significant economic losses due to damage to property, infrastructure, and businesses without any ability to adapt or survive. The costs associated with rebuilding and repairing affected areas can be substantial and can strain local economies and resources. Nonactionable science can lead to the degradation and loss of valuable coastal ecosystems, including wetlands, mangroves, and coral reefs. These ecosystems serve as natural buffers against sea level rise, providing coastal protection, maintaining biodiversity, and supporting important fisheries and tourism industries. Nonactionable science can exacerbate existing social and environmental injustices, disproportionately affecting marginalized and vulnerable communities. Lack of access to actionable information and inadequate consideration of social equity can lead to unequal distribution of resources, exacerbating inequalities and placing disadvantaged communities at greater risk. In many cases, low-income communities and communities of color, who often reside in low-lying areas, face higher risks from sea level rise due to historical patterns of social and environmental injustice. Without actionable science guiding equitable adaptation strategies, these communities may be further marginalized and left without adequate support.

The chapter will reexamine the motivation, questions, current practice and policies, and what science can go from now surrounding the rising sea level. I hope to provide a comprehensive understanding of the contributions of science to addressing sea level rise, while highlighting the importance of interdisciplinary collaboration, policy integration, and stakeholder engagement in achieving sustainable and resilient coastal communities. It will present a collection of real-world case studies and best practices that demonstrate successful applications of science in addressing sea level rise. These examples highlight innovative approaches, community-based initiatives, and international collaborations that have effectively tackled the challenges posed by rising sea levels. It will also explore science-based approaches to adapting to sea level rise. It highlights the importance of coastal management strategies, such as beach nourishment, seawalls, and managed retreat, in reducing the impacts of rising sea levels on human settlements and ecosystems.

2 Motivation and Goal of Actions

Common countering activities to address the challenges posed by rising sea levels include implementing measures such as coastal defense systems, land-use planning, adaptation strategies, and international cooperation. The strong motivation is to best mitigate or adapt to its influences on coastal communities, ecosystems, and infrastructure. The goal is to reduce vulnerability, protect lives and property, and ensure the long-term sustainability of coastal regions.

2.1 Glacier Cycles and Sea Level Rise in Earth History

Glacier cycles mean the periodic expansion and retreat of glaciers over long periods of time (Lambeck and Chappell 2001; Lambeck et al. 2002). These cycles are primarily driven by changes in Earth's climate, specifically variations in temperature and precipitation. Understanding glacier cycles is important for studying past climate change and its implications for sea level rise, water resources, and landscape formation (Gildor and Tziperman 2001). Glacial periods, also known as ice ages, are characterized by the growth and expansion of glaciers. During these periods, temperatures are significantly colder, and ice accumulates in regions where snowfall exceeds melting. This results in the formation of large ice sheets and glaciers in polar and high-altitude regions. Interglacial periods occur between glacial periods and are marked by a general warming of the climate. During these periods, glaciers begin to retreat as melting exceeds accumulation. Interglacial periods are generally shorter in duration compared to glacial periods. The timing and intensity of glacier cycles are influenced by Milankovitch cycles (Bennett 1990), which are changes in Earth's orbit and axial tilt (Lourens 2021). These cycles occur over thousands of years and include eccentricity, obliquity, and precession. These orbital variations impact the amount and distribution of solar radiation reaching the Earth's surface, which in turn affects the climate and glacier dynamics. Glaciers move under the influence of gravity. As snow accumulates and compacts over time, it transforms into ice. The weight of the ice causes it to flow downslope, carving valleys, and eroding the landscape. Glaciers can advance or retreat depending on the balance between accumulation (snowfall) and ablation (melting and sublimation). The North American Cordillera has revealed multiple glacial advances and retreats over the past million years. These cycles have shaped the landscapes of regions such as the Rocky Mountains and the Sierra Nevada. Similarly, the study of ice cores from Antarctica and Greenland has provided detailed records of past climate and glacial fluctuations (Benn and Evans 2014).

Lambeck et al. (2014) studied that there were periods of significant sea level rise caused by the melting of ice sheets and glaciers. One of the most prominent events was the Pleistocene–Holocene transition, which occurred approximately 11,000 years ago (Rosen and Rivera-Collazo 2012). This period marked the end of the last major glacial period and the beginning of the current interglacial period. As the Earth's climate warmed, massive ice sheets that covered large parts of North America, Europe, and Asia began to melt rapidly. This resulted in a substantial increase in global sea levels, which rose by approximately 120 m (400 ft) during the deglaciation period (the past 20,000 years), with an average rate of around 1 cm (0.4 inches) per year (Clark and Mix 2002; Lambeck et al. 2014). Evidence for sea level rise during the Pleistocene–Holocene transition comes from various sources, including sedimentary deposits, coral reefs, and fossilized marine organisms. Scientists analyze sediment cores and drill samples from coastal regions to study the layers of sediment and the types of marine organisms present. Radiocarbon dating and other dating techniques help determine the age of these deposits and provide insight into past sea level changes. One well-studied example of sea level rise in this period is the flooding of the Sundaland region, which includes present-day Indonesia, Malaysia, and parts of Southeast Asia. The melting ice causes sea levels to rise, leading to the submergence of vast coastal areas and the formation of new islands and archipelagos.

2.2 *Where Is Rising?*

Although sea level rise is a global event and impacts coastal regions worldwide, at present the most impacted regions are low-lying islands, such as Pacific Island nations, coastal areas of Bangladesh, cities with coastal exposure like Miami, Venice, and Shanghai. Most small island developing states (SIDS), including nations in the Caribbean, Indian Ocean, and the Pacific, are highly vulnerable to the impacts that extend beyond physical risks to include socioeconomic and cultural implications. For example, the Barbados Coastal Zone Management Unit has been implementing measures such as beach nourishment and coastal protection to mitigate the effects of sea level rise (Mycoo et al. 2012). Besides, the Arctic is also experiencing the effects of global warming at an accelerated rate, resulting in the melting of polar ice and contributing to rising sea levels, and the coastal communities in Alaska, Canada, and Greenland inside the Arctic Circle are witnessing changes in shoreline erosion, permafrost thaw, and increased coastal flooding (AMAP 2017).

2.3 *Who Are Impacted?*

People living in coastal communities are among the most directly affected by sea level rise. For example, the residents of the Sundarbans region in Bangladesh, home to millions of people, are exposed to the threat of displacement and loss of livelihoods due to sea level rise (Pethick and Orford 2013). Socioeconomically disadvantaged communities often face disproportionate impacts from sea level rise. These communities may have limited resources and infrastructure to adapt to changing coastal conditions. For example, marginalized communities in coastal regions of the United States, such as in Louisiana and Florida, are at greater risk due to a combination of economic, social, and environmental factors (Nicholls and Cazenave 2010). In addition, indigenous communities, especially those with cultural and historical ties to coastal areas, are at risk from sea level rise. These communities often have deep connections to the land and face potential displacement, loss of cultural heritage sites, and disruption to traditional practices. For instance, indigenous populations in Alaska, such as the Inupiat and Yupik, are experiencing the impacts of coastal erosion and loss of hunting and fishing grounds (Nelson et al. 2007). Economic sectors dependent on coastal regions, such as tourism, fisheries, and agriculture will take the hit. Coastal tourism destinations, like the Florida Keys in the United States or the Maldives, may face declining visitor numbers due to the degradation of coastal ecosystems and infrastructure (Amelung and Nicholls 2014).

2.4 *What Can We Do About It?*

There are pretty limited things we can do for now, but here are things we can prepare and adapt ourselves to the changes. People can educate themselves and others about the causes and consequences of sea level rise to foster a sense of urgency and understanding. By making sustainable choices in daily life, such as using public transportation, conserving energy, and reducing water consumption, individuals can contribute to mitigating climate change, which is a major driver of sea level rise. Participate in or support local initiatives that aim to protect and restore coastal ecosystems, such as beach clean-ups, dune restoration, and mangrove conservation.

Governments should develop and enforce policies that aim to reduce greenhouse gas emissions and promote renewable energy sources. This can include setting emissions reduction targets, implementing carbon pricing mechanisms, and investing in clean energy infrastructure. Governments can invest in coastal protection measures, such as sea walls, dikes, and beach nourishment, to reduce the impacts of sea level rise. They can also develop comprehensive coastal management plans that account for future sea level rise projections and incorporate nature-based solutions, such as wetland restoration and mangrove conservation. Governments should prioritize the needs of vulnerable communities affected by sea level rise by providing financial assistance, relocation support, and access to resources for adaptation measures.

3 Cutting-Edge Scientific Research and Gaps with Actions

Advancements in renewable energy technologies, such as solar, wind, and hydroelectric power, are promising for reducing greenhouse gas emissions. Researchers are exploring ways to enhance the efficiency, affordability, and scalability of renewable energy systems. For instance, offshore wind farms are being developed in coastal areas to harness strong and consistent winds, providing a sustainable source of electricity without emissions (Bilgili et al. 2011). Carbon dioxide capture and storage (CCS) technologies (Benson and Orr 2008) capture carbon dioxide (CO_2) emissions from power plants and industrial facilities and store them underground, preventing them from entering the atmosphere. Developing low-carbon transportation systems is critical. Research is focused on electric vehicles (EVs) (Hawkins et al. 2012;Sanguesa et al. 2021), advanced biofuels, and improved public transportation networks. For example, the advancement of EV technology and infrastructure has the potential to significantly reduce emissions from the transportation sector. Sustainable land-use practices, including reforestation, afforestation, and forest conservation, help sequester carbon and mitigate climate change. Research is being conducted to understand the carbon sequestration potential of different ecosystems and identify effective land management strategies. Improving energy efficiency in buildings, industries, and appliances is a key mitigation strategy. Research aims to develop technologies and policies that promote energy conservation and reduce emissions. Smart buildings equipped with energy-efficient systems and advanced monitoring technologies can significantly reduce energy consumption.

4 Research Challenges and Trade-Offs in Real-World Application

4.1 Challenges and Issues Preventing Research Progress

Sea level rise involves complex dynamics, including the interaction between ice sheets, oceans, and land masses (Lenaerts et al. 2019). Understanding these processes and their future trajectories involves dealing with uncertainties. Uncertainties arise from factors such as incomplete understanding of ice sheet behavior, potential feedback mechanisms, and regional variations in sea level rise. Addressing and quantifying these uncertainties is important for effective research and decision-making. Meanwhile, gathering accurate and extensive data on sea level rise is essential for effective research. However, data collection can be challenging due to the remoteness of certain regions, limited monitoring infrastructure, and the need for long-term observations. Inconsistent or incomplete data can hinder the accuracy of projections and impede the development of reliable models. Also, bridging the gap between research findings and policy implementation is critical in responding to sea level rise. While scientific research provides valuable insights, translating those findings into actionable policies and practices can be challenging. Effective policy frameworks, governance structures, and stakeholder engagement are necessary to ensure research informs decision-making and drive appropriate actions.

4.2 Gaps Between Scientists and Impacted Communities

There is often a disconnect between the scientific language used by researchers studying sea level rise and the understanding of the broader community. Complex scientific terminology and technical jargon can make it challenging for community members to comprehend and engage with the information. Indigenous and local communities residing in coastal areas possess valuable knowledge and insights about their environments and the impacts of sea level rise. However, this knowledge is not always effectively integrated into scientific research and decision-making processes. There may be a lack of trust between scientists and impacted communities. Historical experiences, power imbalances, and perceived disconnects between scientific findings and lived experiences can contribute to this trust deficit. Also, there are often disparities in the representation and inclusion of marginalized communities in scientific research and decision-making processes related to sea level rise. This lack of diversity can limit the perspectives and experiences considered in addressing the impacts of sea level rise. Meaningful collaboration between scientists and impacted communities is essential for addressing the challenges of sea level rise. However, there is a gap in the extent of collaboration, with limited opportunities for joint decision-making, co-design of research, and shared knowledge production. By acknowledging these gaps and addressing the underlying issues, we can work toward more inclusive and effective strategies for understanding and responding to the impacts of sea level rise.

4.3 Economic, Social, and Political Challenges to Implementing Solutions

Implementing scientific solutions for sea level rise will inevitably face economic, social, and political challenges. It is a global issue that requires cooperation and collaboration across different jurisdictions and countries. Negotiating agreements, sharing responsibilities, and establishing effective governance mechanisms can be challenging in the context of differing political interests. Coordinating and aligning policies at different levels of government (local, regional, national, international) is required for effective sea level rise mitigation and adaptation. However, political fragmentation and differences in priorities can hinder policy coherence and implementation. Rising sea levels can lead to the displacement of communities living in low-lying coastal areas (Noss 2011). Relocation and resettlement efforts can cause social disruptions, including the loss of cultural heritage and community cohesion. Vulnerable communities, such as those with lower socioeconomic status or marginalized groups, may bear a disproportionate burden of the impacts of sea level rise. Addressing social inequities and ensuring that adaptation strategies consider the needs and concerns of all stakeholders is a high-priority task. Implementing large-scale adaptation and mitigation measures to address sea level rise can be financially demanding. The costs of infrastructure upgrades, coastal protection measures, and relocation efforts can be substantial, posing challenges to funding and resource allocation. Economic activities such as coastal development, tourism, and shipping may be impacted by sea level rise mitigation measures. Balancing the economic interests of industries and communities with the need for sustainable solutions can be a complex task.

4.4 *Ethical Considerations During Decision-Making*

Ethical decision-making should prioritize the well-being and rights of coastal communities directly impacted by sea level rise. This involves recognizing the social, cultural, and economic significance of these communities and ensuring that their voices are heard in decision-making processes. Respecting the rights and interests of coastal residents is essential for preserving their livelihoods and ensuring their resilience in the face of sea level rise. Ethical decision-making requires addressing environmental justice concerns. This includes ensuring that vulnerable populations, such as low-income communities and marginalized groups, are not disproportionately burdened by the impacts of sea level rise. Environmental justice principles call for the equitable distribution of environmental risks and benefits, access to resources, and the inclusion of marginalized voices in decision-making processes. Sea level rise may result in the displacement of communities residing in low-lying coastal areas. Ethical considerations involve providing support and resources to affected communities during relocation processes, ensuring that their rights are protected, and minimizing social disruptions. Just and fair approaches to relocation should be pursued, taking into account the cultural, historical, and emotional ties that individuals and communities have to their land.

4.5 *Balancing Short-Term Versus Long-Term Priorities and Trade-Offs*

Balancing short-term and long-term priorities requires considering the environmental impact of response measures. Stakeholders can provide valuable insights into local needs and concerns, ensuring that short-term actions align with long-term goals. Some adaptation strategies, such as hard coastal defenses, may also have negative ecological consequences. It is important to assess the trade-offs between protecting human assets and preserving coastal ecosystems, considering the long-term ecological implications. Participatory approaches, such as community-based planning, facilitate the inclusion of diverse perspectives and help navigate trade-offs. While certain measures may have higher upfront costs, they can provide long-term savings by reducing vulnerability to sea level rise. Cost–benefit analysis and incorporating the concept of discount rates can help evaluate trade-offs between immediate expenses and long-term benefits. Infrastructure decisions, such as constructing buildings, roads, or ports, should account for sea level rise projections and incorporate adaptive design features. Investing in resilient infrastructure can help avoid costly retrofitting or reconstruction in the future. Coastal protection measures such as building sea walls or installing flood gates can provide short-term protection, but long-term plans may need to include managed retreat or ecosystem-based approaches to accommodate rising sea levels sustainably.

5 Ongoing Use Cases and Lessons Learnt

5.1 *Coastal Wetland Restoration*

Coastal wetland restoration involves the creation and enhancement of wetland areas that have been degraded or lost due to human activities. This may include re-establishing tidal flows, planting native vegetation, and creating suitable habitats for wetland species. For example, the California State Coastal Conservancy has implemented numerous projects to restore coastal wetlands, such as the South Bay Salt Pond Restoration Project in San Francisco Bay (Kurth et al. 2022). Barrier islands protect coastlines from storm surges and erosion and the efforts to restore barrier islands involve building dunes, planting vegetation, and replenishing sand to reinforce the island's natural protective features. The Coastal Wetlands Planning, Protection and Restoration Act in the United States has funded projects like the Caminada Headland Beach and Dune Restoration in Louisiana to restore and protect barrier islands (Khalil and Raynie 2015). In addition, living shorelines are environmentally friendly alternatives to hard coastal defenses. These projects aim to stabilize shorelines, reduce erosion, and enhance habitats using natural materials such as oyster reefs, marsh grasses, and submerged aquatic vegetation. The National Oceanic and Atmospheric Administration (NOAA) has supported various living shoreline projects, including the Napatree Point Conservation Area in Rhode Island (Mayo et al. 2015). Meanwhile, tidal marshes also provide essential habitats, protect coastlines from erosion, and help mitigate climate change by sequestering carbon, and restoration efforts mainly focus on re-establishing tidal flows, removing invasive species, and restoring native vegetation (Brockmeyer et al. 2022).

5.2 *Managed Retreat and Land-Use Planning*

Managed retreat means a planned and strategic process of moving communities, buildings, and infrastructure away from coastal areas at risk of sea level rise. It recognizes the long-term challenges posed by rising sea levels and aims to ensure the safety and resilience of communities. Managed retreat can take several forms, including voluntary buyouts, relocation, and the restoration of natural coastal buffers. An example of managed retreat is the Staten Island Bluebelt Program in New York City (Gumb et al. 2008), which focuses on acquiring and converting flood-prone properties into green infrastructure and open space. At the same time, land-use planning refers to the careful assessment and regulation of land development to minimize exposure to sea level rise and protect coastal areas. It involves zoning regulations, building codes, and development restrictions aimed at reducing vulnerability and promoting resilience. Land-use planning considers factors such as elevation, flood risk, and ecological values to guide decisions on where and how development should occur, for example, the Coastal Zone Management Act (Chasis

1985), which encourages states to develop comprehensive coastal management programs to address sea level rise and coastal hazards.

5.3 Nature-Based Infrastructure

This means using natural systems and processes as a means to mitigate the impacts of sea level rise. Wetlands, such as salt marshes and mangroves, act as natural buffers that absorb wave energy and reduce the impacts of storm surges. Restoring and protecting these wetland ecosystems can provide valuable coastal protection. For example, the Tidal Marsh and Barrier Beach Restoration Project in San Francisco Bay, California (Stralberg et al. 2011), aims to restore tidal marshes and enhance natural barriers to protect against rising sea levels. Living shorelines are designed to stabilize and protect coastal areas using natural materials, such as vegetation, oyster reefs, and dunes. They offer a more sustainable alternative to hard structures like seawalls. Another example is the Living Shoreline Project in Chesapeake Bay, Maryland (Davis et al. 2006), which combines the planting of native marsh grasses with the creation of oyster reefs to protect shorelines and enhance habitat. Beach nourishment involves adding sand to eroded beaches to restore their width and elevation. This helps to provide a natural buffer against rising sea levels and storm surges (Fig. 7.1). For instance, the Miami-Dade County Beach Erosion Control and Hurricane Protection Project in Florida uses beach nourishment to protect coastal areas and support the tourism industry.

Fig. 7.1 Law-protected sand dunes on the beach of Assateague Island state park in Maryland

5.4 *Climate-Resilient Building Design*

Building structures on stilts or raised platforms above the anticipated flood level can minimize the risk of damage from storm surges and rising sea levels. For example, in the Netherlands, the Floating Houses project in Maasbommel features homes designed to float during floods, reducing the potential for property damage (Danilescu 2020). Using flood-resistant materials and construction techniques can also help protect buildings from water damage. This includes using corrosion-resistant materials, waterproofing foundations, and designing flood-resistant building envelopes. The Surry Hills Library and Community Centre in Sydney, Australia, incorporates flood-resistant design features to protect against potential flooding in the area (Malighetti 2011). Implementing effective stormwater management strategies can help mitigate the impacts of sea level rise. This includes designing buildings with green roofs, rain gardens, and permeable surfaces to absorb and manage excess water. The Bullitt Center in Seattle, Washington, is an example of a building that incorporates sustainable stormwater management practices (Sojka et al. 2016). Integrating energy-efficient design principles and renewable energy technologies can reduce reliance on fossil fuels and contribute to climate change mitigation efforts. Energy-efficient buildings help reduce greenhouse gas emissions and minimize the carbon footprint associated with energy consumption. The Edge Building in Amsterdam, the Netherlands, is a sustainable office building that incorporates energy-efficient design elements and renewable energy sources (Mehmood et al. 2019). Adapting existing buildings through retrofitting and repurposing can contribute to climate resilience. This involves upgrading structures to withstand changing environmental conditions and sea level rise. The New York City Department of Parks and Recreation's Climate-Resilient Retrofit Guidelines provide guidance for retrofitting existing buildings to enhance resilience to climate impacts (Dolman 2021).

6 Vision of Actionable Science in Mitigating Sea Level Rise

As mentioned above, the current main role of science is to enhance our understanding of the physical processes driving sea level rise, such as thermal expansion, melting glaciers, and ice sheet dynamics, such as studying ocean circulation patterns, climate dynamics, and the interactions between the atmosphere, oceans, and ice. Scientists are continually improving their understanding of feedback mechanisms and interactions between various factors influencing sea level rise. This includes studying the dynamics of ice sheets, ocean currents, and land subsidence. By unraveling these complex interactions, scientists can enhance predictions and develop more effective mitigation strategies. By improving our knowledge of these processes, scientists can develop more accurate projections and inform decision-making. Meanwhile, through modeling and simulation, scientists can predict future sea level rise scenarios based on different greenhouse gas emissions scenarios and climate projections. These predictions provide valuable information for planning and decision-making at various scales. For example, the Intergovernmental Panel on Climate Change (IPCC) produces comprehensive reports that assess the latest scientific understanding of sea level rise and its potential impacts. Science can help develop adaptation strategies to mitigate the impacts of sea level rise. This includes identifying vulnerable areas, assessing risks, and designing and implementing measures to protect coastal communities and ecosystems. For instance, coastal engineers and scientists collaborate to design and construct sea walls, levees, and coastal restoration projects to minimize flooding and erosion risks. Science supports the development and promotion of sustainable practices to reduce greenhouse gas emissions and limit further sea level rise. This includes research on renewable energy sources, energy-efficient technologies, and sustainable land and water management practices. For example, promoting renewable energy adoption can help reduce dependence on fossil fuels, thereby mitigating climate change and sea level rise. Addressing sea level rise requires global cooperation and collaboration among scientists, policymakers, and stakeholders. International efforts, such as the United Nations Framework Convention on Climate Change (UNFCCC) (Kyoto Protocol 1997) and the Paris Agreement (Savaresi 2016), aim to foster collaboration and support collective action to mitigate climate change and its impacts. These agreements provide a framework for sharing knowledge, resources, and best practices in addressing sea level rise. The future potential of science in actively handling sea level rise lies in continued research, technological advancements, and collaborative efforts. By leveraging scientific knowledge, engaging stakeholders, and integrating science into policy and decision-making processes, society can work toward resilient coastal communities, protection of vulnerable ecosystems, and sustainable development in the face of sea level rise.

7 Conclusion

This chapter introduced the various strategies and approaches to address sea level rise and explored the significance of sea level rise, the vulnerable regions and populations, and the potential consequences of inaction. It discussed the role of science in providing valuable insights, data, and modeling to inform decision-making processes. It also walked through several mitigation strategies, such as coastal wetland restoration, managed retreat, land use planning, nature-based infrastructure, and climate-resilient building design. It emphasized the importance of integrating science, technology, and stakeholder collaboration to develop innovative solutions. The challenges and gaps between scientists and impacted communities were also identified, including the need for improved communication, equity considerations, and ethical decision-making.

We also discussed the future potential of science in actively addressing sea level rise, like advancing technology and data collection, improving understanding of feedback mechanisms, integrating climate and sea level rise models, developing innovative engineering and infrastructure solutions, fostering stakeholder engagement, and integrating science into policy and decision-making processes. The vision for the future involves a multidisciplinary and collaborative approach, where science will be central in developing sustainable and resilient coastal communities, protecting vulnerable ecosystems, and informing long-term adaptation and mitigation strategies.

References

Adebisi, Naheem, Abdul-Lateef Balogun, Teh Hee Min, and Abdulwaheed Tella. 2021. Advances in estimating Sea Level Rise: A review of tide gauge, satellite altimetry and spatial data science approaches. *Ocean & Coastal Management* 208: 105632.

AMAP. 2017. *Snow, water, ice and permafrost in the Arctic (SWIPA) 2017*. AMAP.

Amelung, Bas, and Sarah Nicholls. 2014. Implications of climate change for tourism in Australia. *Tourism Management* 41: 228–244.

Benn, Douglas, and David J.A. Evans. 2014. *Glaciers and glaciation*. Routledge.

Bennett, Keith D. 1990. Milankovitch cycles and their effects on species in ecological and evolutionary time. *Paleobiology* 16 (1): 11–21.

Benson, Sally M., and Franklin M. Orr. 2008. Carbon dioxide capture and storage. *MRS Bulletin* 33 (4): 303–305.

Bilgili, Mehmet, Abdulkadir Yasar, and Erdogan Simsek. 2011. Offshore wind power development in Europe and its comparison with onshore counterpart. *Renewable and Sustainable Energy Reviews* 15 (2): 905–915.

Brockmeyer, Ronald E., Melinda Donnelly, Jorge R. Rey, and Douglas B. Carlson. 2022. Manipulating, managing and rehabilitating mangrove-dominated wetlands along Florida's east coast (USA): Balancing mosquito control and ecological values. *Wetlands Ecology and Management* 30: 987–1005.

Cazenave, Anny, and Robert S. Nerem. 2004. Present-day sea level change: Observations and causes. *Reviews of Geophysics* 42 (3): RG3001.

Chasis, Sarah. 1985. The coastal zone management act: A protective mandate. *Natural Resources Journal* 25: 21.

Church, John A., Peter U. Clark, Anny Cazenave, Jonathan M. Gregory, Svetlana Jevrejeva, Anders Levermann, Mark A. Merrifield, et al. 2013. Sea-level rise by 2100. *Science* 342 (6165): 1445–1445.

Clark, Peter U., and Alan C. Mix. 2002. Ice sheets and sea level of the Last Glacial Maximum. *Quaternary Science Reviews* 21 (1–3): 1–7.

Danilescu, Iulia Claudia. 2020. Study on the architectural projects of the floating structures for housing and leisure, along the Danube. *Journal of Danubian Studies and Research* 10 (2): 224–234.

Davis, Jana L.D., Richard L. Takacs, and Robert Schnabel. 2006. Evaluating ecological impacts of living shorelines and shoreline habitat elements: An example from the upper western Chesapeake Bay. In *Management, policy, science, and engineering of nonstructural erosion control in the Chesapeake Bay*, 55–61. Chesapeake Research Consortium.

Dolman, Nanco 2021. Integration of water management and urban design for climate resilient cities. Climate Resilient Urban Areas: Governance, design and development in coastal delta cities, 21–43.

Gildor, Hezi, and Eli Tziperman. 2001. A sea ice climate switch mechanism for the 100-kyr glacial cycles. *Journal of Geophysical Research: Oceans* 106 (C5): 9117–9133.

Gumb, Dana, James Garin, Sandeep Mehrotra, and Brian Henn. 2008. Watershed approach to integrating green and hard infrastructure: New York City's Staten Island Bluebelt. In *Sustainability conference 2008*, 951–958. Water Environment Federation.

Hawkins, Troy R., Ola Moa Gausen, and Anders Hammer Strømman. 2012. Environmental impacts of hybrid and electric vehicles—a review. *The International Journal of Life Cycle Assessment* 17: 997–1014.

Jaschik, Kevin. 2014. Small states and international politics: Climate change, the Maldives and Tuvalu. *International Politics* 51 (2): 272–293.

Keyzer, L.M., Peter M.J. Herman, B.P. Smits, J.D. Pietrzak, R.K. James, Adam S. Candy, R.E.M. Riva, et al. 2020. The potential of coastal ecosystems to mitigate the impact of sea-level rise in shallow tropical bays. *Estuarine, Coastal and Shelf Science* 246: 107050.

Khalil, Syed M., and Richard C. Raynie. 2015. Coastal restoration in Louisiana: An update. *Shore & Beach* 83 (4): 4.

Kurth, Margaret H., Bari N. Greenfeld, Matthew A. Smith, Samuel M. Fielding, Marriah S. Abellera, and Jeffrey K. King. 2022. *Financing natural infrastructure: South Bay Salt Pond Restoration Project, California*. US Army Engineer Research and Development Center [Environmental Laboratory].

Kyoto Protocol. 1997. United Nations framework convention on climate change. *Kyoto Protocol, Kyoto* 19 (8): 1–21.

Lambeck, Kurt, and John Chappell. 2001. Sea level change through the last glacial cycle. *Science* 292 (5517): 679–686.

Lambeck, Kurt, Tezer M. Esat, and Emma-Kate Potter. 2002. Links between climate and sea levels for the past three million years. *Nature* 419 (6903): 199–206.

Lambeck, Kurt, Hélène Rouby, Anthony Purcell, Yiying Sun, and Malcolm Sambridge. 2014. Sea level and global ice volumes from the Last Glacial Maximum to the Holocene. *Proceedings of the National Academy of Sciences 111 (43): 15296–15303.*

Lenaerts, Jan T.M., Brooke Medley, Michiel R. van den Broeke, and Bert Wouters. 2019. Observing and modeling ice sheet surface mass balance. *Reviews of Geophysics* 57 (2): 376–420.

Lourens, Lucas J. 2021. The variation of the Earth's movements (orbital, tilt, and precession) and climate change. In *Climate change*, 583–606. Elsevier.

Malighetti, Laura Elisabetta. 2011. Surry Hills Library and Community Centre, Sydney, Australia, Francis-Jones Morehen Thorp. *ARKETIPO* 52: 80–93.

Masson-Delmotte, Valérie, Panmao Zhai, Hans-Otto Pörtner, Debra Roberts, Jim Skea, Priyadarshi R. Shukla, Anna Pirani, et al. 2018. *Global warming of 1.5°C: An IPCC Special Report on the impacts of global warming of 1.5°C*. No. 5, 43–50. Intergovernmental Panel on Climate Change.

Masson-Delmotte, Valérie, Panmao Zhai, Hans-Otto Pörtner, Debra Roberts, Jim Skea, and Priyadarshi R. Shukla. 2022. Global Warming of 1.5°C IPCC special report on impacts of global warming of 1.5°C above pre-industrial levels in context of strengthening response to climate change, sustainable development, and efforts to eradicate poverty. Cambridge University Press.

Mayo, Thomas W., Peter W.C. Paton, and Peter V. August. 2015. Responses of birds to humans at a coastal barrier beach: Napatree Point, Rhode Island. *Northeastern Naturalist* 22 (3): 501–512.

Mehmood, Muhammad Uzair, Daye Chun, Hyunjoo Han, Gyuyeob Jeon, and Kuan Chen. 2019. A review of the applications of artificial intelligence and big data to buildings for energy-efficiency and a comfortable indoor living environment. *Energy and Buildings* 202: 109383.

Mimura, Nobuo. 2013. Sea-level rise caused by climate change and its implications for society. *Proceedings of the Japan Academy, Series B* 89 (7): 281–301.

Mycoo, Michelle, and Andrew Chadwick. 2012. "Adaptation to climate change: the coastal zone of Barbados." *In Proceedings of the Institution of Civil Engineers-Maritime Engineering, 165 (4): 159–168. Thomas Telford Ltd.*

Nelson, Donald R., W. Neil Adger, and Katrina Brown. 2007. Adaptation to environmental change: Contributions of a resilience framework. *Annual Review of Environment and Resources* 32: 395–419.

Nicholls, Robert J., and Anny Cazenave. 2010. Sea-level rise and its impact on coastal zones. *Science* 328 (5985): 1517–1520.

Nicholls, Robert J., Susan E. Hanson, Jason A. Lowe, Richard A. Warrick, Xianfu Lu, and Antony J. Long. 2014. Sea-level scenarios for evaluating coastal impacts. *Wiley Interdisciplinary Reviews: Climate Change* 5 (1): 129–150.

Noss, Reed F. 2011. Between the devil and the deep blue sea: Florida's unenviable position with respect to sea level rise. *Climatic Change* 107 (1–2): 1–16.

Pethick, John, and Julian D. Orford. 2013. Rapid rise in effective sea-level in southwest Bangladesh: Its causes and contemporary rates. *Global and Planetary Change* 111: 237–245.

Rosen, Arlene M., and Isabel Rivera-Collazo. 2012. Climate change, adaptive cycles, and the persistence of foraging economies during the late Pleistocene/Holocene transition in the Levant. *Proceedings of the National Academy of Sciences* 109 (10): 3640–3645.

Sanguesa, Julio A., Vicente Torres-Sanz, Piedad Garrido, Francisco J. Martinez, and Johann M. Marquez-Barja. 2021. A review on electric vehicles: Technologies and challenges. *Smart Cities* 4 (1): 372–404.

Savaresi, Annalisa. 2016. The Paris agreement: A new beginning? *Journal of Energy & Natural Resources Law* 34 (1): 16–26.

Sojka, Sarah, Tamim Younos, and David Crawford. 2016. Modern urban rainwater harvesting systems: Design, case studies, and impacts. In *Sustainable water management in urban environments*, 209–234. Cham: Springer.

Stralberg, Diana, Matthew Brennan, John C. Callaway, Julian K. Wood, Lisa M. Schile, Dennis Jongsomjit, Maggi Kelly, V. Thomas Parker, and Stephen Crooks. 2011. Evaluating tidal marsh sustainability in the face of sea-level rise: A hybrid modeling approach applied to San Francisco Bay. *PLoS One* 6 (11): e27388.

Chapter 8
Actionable Science for Irrigation

Hui Fang

Contents

H. Fang (✉)
Fairfax County Public School, Fairfax, VA, USA

Z. Sun (ed.), *Actionable Science of Global Environment Change*,
https://doi.org/10.1007/978-3-031-41758-0_8

1 Introduction

Irrigation, which involves the controlled application of water to plants for agricultural purposes, delivers water to the soil to infiltrate and percolate through the soil profile (Bos and Nugteren 1990). The physical properties of the soil, such as texture, structure, and hydraulic conductivity, will determine the movement patterns of water (Rawls et al. 1998). As water moves through the soil, plant roots absorb it through a process called transpiration which is driven by the difference in water potential between the soil and the plant, creating a water uptake pathway that sustains crop growth (Hopmans and Bristow 2002). Irrigation water can also carry dissolved nutrients, such as nitrogen, phosphorus, and potassium, which are essential for plant growth (Haynes 1985). As water infiltrates the soil, it dissolves nutrients present in the soil matrix and makes them available for plant uptake to ensure that crops have access to the necessary nutrients for their growth and development (Clothier and Green 1994). By providing water to the root zone, irrigation replenishes the soil moisture lost through evaporation and transpiration, and adequate soil moisture is essential for sufficient nutrient uptake, enzymatic processes, and overall plant health (Gan et al. 2013). It also helps regulate soil moisture, preventing water stress and promoting optimal crop performance. In certain regions, the irrigation water may contain dissolved salts and with repeated irrigation, these salts can accumulate in the soil, leading to soil salinization (Rengasamy 2006). Proper irrigation management involves applying water in quantities that exceed the plant's evapotranspiration rate, allowing excess water to leach the accumulated salts below the root zone. This process, known as leaching, helps prevent salt buildup and maintains a favorable soil environment for crop growth (Tukey Jr 1970).

Irrigation has been in practice for thousands of years, and it enables farmers to grow crops in areas that would otherwise be unsuitable for agriculture, increasing crop yields and helping to meet the growing demand for food (Tilman et al. 2002). According to the Food and Agriculture Organization (FAO), irrigation accounts for approximately 70% of global freshwater withdrawals every year (Siebert et al. 2010). With climate change affecting rainfall patterns and freshwater availability, efficient and effective irrigation practices are a high priority for ensuring global food security (Misra 2014). Science is the key to improve irrigation practices, from selecting appropriate crops, to designing efficient irrigation systems, to optimizing water usage. With the help of advanced technologies and data analysis, actionable science can provide farmers with valuable insights into soil moisture levels, crop water requirements, and other critical parameters, enabling them to make more informed decisions and manage their resources more sustainably (Evans and Sadler 2008). The increasing global population results in a growing demand for food, and irrigation is a critical factor in meeting this demand (Mancosu et al. 2015). However, improper irrigation practices can lead to water wastage, soil degradation, and decreased crop yield, resulting in significant economic losses for farmers and food shortages for the population (Pimentel et al. 1995). Science can help address these challenges by providing innovative solutions and strategies to optimize irrigation practices, for example, precision agriculture (Dobermann et al. 2004).

In this chapter, we will explore the various ways actionable science can improve irrigation practices, including case studies and examples of successful implementation. We will also discuss the challenges and limitations of actually applying these techniques in the fields and explore future research directions. This chapter also provides an overview of the current state of the art in actionable science for irrigation and point out the promising directions for further research and innovation to realize sustainable agriculture and protect global food security.

2 Current Irrigation Practice

2.1 Types of Irrigation Systems

Flood Irrigation

In flood irrigation, water is applied to the fields by flooding the entire surface of the soil (Mitchell and Van Genuchten 1993). Water is delivered to the fields via canals, ditches, or pipes and is allowed to flow across the fields, covering the soil surface and infiltrating into the soil. This method is typically used for crops such as rice (Westcott and Vines 1986) that require large amounts of water. However, it can be less efficient than other methods of irrigation and can result in soil erosion, as the water may cause soil to be washed away or deposited in other areas of the field (Patel et al. 2010).

For example, Fresno County in San Joaquin Valley, California, occasionally adopts flood irrigation in its history to fight with the semi-arid climate with hot and dry summers and mild winters (Alexander et al. 1990; Griffith et al. 2016; Schmidt and Sherman 1987). The primary water source for flood irrigation in this region is the Central Valley Project (Becker et al. 1976; De Roos 2000), which collects water from the Sierra Nevada mountains and stores it in reservoirs like the San Luis Reservoir (Carle 2015). Then, the water is delivered to Fresno County through a network of canals and distribution systems managed by the Fresno Irrigation District (FID) (Hopkins and Stretch 2009). Before flooding, the fields are prepared by leveling the land and constructing borders or bunds around the field perimeter to contain the water (Bachand et al. 2014). The borders help prevent water from flowing outside the intended area. The FID also coordinates with farmers to schedule water releases based on their water rights and the crop water requirements (Stretch and Mowry 2010). When it is time to irrigate, water is allowed to flow into the field by opening gates or using other irrigation structures. The water spreads across the field, covering it evenly and saturating the soil. In common practice, the level of flooding is carefully monitored to avoid over-irrigation and minimize water loss due to deep percolation (Bouman 2007). After the flooding, a soaking period is provided to allow the water to infiltrate the soil and reach the root zone of the crops (Bouman and Tuong 2001). This period can vary depending on factors such as soil type, crop water requirements, and evaporation rates. Once the desired soaking period is completed, excess water is drained or allowed to percolate back into the groundwater table. Drainage channels or outlets are strategically placed to remove excess water from the fields and prevent waterlogging (Robinson and Rycroft 1999). If everything goes well, the flooded fields will provide a favorable environment for crop growth, as the water helps nourish the plants and replenish the soil moisture (Shaxson and Barber 2003).

Sprinkler Irrigation

Sprinkler irrigation (Pair 1969) distributes water over the field in the form of a spray or sprinkler, simulating rainfall. It uses sprinkler heads that are connected to a network of pipes and valves and rotate them or have a fixed position and release water in a controlled manner, distributing it over the crops (Lyle and Bordovsky 1983).

Let us take a look at Maricopa County, Arizona, which is known for its arid climate and extensive agricultural activities. Due to the desert climate with hot summers and mild winters, the primary water sources for sprinkler irrigation in Maricopa County include surface water from the Central Arizona Project (CAP) canal system (Hanemann 2002), which delivers water from the Colorado River, and groundwater extracted from aquifers through wells. Before installing the sprinkler system, farmers usually prepare the fields by clearing debris, leveling the land, and installing necessary infrastructure such as pumps, filters, and irrigation lines. Sprinkler heads are strategically placed throughout the field, considering factors like crop spacing and water distribution uniformity requirements (Hargreaves and Merkley 1998). It operates based on a predetermined schedule and can be controlled manually or automated using timers and sensors. It can be adjusted to deliver the required amount of water based on crop type, growth stage, and environmental conditions. When the sprinkler system is activated, pressurized water is distributed through the sprinkler heads and the water is sprayed into the air in a controlled pattern and falls onto the crop canopy, providing irrigation (Zazueta et al. 1994). Also, the sprinkler system is designed to deliver water at a specific application rate, typically measured in inches per hour. This rate is determined based on factors such as soil infiltration rate, evapotranspiration rates, and crop water requirements (Amosson et al. 2002). Farmers in Maricopa County follow irrigation scheduling techniques based on factors like crop type, soil moisture monitoring, weather conditions, and plant growth stage (Eakin et al. 2016) to provide adequate water to the crops while avoiding over-irrigation and water wastage. Besides, regular maintenance of the sprinkler system is required to ensure its proper functioning, which includes checking for leaks, clogs, and damaged components, as well as adjusting the system to account for changes in crop needs and environmental conditions (Mutchek and Williams 2014).

Sprinkler irrigation allows for efficient and uniform water distribution, reducing water loss due to evaporation and minimizing runoff. It is suitable for a variety of crops, including grains, vegetables, and fruits. The flexibility of sprinkler heads allows farmers to adjust the irrigation patterns based on the crop's water requirements and growth stage. However, it is worth noting that sprinkler irrigation requires careful system design and maintenance (Louie and Selker 2000). Factors such as wind conditions, nozzle selection, and water pressure need to be considered to ensure optimal water distribution and minimize water waste.

Drip Irrigation

Drip irrigation directly delivers water to the plant's root zone in a slow, steady, and precise manner via a network of tubes or pipes with emitters or drippers that release water drop by drop at or near the plant's base (Camp 1998). Drip irrigation is highly efficient because it delivers water directly to the plants, minimizing evaporation and reducing water loss through runoff. It allows for precise control of water application, ensuring that each plant receives the right amount of water. Drip irrigation is commonly used for various crops, including tomatoes, fruits, vegetables, and even in landscaping applications. It is especially beneficial in arid regions with limited water availability. While drip irrigation requires careful planning and maintenance to avoid clogging of emitters and ensure proper water distribution, it is known for its water efficiency and ability to conserve resources. It also helps reduce weed growth and minimize soil erosion compared to other irrigation methods.

For example, Yolo County, California, uses drip irrigation (Johnstone et al. 2005; Mehta et al. 2013) and diverted waters from the close-by rivers into canals, which are managed by irrigation districts such as the Yolo County Flood Control and Water Conservation District (http://www.ycfcwcd.org). Farmers work with irrigation experts to design the drip irrigation system based on factors such as crop water requirements, soil characteristics, and field layout (Shock 2006). The design includes the selection and placement of drip lines, emitters, filters, and pressure regulators. Before installing the drip irrigation system, farmers prepare the fields by clearing vegetation, leveling the land, and ensuring proper soil drainage. Filtration systems are also installed in advance to prevent clogging of the drip emitters (Capra and Scicolone 2004). Then, drip lines are laid out along the rows or beds where the crops are planted. The drip lines consist of flexible tubes with evenly spaced emitters or drippers that deliver water directly to the root zone of the plants. Also, pressure regulators and filters are installed to ensure uniform water distribution and prevent damage to the emitters. The system is configured with control valves and tubing connectors to create zones that allow for precise control of water delivery.

When the drip irrigation system is activated, water flows from the water source through the main supply line and into the drip lines. Emitters release water slowly and directly to the base of each plant, delivering water at a controlled rate. Farmers can determine the appropriate irrigation schedule based on factors such as crop water requirements, weather conditions, and soil moisture monitoring (Thompson et al. 2007). This helps ensure that plants receive adequate water while avoiding over-irrigation. On the other hand, regular maintenance is necessary to keep the drip irrigation system functioning optimally. Farmers have to monitor the system for leaks, clogs, or damaged emitters, and replace or clean them as needed. They also monitor soil moisture levels to fine-tune irrigation schedules. Some drip irrigation systems also incorporate fertigation to allow the simultaneous application of water and fertilizers (Singandhupe et al. 2003).

Deficit Irrigation

Deficit irrigation reduces the amount of water applied to crops below the full crop water requirement (Fereres and Soriano 2007). Deficit irrigation involves intentionally applying less water than the crop's full water requirement. This approach introduces a certain level of risk, as water stress can potentially impact crop yield and quality for the return of more efficient water use. For example, in California's wine grape production, deficit irrigation is often employed during certain growth stages, such as berry development, to optimize grape quality and concentrate flavors (Permanhani et al. 2016). By carefully managing water inputs, growers can influence vine growth and grape composition, achieving desired characteristics in the final wine product. Scientific research and field trials have provided insights into the appropriate timing and extent of deficit irrigation for different grape varieties, helping wineries produce high-quality wines with less water.

However, there are several challenges associated with implementing it in the real world. Different crops have varying sensitivity to water stress during different growth stages. Determining the optimal timing and extent of deficit irrigation requires a deep understanding of the crop's water requirements, growth patterns, and response to water stress. It can be challenging to accurately assess the crop's tolerance to water deficit and ensure that yield and quality are not compromised. Also, implementing deficit irrigation practices requires precise water management decisions. Determining the appropriate deficit level, irrigation timing, and duration requires accurate monitoring of soil moisture, weather conditions, and crop development stages. Meantime, deficit irrigation practices need to be adapted to the specific agro-climatic conditions, soil types, and crop varieties of each region. Factors such as evapotranspiration rates, rainfall patterns, and soil characteristics vary across locations, making it challenging to generalize deficit irrigation strategies. Site-specific adaptation is also necessary to optimize the benefits of deficit irrigation in different agricultural settings. Lack of awareness, training, and access to technical expertise may hinder the adoption and proper implementation of deficit irrigation techniques. The economic viability of deficit irrigation practices depends on various factors, including crop value, market conditions, and production costs. Assessing the economic implications and ensuring a favorable cost-benefit ratio is difficult for widespread adoption. In some regions, water rights and regulatory frameworks may pose constraints on implementing deficit irrigation (Linker et al. 2016). Water allocation policies, permits, and restrictions may limit the flexibility of farmers to adopt deficit irrigation practices. Addressing policy barriers and establishing supportive frameworks can encourage the adoption of deficit irrigation and incentivize water-efficient practices.

2.2 *Parties Involved in Irrigation Decisions and Activities in the United States*

There are several parties and activities involved in a typical irrigation decision and practice in the United States. Farmers of course are the primary party responsible for making irrigation decisions and implementing irrigation practices on their land. There are also irrigation companies who manage the water distribution system, including canals, pipelines, and water delivery infrastructure (Shah and Bhattacharya 1993). They allocate water resources to farmers based on water rights and contracts. Water management agencies like EPA (Environment Protection Agency) (Solley et al. 1998) oversee water allocation and usage regulations, ensuring compliance with local, state, and federal laws. They may set water usage limits and promote water conservation practices. Local organizations or committee boards (Caponera and Nanni 2019) working for water conservation and promoting water-saving practices will also provide education on efficient irrigation techniques, and offer incentives for implementing water conservation measures.

Meanwhile, research institutions will conduct studies and research on irrigation techniques, crop water requirements, and water-use efficiency and their findings contribute to the development of best practices and guidelines for irrigation decision-making. Also, farmers often need to involve agricultural consultants who are independent professionals who offer expertise on irrigation practices, system design, and water management strategies tailored to specific crops and farming operations. Within the local government, especially those states with a large portion of agriculture in their GDP (Gross Domestic Product), there will be irrigation monitoring and enforcement departments to ensure compliance with water usage regulations, conduct inspections, and monitor water diversion and distribution systems (Ma and Ortolano 2000). For those water conservation districts, special districts will be established to manage and conserve water resources within a specific geographical area. They may also provide funding, technical support, and educational resources to farmers for implementing efficient irrigation practices. These parties and activities will be weaved together and collectively contribute to effective irrigation decision-making and practices in the United States, to optimize water usage, sustain crop productivity, and conserve water resources.

2.3 What Does a Day of Farmers Look Like During the Irrigation Season?

Farmers often start their day very early to make the most of daylight hours. They may begin by checking weather forecasts and monitoring irrigation schedules, and typically visit their fields to assess crop health, soil moisture levels, and overall irrigation needs (Wang and Cai 2009). They may walk or drive through the fields, observing plant growth, checking for signs of pests or diseases, and evaluating the effectiveness of previous irrigation applications. Farmers also allocate time for maintaining and inspecting their irrigation systems. This includes checking irrigation equipment, such as pumps, pipes, valves, and sprinklers, to ensure they are functioning properly. Repairs or adjustments may be made as necessary. Based on the data collected from soil moisture sensors, weather information, and field observations, farmers make decisions about irrigation scheduling and application rates. They calculate the water requirements of the crops, taking into account factors like evapotranspiration rates and soil moisture levels. If the farm operates under a water allocation system, farmers may need to manage their water resources efficiently. They allocate water to different fields of crops based on priority and water availability. This involves monitoring water usage, tracking water levels in reservoirs or canals, and complying with any regulations or restrictions in place.

They will operate the irrigation equipment by starting and stopping pumps, adjusting sprinkler settings, or managing automated irrigation systems to ensure that water is distributed evenly across the fields (Mermoud et al. 2005). Farmers usually maintain their detailed records of irrigation activities, such as the amount of water applied, irrigation start and stop times, and any observations or adjustments made. These records help in monitoring water usage, evaluating irrigation effectiveness, and planning for future irrigation cycles. In addition, farmers need to conduct other tasks such as crop planting, pest control, fertilization, harvesting, equipment maintenance, and managing farm personnel. Throughout the day, farmers keep an eye on weather conditions, soil moisture levels, and any changes in crop health. They may utilize technology, such as remote monitoring systems or mobile applications, to stay updated on real-time data and receive alerts if any irrigation issues arise. Towards the end of the day, farmers will review the day's activities, assess the effectiveness of irrigation, and plan for the next day's tasks.

2.4 How Does the Government Manage Irrigation? How Do Water Managers Allocate Water?

The governments establish regulatory frameworks, policies, and institutions to oversee water management practices. In many regions, the government allocates water rights and permits to users, including agricultural entities. These rights grant users the legal entitlement to a certain volume or flow of water for specific purposes, such as irrigation. Water rights may be based on seniority, beneficial use, or other allocation principles. Water managers, often operating under the authority of government agencies, develop water allocation plans (Molden 2013). These plans outline the distribution of available water resources among different user groups, including farmers. Allocation decisions consider factors such as water availability, demand, environmental considerations, and social priorities. Water managers employ various methods to allocate water among users. These methods can include predetermined allocations, proportional sharing based on water rights, rotational schedules, priority systems, or market-based mechanisms. The specific method used depends on local circumstances and water management goals. Water managers implement monitoring systems to track water usage, measure water flows, and assess water availability. This involves installing water meters, monitoring river/stream flows, and collecting data on groundwater levels. Accurate measurement helps in enforcing water rights, managing water allocations, and detecting unauthorized or excessive water use.

Governments may establish pricing mechanisms and incentives to encourage efficient water use (Dinar et al. 1997). This can include tiered pricing structures that charge higher rates for excessive water use or offering incentives for implementing water-efficient technologies and practices. The aim is to promote responsible water use and discourage wasteful practices. Governments often implement water conservation programs and initiatives to promote efficient irrigation practices. These programs may include educational campaigns, subsidies for water-saving technologies, funding for irrigation system upgrades, and support for research and development of water-efficient farming techniques. Government agencies engage with stakeholders, including farmers, water user associations, and environmental groups, to ensure a participatory approach to water management (Jonsson 2005). Consultation processes allow for input from various stakeholders, consideration of diverse perspectives, and the development of collaborative solutions. Governments enforce water regulations and allocate resources to monitor compliance with water allocation rules and regulations. Penalties may be imposed on those who violate water rights, exceed allocations, or engage in unauthorized water use. Enforcement actions are aimed at ensuring fairness, equity, and sustainability in water allocation practices.

3 Cutting-Edge Research for Irrigation and Struggles to Be Actionable

3.1 Artificial Intelligence (AI) and Machine Learning

AI and machine learning algorithms can analyze vast amounts of data collected from various sources (Sun et al. 2020), such as soil moisture sensors, weather forecasts, and crop characteristics. By processing this information, the algorithms can identify patterns, correlations, and trends to develop optimized irrigation strategies.

AI has great potential to enable precision irrigation by integrating real-time data with irrigation systems (Vianny et al. 2022). AI systems can continuously monitor and adjust irrigation based on dynamic factors, such as soil moisture levels, plant water stress, and weather conditions (Abioye et al. 2020). AI-based decision support systems also provide farmers with actionable insights and recommendations for irrigation management. By analyzing data from multiple sources and employing machine learning algorithms (Sun et al. 2019a, b), these systems can suggest optimal irrigation schedules, water application rates, and irrigation techniques tailored to specific crop types, soil conditions, and weather patterns. For instance, a study conducted in Spain demonstrated the application of AI algorithms for optimizing irrigation scheduling in olive orchards (Linaza et al. 2021). The researchers employed machine learning techniques to analyze data from soil moisture sensors and weather forecasts and automatically recommend irrigation schedules that optimized water use and maintained crop yield. The study showed significant water savings while improving olive tree growth and productivity. However, AI faces challenges in becoming actionable for irrigation in the United States due to several factors. AI models require high-quality data for accurate predictions and recommendations. In the case of irrigation, data on soil moisture, weather conditions, crop types, and land management practices must be reliable and readily available. In some regions, data may be scarce, inconsistent, or of low quality, making it challenging to develop AI models that can provide actionable insights. AI tools need to be compatible with existing irrigation infrastructure and technologies, which can vary widely. Ensuring that AI systems can integrate with different hardware and software platforms is still work in progress for practical implementation. Meanwhile, the adoption of AI technologies for irrigation may involve initial costs for hardware, software, and training. Ensuring affordability and accessibility for a broad range of farmers, including smallholders, is also crucial for widespread adoption. AI recommendations must consider environmental sustainability and water conservation goals. Striking a balance between maximizing crop yields and minimizing water usage and environmental impact can be a complex task for AI too.

3.2 Unmanned Aerial Vehicles (UAVs)

UAVs, commonly known as drones, have gained significant attention in agriculture, particularly for monitoring crop health, detecting water stress, and optimizing irrigation practices (Matese et al. 2018). UAVs equipped with various remote sensing technologies (Sun et al. 2014), such as multispectral or thermal cameras, LiDAR (Light Detection and Ranging) (Christiansen et al. 2017), or hyperspectral sensors, can capture high-resolution imagery and collect data on crop health parameters, including water stress indicators. These sensors capture reflected or emitted energy from crops, providing valuable information on vegetation vigor, leaf temperature, and water content. UAVs also enable farmers to gather detailed field information quickly and accurately and create irrigation prescriptions tailored to specific field conditions.

For example, researchers at the University of California, Davis conducted a study using UAVs to monitor crop water stress and guide precision irrigation in vineyards (Tang et al. 2022). The study utilized multispectral and thermal cameras mounted on UAVs to capture high-resolution imagery. The imagery was processed to derive vegetation indices and crop water stress indicators. This information helped identify areas of the vineyard experiencing water stress and guided irrigation interventions to target those areas specifically. The study demonstrated the potential of UAVs for optimizing irrigation practices and improving water management in vineyards. However, the operation of UAVs is subject to strict regulations by the federal aviation administration (FAA). Obtaining the necessary licenses and permissions for agricultural drone use can be a time-consuming and bureaucratic process in some states. Farmers and service providers must adhere to safety and airspace regulations, which can limit the flexibility and immediacy of drone-based irrigation monitoring. Most agricultural drones have limited flight endurance, typically ranging from 30 minutes to a few hours, depending on the model. This limited flight time may not cover the entire field or growing season, requiring multiple flights and battery changes, which can be inconvenient and time-consuming. Meanwhile, the costs could be a problem. The initial cost of purchasing drones and associated sensors, as well as ongoing maintenance and repair expenses, can be a significant financial burden for farmers, particularly smallholders. Additionally, training personnel to operate and maintain UAVs adds to the overall cost.

3.3 Nanotechnology for Water Delivery

Nanotechnology holds potential for precise water delivery to plant roots through nanoscale channels or coatings (Agrawal and Rathore 2014). Nanotechnology enables the design and fabrication of materials with nanoscale structures, such as carbon nanotubes or nanoscale coatings, that can facilitate water transport (Noy et al. 2007). These structures can create channels or coatings that effectively control the movement of water, allowing precise delivery to plant roots. For example, researchers at the Massachusetts Institute of Technology (MIT) have explored the use of carbon nanotubes as water transporters within the plant xylem, the tissue responsible for water transport from roots to leaves (Lew et al. 2020). They investigated the potential of carbon nanotubes to facilitate more efficient water uptake by plant roots, leading to reduced irrigation needs. But nanotechnology is a relatively new field, and there is still much to learn about the long-term effects and environmental impacts of nanomaterials in agricultural settings. Farmers and regulators may have concerns about the safety and potential risks associated with the use of nanotechnology in irrigation. The introduction of nanotechnology into agriculture is subject to regulatory approval in many countries. Ensuring that nanomaterials meet safety and environmental standards can be a lengthy and rigorous process, delaying their adoption for agricultural use. Nanoscale materials are typically designed for specific and targeted applications, making it difficult to apply them uniformly across extensive farmland. More importantly, nanomaterials can potentially accumulate in soil and water systems, raising environmental concerns. Understanding the ecological impact of nanotechnology in agriculture is essential for responsible adoption.

3.4 Soil Amendments

Soil amendments are the materials added to the soil to enhance its properties (Park et al. 2011). In the context of water retention, researchers have focused on incorporating hydrogels or superabsorbent polymers into the soil. These materials have the ability to absorb and retain large amounts of water, forming gel-like structures that hold moisture within the soil profile. This improves the soil's water holding capacity, reducing the frequency and volume of irrigation required. The retained water becomes available to plant roots, enhancing plant water availability, especially during periods of drought or limited water supply. By enhancing the water holding capacity of the soil, hydrogels or superabsorbent polymers can also minimize water losses through evaporation. The retained water remains within the soil, reducing the rate at which moisture is lost to the atmosphere. This allows for more efficient water use, as a larger proportion of applied water is utilized by the plants rather than being lost to evaporation. For example, a study conducted by Saha et al. (2020) investigated the effects of hydrogel amendment on water retention and plant growth in sandy soils. The research demonstrated that the addition of hydrogels significantly improved soil water holding capacity, resulting in enhanced plant growth and reduced irrigation requirements. The study highlights the potential of soil amendments like hydrogels in improving water availability and plant performance.

3.5 Nanostructured Materials

Nanotechnology involves manipulating materials and structures at the nanoscale (typically at the size of one billionth of a meter) to create new properties and functionalities (Nasrollahzadeh et al. 2019). In the context of irrigation, nanotechnology offers opportunities to develop materials with enhanced water-related characteristics, such as improved water retention, reduced evaporation, and controlled water movement (Lowry et al. 2019). Nanoscale particles can be incorporated into soil or coatings to modify the behavior of water in the irrigation system. These materials can alter the physical properties of the soil or create barriers that regulate water movement, leading to improved water management and increased water-use efficiency (Mauter et al. 2018). By adjusting the flow and distribution of water, nanotechnology can help maintain optimal soil moisture levels for plant growth and reduce water losses due to deep percolation or surface runoff. Nanomaterials can also be used to develop coatings or films that reduce evaporation from soil or water surfaces. These coatings create a barrier that limits the escape of water vapor, thereby reducing water losses and improving overall water-use efficiency. For instance, a study by Dimkpa and Bindraban (2017) explored the use of nanoscale hydroxyapatite particles to modify the properties of sandy soil and demonstrated that incorporating hydroxyapatite nanoparticles can increase the water holding capacity and improved soil water retention, leading to enhanced plant growth and water-use efficiency.

3.6 Desalination Technologies

Desalination is the process of removing salts and other impurities from saline water, making it suitable for irrigation purposes. Desalination technologies have seen significant advancements, driven by progress in physics and chemistry, allowing for the production of freshwater from seawater or brackish water sources. Reverse osmosis is a widely used desalination technology that utilizes a semi-permeable membrane to separate salts and impurities from water. Under pressure, water is forced through the membrane, leaving behind the dissolved salts and contaminants. The resulting freshwater can be used for various applications, including irrigation. Membrane distillation is an emerging desalination process that utilizes a hydrophobic membrane to separate water from saline solutions. The membrane allows water vapor to pass through while preventing the passage of salts and impurities. By creating a vapor-pressure difference, freshwater is generated through condensation, leaving behind concentrated brine. Advancements in physics and chemistry have enabled the development of more efficient and cost-effective desalination processes. These technologies have witnessed improvements in energy efficiency, membrane durability, and overall system performance, making desalination a viable option for providing irrigation water in regions facing water scarcity. Zein et al. (2023) investigated the performance of a solar-powered reverse osmosis desalination system for irrigation water supply in a remote agricultural area. The research proved the feasibility and effectiveness of using renewable energy sources to power desalination processes, providing a sustainable solution for irrigation water production.

3.7 Water Treatment and Recycling

Innovative approaches in water treatment and recycling have been developed to minimize water waste and improve the quality of irrigation water. Physicochemical processes, including coagulation, filtration, and advanced oxidation, are commonly employed to remove contaminants and pathogens from water sources. Coagulation (Palta et al. 2014) is a process in which chemicals called coagulants are added to water to destabilize and aggregate suspended particles, such as clay, silt, and organic matter. The formed aggregates, called flocs, can be easily removed through sedimentation or filtration processes. Coagulation is an effective method for reducing turbidity and removing particulate matter from water. Filtration involves passing water through a porous medium, such as sand, activated carbon, or membranes, to remove suspended particles, microorganisms, and other contaminants. Different types of filters, including rapid sand filters, multimedia filters, and membrane filters, are used in water treatment systems. Filtration helps improve water clarity and removes a wide range of impurities. Advanced oxidation processes utilize powerful oxidants, such as ozone, ultraviolet (UV) light, and hydrogen peroxide, to break down and remove organic pollutants, pesticides, and pathogens from water. These processes generate highly reactive hydroxyl radicals that can effectively degrade and transform organic compounds into harmless byproducts. Norton-Brandão et al. (2013) investigated the use of coagulation, filtration, and UV disinfection for treating wastewater to be used for agricultural irrigation. The research demonstrated the effectiveness of these physicochemical processes in removing suspended solids, pathogens, and contaminants, ensuring the safety and quality of the irrigation water.

3.8 *Vertical Farming*

Vertical farming is an innovative approach to agriculture that involves growing crops in vertically stacked layers or on vertically inclined surfaces using artificial lighting. Vertical farming systems create controlled environments where temperature, humidity, lighting, and nutrient levels can be precisely regulated. This allows for optimal growing conditions and maximizes crop productivity throughout the year, independent of external climate variations. Vertical farming methods, such as hydroponics and aeroponics, utilize water in a highly efficient manner. Hydroponic systems circulate a nutrient-rich solution to provide plants with the necessary nutrients, while aeroponic systems mist the plant roots with a nutrient solution. These systems typically use significantly less water compared to traditional soil-based farming methods, as water is recirculated within the system, reducing water waste and evaporation. Vertical farming relies on artificial lighting systems, such as LED lights, to provide the necessary light energy for plant growth (Wong et al. 2020). LED lights can be tailored to emit specific wavelengths of light that are optimal for plant photosynthesis, resulting in energy-efficient lighting and targeted crop growth. For example, AeroFarms and Plenty are two prominent companies that have implemented vertical farming systems (O'sullivan et al. 2019). Both operate indoor vertical farms that utilize aeroponic systems and LED lighting to grow leafy greens and herbs in densely stacked trays.

3.9 *Aquaponics*

Aquaponics is a sustainable farming method that combines aquaculture (fish farming) with hydroponics (soilless plant cultivation) (Savidov et al. 2005). The system utilizes the waste produced by fish to provide nutrients for plants, while the plants filter the water, creating a symbiotic relationship. Aquaponics systems can be implemented in urban areas and are known for their high water and land use efficiency. Companies like Growing Power (Proksch et al. 2019) have pioneered the use of aquaponics in farming. The main hurdle for its actionableness is the initial cost. Setting up an aquaponics system can require a significant initial investment in infrastructure, including tanks, pumps, filtration systems, and grow beds. The costs associated with building and maintaining the system can be a barrier for many farmers, particularly small-scale or resource-constrained operations.

3.10 Genetic Engineering and Biotechnology

Advances in genetic engineering and biotechnology offer opportunities to develop crop varieties with enhanced traits, such as drought tolerance, disease resistance, and increased nutrient content (Bhalla 2006). These technologies aim to develop climate-resilient crops that can thrive under changing environmental conditions with limited irrigation. For example, genetically modified (GM) crops like drought-tolerant maize (corn) have been developed to withstand water scarcity (Kumar et al. 2020). Organizations like the International Maize and Wheat Improvement Center (CIMMYT) (Singh 1988) have been actively involved in research on climate-resilient crop varieties. These genetic modifications can raise regulatory and safety concerns, both in terms of environmental impact and potential health risks. Navigating complex regulatory frameworks can be a significant barrier to the adoption of genetically modified crops for irrigation. Importantly, genetically modified organisms (GMOs) can face resistance from the public and consumer groups who have concerns about the safety of such crops. Negative public perception can lead to market rejection and reluctance among farmers to adopt genetically engineered crops for irrigation. In addition, over time, pests and pathogens may develop resistance to genetically modified crops, reducing their effectiveness and requiring ongoing research and development efforts to stay ahead of evolving threats.

4 Successful Use Cases of Science with High Actionableness

4.1 Improving Irrigation Efficiency Using Remote Sensing and Soil Moisture Data

By combining remote sensing and soil moisture data, stakeholders can better understand crop water requirements and make informed decisions about when and how much to irrigate (Deines et al. 2021). The Yakima Basin Agricultural Water Enhancement Program in Washington State (Vano et al. 2010) is a collaborative effort between the Natural Resources Conservation Service (NRCS) (Schaefer et al. 2007), the Washington State Department of Ecology, and local farmers, and aims to improve water-use efficiency and agricultural sustainability in the Yakima Basin region (Fig. 8.1).

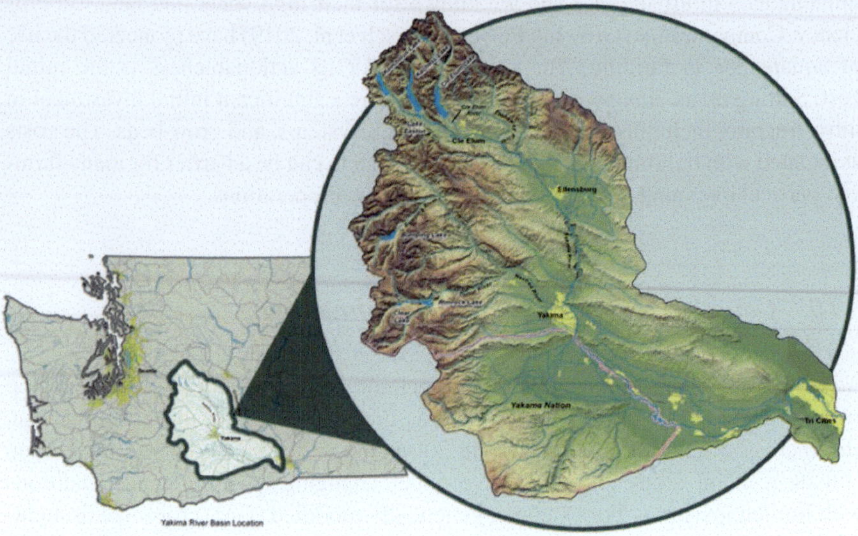

Fig. 8.1 Yakima River Basin. (Image courtesy: https://apps.ecology.wa.gov/publications/documents/1512003.pdf)

By analyzing the satellite imagery and aerial survey data, the program assessed the health of the crops, such as examining vegetation indices, which indicate the overall health and vigor of the plants. Unhealthy or stressed areas can be identified, enabling targeted interventions. Satellite imagery and aerial surveys also help in monitoring water stress levels in the crops (Sheffield et al. 2018). By analyzing indicators like canopy temperature and infrared reflectance, the program identified areas experiencing water stress, which require additional irrigation or water management strategies (Vano et al. 2010). Soil moisture measurements provide quantitative data on the amount of moisture present in the soil at different depths. The sensors continuously monitor and collect soil moisture data over time. The measurements can be logged manually or transmitted wirelessly to a central data management system. Then, the collected soil moisture data is analyzed and interpreted to understand the moisture distribution within the soil profile. This information helps in determining the overall water availability and moisture status in different soil layers. Irrigation scheduling recommendations, generated based on the collected data, are customized to each farmer's specific needs and take into account factors like crop water requirements, spatial variability in soil moisture, and other relevant parameters. Web-based platforms or mobile applications serve as decision support tools in the program. Farmers can access these tools, typically through their devices or computers, to receive the tailored irrigation scheduling recommendations. The decision support tools provide real-time updates on crop health, soil moisture conditions, and irrigation recommendations. The farmers can align their irrigation activities with the specific needs of their crops and the moisture conditions in the soil, thus improving irrigation efficiency and conserving water resources.

4.2 *Managing Irrigation During Drought Using Climate Forecasting and Soil Moisture Monitoring*

Managing irrigation during drought using climate forecasting and soil moisture monitoring involves using weather forecasts and soil moisture data to make informed decisions about irrigation during periods of drought. This can help to conserve water resources and maintain crop yields. For example, the Murray-Darling Basin in Australia, one of the country's largest agricultural regions, has experienced prolonged drought periods in the past (Leblanc et al. 2012). To manage irrigation during such drought conditions, the Murray-Darling Basin Authority (MDBA) implemented a program that combines climate forecasting and soil moisture monitoring (Grafton 2017). The program utilizes climate forecasting model results which take into account various factors such as atmospheric pressure, temperature, humidity, and oceanic conditions to generate forecasts. It also gathers data from meteorological stations strategically located throughout the Murray-Darling Basin. By analyzing these data together, the program can predict rainfall patterns and estimate future weather conditions. Relying on climate forecasting and soil moisture monitoring data, water managers and farmers can gain valuable information to guide their decision-making process. If forecasts indicate periods of increased rainfall, they can delay or reduce irrigation to avoid overwatering. Or, if dry conditions are anticipated or soil moisture levels are low, they can schedule irrigation to supplement the water needs of the crops. These informed decisions can lead to improved crop health and yield. Crops will receive the right amount of water at the right time, minimizing water stress and maximizing productivity.

4.3 *Precision Irrigation for Water and Energy Savings*

Precision irrigation is a water management approach that uses technology to apply water only where and when it is needed. This approach can help to conserve water resources, reduce energy use, and improve crop yields. For example, the FieldNET Precision Irrigation system can enable farmers to optimize water and energy usage while improving crop yields (Saiz-Rubio and Rovira-Más 2020; Cohen et al. 2021). One successful implementation of this system can be found in the cornfields of Nebraska, USA. Similarly in the fields, soil moisture sensors are strategically placed in the fields to gather accurate and representative data. They are typically distributed across various locations within the fields to capture the spatial variability of moisture content.

Besides soil moisture, ET (evapotranspiration) monitoring technology is integrated into the precision irrigation system (Abioye et al. 2020). It measures the rate of water loss from the soil through evaporation and the transpiration of plants. This technology provides valuable information on the water needs of crops and typically involves advanced irrigation methods such as drip irrigation or center pivot systems

Fig. 8.2 Standard central pivot facilities in the soybean and wheat fields in Nebraska

(Fig. 8.2). In Lindsay precision irrigation system (Bhatti et al. 2020), ET technology is integrated to enhance irrigation management. With the information obtained from ET monitoring and weather data, farmers can make informed decisions to adjust their irrigation practices accordingly. They can optimize the timing, duration, and application rates of irrigation to match the actual water requirements of the crops.

Precision irrigation system incorporates Variable Rate Irrigation (VRI) technology, which allows for the application of different amounts of water across the field based on crop variability, soil conditions, and water requirements. VRI takes into account factors like crop evapotranspiration rates, plant water stress levels, and desired moisture levels in the root zone to deliver water precisely where and when it is needed. It eliminates the irrigation of non-cropped areas or areas with excess moisture, reducing unnecessary water usage and potential environmental impacts.

5 Suggestions for Improving Actionableness

5.1 Suggestions for Scientists

Agricultural scientists should actively engage and collaborate with farmers, irrigation practitioners, water managers, and policymakers throughout the research process (Gonsalves 2005). Involving stakeholders in study design, data collection, and interpretation helps ensure that research findings are relevant, practical, and applicable to real-world irrigation challenges. Conduct field validation trials to assess the performance and feasibility of research outcomes in real agricultural settings (Jones et al. 1998). Demonstrating the effectiveness of new irrigation technologies, practices, or management approaches under practical conditions can enhance their adoption and implementation. Also, conduct long-term studies to capture the effects and sustainability of irrigation interventions over extended periods as longitudinal research provides valuable insights into the long-term impacts. Assess the potential benefits, such as increased crop yield, water savings, and reduced energy costs, in relation to the investment required. This analysis can help stakeholders evaluate the investment of new irrigation technologies or practices. Try to develop practical guidelines and protocols based on research findings and convert scientific knowledge into actionable recommendations that can be easily understood and implemented by farmers, irrigation practitioners, and policymakers. Also, scientists need to provide clear step-by-step instructions, decision-support tools, or software platforms to facilitate the practical application of research outcomes. Educational resources, workshops, and training programs will be very helpful to increase awareness and knowledge among farmers and irrigation professionals. Meanwhile, stay engaged with policymakers and relevant institutions to advocate for policies that support sustainable irrigation practices, and share research findings, data, and evidence-based recommendations to influence policy development and encourage the adoption of water-efficient irrigation technologies and management strategies. It is part of scientists' obligation to establish mechanisms for continuous monitoring, evaluation, and feedback on the implementation of research outcomes. After the research is in practice, scientists need to pay attention to collect data on the adoption, performance, and challenges faced by stakeholders when applying recommended irrigation practices. This feedback loop helps refine and improve research interventions, ensuring ongoing actionability.

5.2 *Suggestions for Farmers*

Farmers should actively seek out education and training programs related to irrigation management. By increasing their knowledge and understanding of irrigation principles, technologies, and best practices, farmers can make informed decisions and implement effective irrigation strategies on their farms. Participate in on-farm demonstrations organized by agricultural extension services, research institutions, or industry experts. These demonstrations provide an opportunity to witness and learn firsthand about the practical implementation of irrigation technologies and management techniques. They should consider investing in soil moisture monitoring tools and sensors, as regularly monitoring soil moisture helps farmers make informed decisions regarding irrigation scheduling, ensuring that water is applied when and where it is needed, leading to improved water use efficiency and crop performance. Develop and follow a well-defined irrigation schedule based on crop water requirements, weather conditions, and soil characteristics. Consult with irrigation experts or use available online tools and mobile applications that provide customized irrigation recommendations based on local conditions and crop type. Also, consider the adoption of efficient irrigation technologies such as drip irrigation, micro-sprinklers, or precision irrigation systems. These technologies help deliver water directly to the root zone, reducing water losses due to evaporation or runoff and maximizing water use efficiency. Farmers usually need to inspect and maintain irrigation equipment to ensure optimal performance, and meantime check for leaks, clogs, or damaged components that can affect the distribution of water. Proper maintenance helps prevent water wastage and ensures the longevity of irrigation infrastructure. Stay engaged with fellow farmers, participate in farmer networks, and attend community meetings or workshops focused on irrigation management. Utilize available data and information sources from the science communities and government, and access weather forecasts, soil moisture data, and crop water requirement calculators to make data-driven decisions on irrigation timing and application rates.

5.3 Suggestions for Government and Water Managers

It is recommended to adopt an integrated approach to water management that considers various stakeholders, including farmers, industries, and environmental concerns. Implement fair and transparent water allocation and pricing policies that encourage efficient water use and incentivize conservation practices. Use tools such as water rights systems, water trading mechanisms, and tiered pricing structures to promote responsible irrigation practices and encourage farmers to adopt water-saving technologies. Governments can improve or expand their financial incentives, grants, or subsidies to farmers for adopting efficient irrigation technologies and practices. These incentives can help offset the initial investment costs and encourage farmers to upgrade their irrigation systems, thereby improving water use efficiency and reducing water losses. Also, it is important to offer technical assistance, training programs, and capacity-building initiatives to farmers and irrigation professionals. Establish clear regulatory frameworks for irrigation management, including water usage limits, reporting requirements, and enforcement mechanisms. Government agencies are obligated to invest in research and development initiatives that focus on irrigation technologies, water management strategies, and climate adaptation measures, and support research institutions, universities, and agricultural extension services to develop innovative solutions and provide evidence-based recommendations for irrigation management. Make public data platforms accessible to farmers and water managers, allowing them to make informed decisions regarding irrigation planning and water allocation. Establish fundamental monitoring systems to assess the effectiveness of irrigation management strategies and programs and regularly evaluate the impact of interventions, identify areas for improvement, and adapt policies and practices accordingly.

6 Conclusion

This chapter overviewed the current practice of irrigation and the existing challenges. For example, the availability of reliable and accurate data is essential for implementing effective irrigation management practices, which can be challenging to get especially in developing countries or remote areas. Also, different data sources may not be compatible with each other, making integration and analysis difficult. Meanwhile, the irrigation management practices often require the use of advanced technology, such as sensors, remote sensing tools, and modeling software, which are inaccessible to many farmers, or they may lack the technical expertise to use effectively. It can also be unaffordable, especially for small-scale farmers. To address these challenges, various strategies should be employed, such as making data accessible, reliable, and compatible across different sources, providing support to make technology more usable to farmers via subsidies, free training, and technical assistance, developing affordable technology options, or providing financing or incentives for farmers to adopt these practices without breaking the bank or worrying about the invest-return issues.

Future directions are focused on addressing the challenges of water scarcity, increasing agricultural productivity, and reducing environmental impacts. Smart irrigation systems (Sun and Di 2021) use real-time data to determine irrigation needs and apply water precisely, reducing water waste and increasing efficiency. The integration of artificial intelligence, machine learning (Sun et al. 2019a, b), and the Internet of Things (IoT) technologies can further optimize irrigation management. Climate change is expected to increase the frequency and intensity of extreme weather events like droughts and floods (Kumar et al. 2018). As water scarcity becomes increasingly severe, finding alternative water sources for irrigation like treated wastewater or desalinated water will become more important (DeNicola et al. 2015). Developing technologies that can treat and use alternative water sources effectively is a promising area of research. Also, precision agriculture will continue to be a popular research topic in the coming years. There are many other promising directions in the fundamental research of the soil, seed, water, and environment which could greatly enhance our ability to produce adequate food while controlling the water use in irrigation to sustainably secure food supply without damaging the environment and climate. This chapter aims to establish the groundwork for the ongoing research and hope to facilitate a seamless transition from laboratory concepts to practical field solutions, making it easier to implement advanced irrigation practices in the future .

References

Abioye, Emmanuel Abiodun, Mohammad Shukri Zainal Abidin, Mohd Saiful Azimi Mahmud, Salinda Buyamin, Mohamad Hafis Izran Ishak, Muhammad Khairie Idham Abd Rahman, Abdulrahaman Okino Otuoze, Patrick Onotu, and Muhammad Shahrul Azwan Ramli. 2020. A review on monitoring and advanced control strategies for precision irrigation. *Computers and Electronics in Agriculture* 173: 105441.

Agrawal, Shweta, and Pragya Rathore. 2014. Nanotechnology pros and cons to agriculture: A review. *International Journal of Current Microbiology and Applied Sciences* 3 (3): 43–55.

Alexander, Barton Stone, George Henry Mendell, and George Davidson. 1990. *Engineers and irrigation: Report of the Board of Commissioners on the irrigation of the San Joaquin, Tulare, and Sacramento Valleys of the State of California, 1873*, No. 1–39. Vol. 870. Office of History US Army Corps of Engineers.

Amosson, Stephen H., Leon New, Lal Almas, Fran Bretz, and Thomas Marek. 2002. *Economics of irrigation systems*. Texas FARMER Collection.

Bachand, Philip A.M., Sujoy B. Roy, Joe Choperena, Don Cameron, and William R. Horwath. 2014. Implications of using on-farm flood flow capture to recharge groundwater and mitigate flood risks along the Kings River, CA. *Environmental Science & Technology* 48 (23): 13601–13609.

Becker, Leonard, Donald Sparks, Dan M. Fults, and William W.G. Yeh. 1976. Operations models for central valley project. *Journal of the Water Resources Planning and Management Division* 102 (1): 101–115.

Bhalla, Prem L. 2006. Genetic engineering of wheat–current challenges and opportunities. *Trends in Biotechnology* 24 (7): 305–311.

Bhatti, Sandeep, Derek M. Heeren, J. Burdette Barker, Christopher M.U. Neale, Wayne E. Woldt, Mitchell S. Maguire, and Daran R. Rudnick. 2020. Site-specific irrigation management in a sub-humid climate using a spatial evapotranspiration model with satellite and airborne imagery. *Agricultural Water Management* 230: 105950.

Bos, Marinus Gijsberthus, and J. Nugteren. 1990. *On irrigation efficiencies*, No. 19. ILRI.

Bouman, B. A. M. 2007. Water management in irrigated rice: coping with water scarcity. *International Rice Research Institute*.

Bouman, B. M., and To Phuc Tuong. 2001. Field water management to save water and increase its productivity in irrigated lowland rice. *Agricultural Water Management* 49 (1): 11–30.

Camp, C.R. 1998. Subsurface drip irrigation: A review. *Transactions of the ASAE* 41 (5): 1353.

Caponera, Dante A., and Marcella Nanni. 2019. *Principles of water law and administration: National and international*. Routledge.

Capra, A., and B. Scicolone. 2004. Emitter and filter tests for wastewater reuse by drip irrigation. *Agricultural Water Management* 68 (2): 135–149.

Carle, David. 2015. *Introduction to water in California*. Univ of California Press.

Christiansen, Martin Peter, Morten Stigaard Laursen, Rasmus Nyholm Jørgensen, Søren Skovsen, and René Gislum. 2017. Designing and testing a UAV mapping system for agricultural field surveying. *Sensors* 17 (12): 2703.

Clothier, Brent E., and Steven R. Green. 1994. Rootzone processes and the efficient use of irrigation water. *Agricultural Water Management* 25 (1): 1–12.

Cohen, Yafit, George Vellidis, Carlos Campillo, Vasileios Liakos, Nitsan Graff, Yehoshua Saranga, John L. Snider, Jaume Casadesús, Sandra Millán, and Maria del Henar Prieto. 2021. Applications of sensing to precision irrigation. In *Sensing approaches for precision agriculture*, 301–332. Cham: Springer.

De Roos, Robert W. 2000. *The thirsty land: The story of the central Valley project*. Beard Books.

Deines, Jillian M., Anthony D. Kendall, James J. Butler, Bruno Basso, and David W. Hyndman. 2021. Combining remote sensing and crop models to assess the sustainability of stakeholder-driven groundwater management in the US High Plains aquifer. *Water Resources Research* 57 (3): e2020WR027756.

DeNicola, Erica, Omar S. Aburizaiza, Azhar Siddique, Haider Khwaja, and David O. Carpenter. 2015. Climate change and water scarcity: The case of Saudi Arabia. *Annals of Global Health* 81 (3): 342–353.

Dimkpa, Christian O., and Prem S. Bindraban. 2017. Nanofertilizers: New products for the industry? *Journal of Agricultural and Food Chemistry* 66 (26): 6462–6473.

Dinar, Ariel, Mark W. Rosegrant, and Ruth Suseela Meinzen-Dick. 1997. *Water allocation mechanisms: Principles and examples*, No. 1779. World Bank Publications.

Dobermann, Achim, Simon Blackmore, Simon E. Cook, and Viacheslav I. Adamchuk. 2004. Precision farming: Challenges and future directions. In *Proceedings of the 4th international crop science congress*, Vol. 26. Brisbane, Australia.

Eakin, Hallie, Abigail York, Rimjhim Aggarwal, Summer Waters, Jessica Welch, Cathy Rubiños, Skaidra Smith-Heisters, Chrissie Bausch, and John M. Anderies. 2016. Cognitive and institutional influences on farmers' adaptive capacity: Insights into barriers and opportunities for transformative change in Central Arizona. *Regional Environmental Change* 16: 801–814.

Evans, Robert G., and E. John Sadler. 2008. Methods and technologies to improve efficiency of water use. *Water Resources Research* 44 (7). https://doi.org/10.1029/2007WR006200.

Fereres, Elias, and María Auxiliadora Soriano. 2007. Deficit irrigation for reducing agricultural water use. *Journal of Experimental Botany* 58 (2): 147–159.

Gan, Yantai, Kadambot H.M. Siddique, Neil C. Turner, Xiao-Gang Li, Jun-Yi Niu, Chao Yang, Liping Liu, and Qiang Chai. 2013. Ridge-furrow mulching systems—An innovative technique for boosting crop productivity in semiarid rain-fed environments. *Advances in Agronomy* 118: 429–476.

Gonsalves, Julian F. 2005. *Participatory research and development for sustainable agriculture and natural resource management: A sourcebook*. Vol. 1. IDRC.

Grafton, R. Quentin. 2017. Water reform and planning in the Murray–Darling Basin, Australia. *Water Economics and Policy* 3 (03): 1702001.

Griffith, Glenn E., James M. Omernik, David W. Smith, Terry D. Cook, E.D. Tallyn, Kendra Moseley, and Colleen B. Johnson. 2016. *Ecoregions of California*. US Geological Survey Open-File Report 1021.

Hanemann, W. Michael. 2002. *The central Arizona project*. California Agricultural Experiment Station, Giannini Foundation of Agricultural Economics.

Hargreaves, George H., and Gary P. Merkley. 1998. *Irrigation fundamentals: An applied technology text for teaching irrigation at the intermediate level*. Water Resources Publication.

Haynes, R.J. 1985. Principles of fertilizer use for trickle irrigated crops. *Fertilizer Research* 6: 235–255.

Hopkins, Randy, and Bill Stretch. 2009. *Developing new water supplies in Fresno Irrigation district the Waldron banking facilities*. PhD dissertation, Colorado State University: Libraries.

Hopmans, Jan W., and Keith L. Bristow. 2002. Current capabilities and future needs of root water and nutrient uptake modeling. *Advances in Agronomy* 77: 103–183.

Johnstone, P.R., T.K. Hartz, M. LeStrange, J.J. Nunez, and E.M. Miyao. 2005. Managing fruit soluble solids with late-season deficit irrigation in drip-irrigated processing tomato production. *HortScience* 40 (6): 1857–1861.

Jones, J.W., G.Y. Tsuji, G. Hoogenboom, L.A. Hunt, P.K. Thornton, P.W. Wilkens, D.T. Imamura, W.T. Bowen, and U. Singh. 1998. Decision support system for agrotechnology transfer: DSSAT v3. In *Understanding options for agricultural production*, 157–177. Springer.

Jonsson, Anna. 2005. Public participation in water resources management: Stakeholder voices on degree, scale, potential, and methods in future water management. *Ambio: A Journal of the Human Environment* 34 (7): 495–500.

Kumar, Praveen, Jayanti Tokas, Naresh Kumar, Manohar Lal, and H.R. Singal. 2018. Climate change consequences and its impact on agriculture and food security. *International Journal of Chemical Studies* 6 (6): 124–133.

Kumar, Krishan, Geetika Gambhir, Abhishek Dass, Amit Kumar Tripathi, Alla Singh, Abhishek Kumar Jha, Pranjal Yadava, Mukesh Choudhary, and Sujay Rakshit. 2020. Genetically modified crops: Current status and future prospects. *Planta* 251: 1–27.

Leblanc, Marc, Sarah Tweed, Albert Van Dijk, and Bertrand Timbal. 2012. A review of historic and future hydrological changes in the Murray-Darling Basin. *Global and Planetary Change* 80: 226–246.

Lew, Tedrick Thomas Salim, Volodymyr B. Koman, Pavlo Gordiichuk, Minkyung Park, and Michael S. Strano. 2020. The emergence of plant nanobionics and living plants as technology. *Advanced Materials Technologies* 5 (3): 1900657.

Linaza, Maria Teresa, Jorge Posada, Jürgen Bund, Peter Eisert, Marco Quartulli, Jürgen Döllner, Alain Pagani, et al. 2021. Data-driven artificial intelligence applications for sustainable precision agriculture. *Agronomy* 11 (6): 1227.

Linker, Raphael, Ilya Ioslovich, Georgios Sylaios, Finn Plauborg, and Adriano Battilani. 2016. Optimal model-based deficit irrigation scheduling using AquaCrop: A simulation study with cotton, potato and tomato. *Agricultural Water Management* 163: 236–243.

Louie, Michael J., and John S. Selker. 2000. Sprinkler head maintenance effects on water application uniformity. *Journal of Irrigation and Drainage Engineering* 126 (3): 142–148.

Lowry, Gregory V., Astrid Avellan, and Leanne M. Gilbertson. 2019. Opportunities and challenges for nanotechnology in the agri-tech revolution. *Nature Nanotechnology* 14 (6): 517–522.

Lyle, W.M., and J.P. Bordovsky. 1983. LEPA irrigation system evaluation. *Transactions of the ASAE* 26 (3): 776–0781.

Ma, Xiaoying, and Leonard Ortolano. 2000. *Environmental regulation in China: Institutions, enforcement, and compliance*. Rowman & Littlefield Publishers.

Mancosu, Noemi, Richard L. Snyder, Gavriil Kyriakakis, and Donatella Spano. 2015. Water scarcity and future challenges for food production. *Water* 7 (3): 975–992.

Matese, Alessandro, Rita Baraldi, Andrea Berton, Carla Cesaraccio, Salvatore Filippo Di Gennaro, Pierpaolo Duce, Osvaldo Facini, Massimiliano Giuseppe Mameli, Alessandra Piga, and Alessandro Zaldei. 2018. Estimation of water stress in grapevines using proximal and remote sensing methods. *Remote Sensing* 10 (1): 114.

Mauter, Meagan S., Ines Zucker, Francois Perreault, Jay R. Werber, Jae-Hong Kim, and Menachem Elimelech. 2018. The role of nanotechnology in tackling global water challenges. *Nature Sustainability* 1 (4): 166–175.

Mehta, Vishal K., Van R. Haden, Brian A. Joyce, David R. Purkey, and Louise E. Jackson. 2013. Irrigation demand and supply, given projections of climate and land-use change, in Yolo County, California. *Agricultural Water Management* 117: 70–82.

Mermoud, A., T.D. Tamini, and H. Yacouba. 2005. Impacts of different irrigation schedules on the water balance components of an onion crop in a semi-arid zone. *Agricultural Water Management* 77 (1–3): 282–295.

Misra, Anil Kumar. 2014. Climate change and challenges of water and food security. *International Journal of Sustainable Built Environment* 3 (1): 153–165.

Mitchell, Alan R., and M.Th. Van Genuchten. 1993. Flood irrigation of a cracked soil. *Soil Science Society of America Journal* 57 (2): 490–497.

Molden, David, ed. 2013. *Water for food water for life: A comprehensive assessment of water management in agriculture*. Routledge.

Mutchek, Michele, and Eric Williams. 2014. Moving towards sustainable and resilient smart water grids. *Challenges* 5 (1): 123–137.

Nasrollahzadeh, Mahmoud, S. Mohammad Sajadi, Mohaddeseh Sajjadi, and Zahra Issaabadi. 2019. An introduction to nanotechnology. In *Interface science and technology*, vol. 28, 1–27. Elsevier.

Norton-Brandão, Diana, Sigrid M. Scherrenberg, and Jules B. van Lier. 2013. Reclamation of used urban waters for irrigation purposes–a review of treatment technologies. *Journal of Environmental Management* 122: 85–98.

Noy, Aleksandr, Hyung Gyu Park, Francesco Fornasiero, Jason K. Holt, Costas P. Grigoropoulos, and Olgica Bakajin. 2007. Nanofluidics in carbon nanotubes. *Nano Today* 2 (6): 22–29.

O'sullivan, C. A., G. D. Bonnett, C. L. McIntyre, Z. Hochman, and A. P. Wasson. 2019. Strategies to improve the productivity, product diversity and profitability of urban agriculture. *Agricultural Systems* 174: 133–144.

Pair, Claude H. 1969. *Sprinkler irrigation*. Sprinkler Irrigation Association.

Palta, Sanjeev, Richa Saroa, and Anshu Palta. 2014. Overview of the coagulation system. *Indian Journal of Anaesthesia* 58 (5): 515.

Park, Jin Hee, Dane Lamb, Periyasamy Paneerselvam, Girish Choppala, Nanthi Bolan, and Jae-Woo Chung. 2011. Role of organic amendments on enhanced bioremediation of heavy metal (loid) contaminated soils. *Journal of Hazardous Materials* 185 (2–3): 549–574.

Patel, D.P., Das Anup, G.C. Munda, P.K. Ghosh, Juri Sandhya Bordoloi, and Manoj Kumar. 2010. Evaluation of yield and physiological attributes of high-yielding rice varieties under aerobic and flood-irrigated management practices in mid-hills ecosystem. *Agricultural Water Management* 97 (9): 1269–1276.

Permanhani, M., J. Miguel Costa, M.A.F. Conceição, R.T. De Souza, M.A.S. Vasconcellos, and M.M. Chaves. 2016. Deficit irrigation in table grape: Eco-physiological basis and potential use to save water and improve quality. *Theoretical and Experimental Plant Physiology* 28: 85–108.

Pimentel, David, Celia Harvey, Pradnja Resosudarmo, K. Sinclair, D. Kurz, M. McNair, S. Crist, et al. 1995. Environmental and economic costs of soil erosion and conservation benefits. *Science* 267 (5201): 1117–1123.

Proksch, Gundula, Alex Ianchenko, and Benz Kotzen. 2019. Aquaponics in the built environment. In *Aquaponics food production systems: Combined aquaculture and hydroponic production technologies for the future*, 523–558. Springer.

Rawls, W.J., D. Gimenez, and R. Grossman. 1998. Use of soil texture, bulk density, and slope of the water retention curve to predict saturated hydraulic conductivity. *Transactions of the ASAE* 41 (4): 983.

Rengasamy, Pichu. 2006. World salinization with emphasis on Australia. *Journal of Experimental Botany* 57 (5): 1017–1023.

Robinson, M., and D.W. Rycroft. 1999. The impact of drainage on streamflow. *Agricultural Drainage* 38: 767–800.

Saha, Abhisekh, Sreedeep Sekharan, and Uttam Manna. 2020. Superabsorbent hydrogel (SAH) as a soil amendment for drought management: A review. *Soil and Tillage Research* 204: 104736.

Saiz-Rubio, Verónica, and Francisco Rovira-Más. 2020. From smart farming towards agriculture 5.0: A review on crop data management. *Agronomy* 10 (2): 207.

Savidov, N. A., E. Hutchings, and J. E. Rakocy. 2005. Fish and plant production in a recirculating aquaponic system: a new approach to sustainable agriculture in Canada. In International Conference and Exhibition on Soilless Culture: ICESC 742, pp. 209–221.

Schaefer, Garry L., Michael H. Cosh, and Thomas J. Jackson. 2007. The USDA natural resources conservation service soil climate analysis network (SCAN). *Journal of Atmospheric and Oceanic Technology* 24 (12): 2073–2077.

Schmidt, Kenneth D., and Irving Sherman. 1987. Effect of irrigation on groundwater quality in California. *Journal of Irrigation and Drainage Engineering* 113 (1): 16–29.

Shah, Tushaar, and Saumindra Bhattacharya. Farmer organisations for lift irrigation: Irrigation companies and tubewell cooperatives of Gujarat. ODI Irrigation Management Network, 1993.

Shaxson, Francis, and Richard Barber. 2003. *Optimizing soil moisture for plant production: The significance of soil porosity*. Rome: UN-FAO.

Sheffield, J., Eric F. Wood, M. Pan, H. Beck, G. Coccia, A. Serrat-Capdevila, and K. Verbist. 2018. Satellite remote sensing for water resources management: Potential for supporting sustainable development in data-poor regions. *Water Resources Research* 54 (12): 9724–9758.

Shock, Clinton C. "Drip irrigation: An introduction." (2006).

Siebert, Stefan, Jacob Burke, Jean-Marc Faures, Karen Frenken, Jippe Hoogeveen, Petra Döll, and Felix Theodor Portmann. 2010. Groundwater use for irrigation–a global inventory. *Hydrology and Earth System Sciences* 14 (10): 1863–1880.

Singh, R.P. 1988. "International Maize and Wheat Improvement Center (CIMMYT)." Pathogenicity variations of Puccinia recondita f. sp. tritici and P. graminis f. sp. tritici in wheat-growing areas of Mexico during 1988 and 1989. *Plant Disease* 75: 790–794.

Singandhupe, R. B., G. G. S. N. Rao, N. G. Patil, and P. S. Brahmanand. 2003. Fertigation studies and irrigation scheduling in drip irrigation system in tomato crop (Lycopersicon esculentum L.). *European Journal of Agronomy* 19 (2): 327–340.

Solley, Wayne B., Robert R. Pierce, and Howard A. Perlman. 1998. *Estimated use of water in the United States in 1995*. Vol. 1200. US Geological Survey.

Stretch, Bill, and David Mowry. 2010. *Fresno Irrigation District's three-year maintenance and facility upgrade plan*. PhD dissertation, Colorado State University: Libraries.

Sun, Ziheng, and Liping Di. 2021. A review of smart irrigation decision support systems. In *2021 9th international conference on agro-geoinformatics (Agro-Geoinformatics)*, 1–4. IEEE.

Sun, Ziheng, Chunming Peng, Meixia Deng, Aijun Chen, Peng Yue, Hui Fang, and Liping Di. 2014. Automation of customized and near-real-time vegetation condition index generation through cyberinfrastructure-based geoprocessing workflows. *IEEE Journal of Selected Topics in Applied Earth Observations and Remote Sensing* 7 (11): 4512–4522.

Sun, Jie, Liping Di, Ziheng Sun, Yonglin Shen, and Zulong Lai. 2019a. County-level soybean yield prediction using deep CNN-LSTM model. *Sensors* 19 (20): 4363.

Sun, Ziheng, Liping Di, and Hui Fang. 2019b. Using long short-term memory recurrent neural network in land cover classification on Landsat and Cropland data layer time series. *International Journal of Remote Sensing* 40 (2): 593–614.

Sun, Ziheng, Liping Di, Hui Fang, and Annie Burgess. 2020. Deep learning classification for crop types in North Dakota. *IEEE Journal of Selected Topics in Applied Earth Observations and Remote Sensing* 13: 2200–2213.

Tang, Zhehan, Yufang Jin, Maria Mar Alsina, Andrew J. McElrone, Nicolas Bambach, and William P. Kustas. 2022. Vine water status mapping with multispectral UAV imagery and machine learning. *Irrigation Science* 40 (4–5): 715–730.

Thompson, R.B., M. Gallardo, L.C. Valdez, and M.D. Fernández. 2007. Using plant water status to define threshold values for irrigation management of vegetable crops using soil moisture sensors. *Agricultural Water Management* 88 (1–3): 147–158.

Tilman, David, Kenneth G. Cassman, Pamela A. Matson, Rosamond Naylor, and Stephen Polasky. 2002. Agricultural sustainability and intensive production practices. *Nature* 418 (6898): 671–677.

Tukey, H.B., Jr. 1970. The leaching of substances from plants. *Annual Review of Plant Physiology* 21 (1): 305–324.

Vano, Julie A., Michael J. Scott, Nathalie Voisin, Claudio O. Stöckle, Alan F. Hamlet, Kristian E.B. Mickelson, Marketa McGuire Elsner, and Dennis P. Lettenmaier. 2010. Climate change impacts on water management and irrigated agriculture in the Yakima River Basin, Washington, USA. *Climatic Change* 102: 287–317.

Vianny, D. Maria Manuel, A. John, Senthil Kumar Mohan, Aliza Sarlan, and Ali Ahmadian. 2022. Water optimization technique for precision irrigation system using IoT and machine learning. *Sustainable Energy Technologies and Assessments* 52: 102307.

Wang, Dingbao, and Ximing Cai. 2009. Irrigation scheduling—Role of weather forecasting and farmers' behavior. *Journal of Water Resources Planning and Management* 135 (5): 364–372.

Westcott, M.P., and K.W. Vines. 1986. A comparison of sprinkler and flood irrigation for Rice 1. *Agronomy Journal* 78 (4): 637–640.

Wong, Chui Eng, Zhi Wei Norman Teo, Lisha Shen, and Yu. Hao. 2020. Seeing the lights for leafy greens in indoor vertical farming. *Trends in Food Science & Technology* 106: 48–63.

Zazueta, Fedro S., Allen G. Smajstrla, and Gary A. Clark. 1994. *Irrigation system controllers*. University of Florida Cooperative Extension Service, Institute of Food and Agriculture Sciences, EDIS.

Zein, Adnan, Sami Karaki, and Mahmoud Al-Hindi. 2023. Analysis of variable reverse osmosis operation powered by solar energy. *Renewable Energy* 208: 385–398.

Chapter 9
Actionable Science for Snow Monitoring and Response

Gokul Prathin Asamani and Ziheng Sun

Contents

G. P. Asamani · Z. Sun (✉)
Center for Spatial Information Science and Systems, Department of Geography and Geoinformation Science, George Mason University, Fairfax, VA, USA
e-mail: zsun@gmu.edu

Z. Sun (ed.), *Actionable Science of Global Environment Change*,
https://doi.org/10.1007/978-3-031-41758-0_9

1 Introduction

Snow is essential for the sustainability and survival of human society, and abnormal snow conditions can have catastrophic consequences (Alldredge and Silver 1988). Snow serves as a natural reservoir of water, particularly in mountainous and polar regions. It accumulates during winter and melts gradually in spring, releasing water into rivers and streams. For instance, in the Western United States, the Sierra Nevada snowpack provides vital water resources for millions of people and extensive agricultural operations in California's Central Valley (Serreze et al. 1999). Snow also serves as a catalyst for winter tourism and recreational activities, attracting visitors and supporting local economies in snow-dependent regions. In addition to its economic importance, snow information also helps regulate water runoff and mitigate flood risks during snowmelt. By gradually releasing water, snow also acts as a natural buffer, reducing the likelihood of sudden and overwhelming flooding events. From a broader perspective, snow can profoundly influence ecosystem dynamics and agricultural productivity (Rasmussen et al. 2012), because its insulating properties shield vegetation, crops, and soil from extreme cold temperatures during winter, allowing for their survival and providing crucial moisture when the snow gradually melts.

Despite its importance, the knowledge and research surrounding snow science have not always translated into actionable solutions for people having their livelihoods relying on snow. While scientific research provides invaluable insights into aspects of snow and its impacts, there is often a gap between scientific findings and their practical application. Challenges like limited resources, technological limitations, and communication barriers can impede the practical application of snow science. This chapter aims to explore the reasons behind this gap and shed light on the potential for making snow science more actionable. Scientific research and NOAA snow monitoring are indispensable in comprehending the aspects of snow and its impacts. The translation of scientific research into actionable information often requires simplification and practical adaptation. It involves distilling the complex findings into practical guidelines and recommendations that can be easily understood and implemented by stakeholders, such as policymakers, land managers, and the general public.

This chapter aims to highlight the importance of making snow science more actionable. It will introduce the advancements in research methods, technological innovations, and communication strategies that can facilitate the practical application of snow science findings. Eventually, we hope the chapter could inspire researchers, policymakers, and practitioners to collaborate and develop more actionable solutions that leverage the immense value of snow science for the benefit of society and the environment.

2 Current Practice in Snow Monitoring and Decision Making

Snow monitoring relies on advanced tools and techniques that employ a range of innovative approaches to gather detailed information about snowpack characteristics. One such technique is LiDAR (Light Detection and Ranging), which employs laser pulses to measure distances and create highly precise three-dimensional maps of the snowpack (Deems et al. 2013). LiDAR systems usually consist of a laser emitter, a scanning mechanism, and a receiver. The emitter releases short pulses of laser light that travel toward the ground surface. When these pulses encounter an object, such as a snowflake or the ground, they bounce back and are detected by the receiver. By measuring the time it takes for the laser pulse to travel to the object and return, LiDAR systems can calculate the distance and create a detailed representation of the snowpack's surface. LiDAR technology is particularly valuable in snow monitoring due to its ability to capture detailed snow depth measurements across vast areas. By flying an aircraft equipped with LiDAR sensors over a snow-covered region, researchers can collect data on the snowpack's thickness and its vertical and horizontal distribution. This data is essential for assessing snow accumulation, understanding snow water equivalent (SWE) (Jonas et al. 2009), and determining potential risks associated with snowmelt and avalanches. The high-resolution three-dimensional maps generated by LiDAR enable researchers to identify and analyze complex snow structures and features. These structures include cornices, snowdrifts, and layers within the snowpack, which are important factors in determining avalanche hazards and slope stability. LiDAR also allows for the estimation of parameters such as snow density, which is essential for hydrological modeling and assessing water resources associated with snowmelt (Deems et al. 2013).

Satellite-based sensors provide high-resolution images of snow-covered areas, enabling researchers to analyze spatial patterns and changes in snow extent over time. By integrating satellite observations with ground-based measurements and weather data, comprehensive snow monitoring systems such as the Snow Data Assimilation System (SNODAS) (Barrett 2003) have been developed. SNODAS incorporates satellite observations from sensors like Moderate Resolution Imaging Spectroradiometer (MODIS), which capture high-resolution images of snow-covered areas. These images provide valuable information about the extent and distribution of snow cover (Fig. 9.1). SNODAS also utilizes ground-based measurements collected from weather stations and snow telemetry (SNOTEL) sites (Serreze et al. 1999), which provide vital data on snow depth, snow water equivalent (SWE) (Jonas et al. 2009), and other snowpack properties. Weather data, including temperature, humidity, wind speed, and precipitation, are also integrated into the system. The assimilation process within SNODAS involves integrating these diverse data sources into a unified model framework. This involves accounting for various factors such as elevation, aspect, slope, vegetation, and land cover characteristics, which can influence snow accumulation and melt patterns. Through sophisticated algorithms and data assimilation techniques, SNODAS combines the observed and modeled data to generate accurate snow cover and snowmelt information at high spatial and temporal resolutions.

Decision making in response to snowstorms involves multiple stakeholders, such as meteorologists, emergency management agencies, transportation authorities, and local governments. The process typically includes meteorologists using weather models and satellite data to predict the occurrence and severity of snowstorms. These forecasts help in alerting relevant authorities and preparing for potential impacts. Once a snowstorm is forecasted, authorities assess the potential risks and impacts, such as road closures, power outages, and disruptions to transportation and infrastructure. Based on the risk assessment, emergency management agencies and local governments develop response plans, mobilize resources, and communicate with the public to ensure preparedness. During a snowstorm, decision makers coordinate response efforts, such as snow removal, prioritizing critical services, and ensuring public safety. After the storm, recovery operations take place, including damage assessment, infrastructure repairs, and restoration of services (Fuchs and Bründl 2005). Compared to snow monitoring, snowstorm forecasting is a more complex task, and forecast accuracy can vary depending on various factors such as atmospheric conditions, topography, and storm dynamics (Mitchell et al. 1998). Effective coordination among multiple stakeholders is crucial for a timely response. Delays can occur if there is a lack of communication or coordination between meteorological agencies, emergency management organizations, and local governments. Developing robust early warning systems that can detect the onset of a snowstorm with greater precision can help in initiating response actions earlier, reducing the

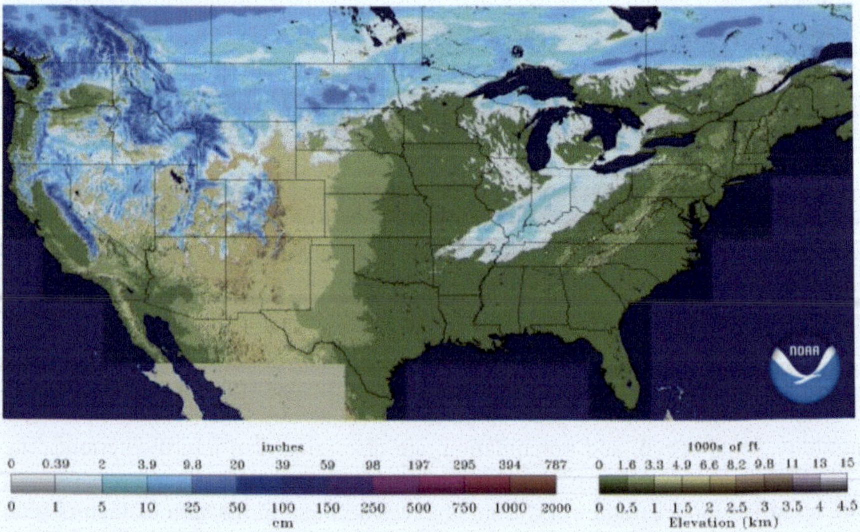

Fig. 9.1 Sample Snow Depth Output from SNODAS for January 15, 2015, United States, in addition to extending well into Canada as well as outlines the coast and contains parts of Mexico. (Image courtesy: NOHRSC)

time lag. Current snowstorm warning systems rely on a combination of weather radar, satellite imagery, weather stations, and meteorological models. Radar systems measure precipitation characteristics, while satellites provide valuable data on cloud cover and temperature (Zrnic et al. 2007). Ground-based weather stations offer real-time observations of temperature, humidity, and wind (Karabatić et al. 2011). These data sources are integrated into meteorological models to generate forecasts. To enhance these systems, radar technology can be improved by increasing resolution and incorporating dual-polarization capabilities to better identify precipitation types and estimate snowfall rates. Other potential upgrades include phased array radar which can enable rapid scanning of larger areas, and multi-frequency radar which can mitigate attenuation issues (Zrnic et al. 2007).

2.1 *How Is Snow Monitored in Operation?*

One of the primary methods used in snow monitoring is remote sensing. Satellites capture data in different spectral bands, enabling scientists to detect and analyze snow cover presence, extent, and changes. Several open-accessible satellites can fulfill this task, like MODIS and Visible Infrared Imaging Radiometer Suite (VIIRS) on terra and aqua. These images, combined with data from ground-based observations and weather stations, form the basis for monitoring and understanding snow dynamics. There are many scientific ground-based observation station networks such as SNOTEL which deploy sensors such as snow pillows and snow depth sensors to gather data (Yang et al. 2023). Located in various mountainous regions, these stations provide real-time measurements of snowpack characteristics. In Colorado, the SNOTEL network encompasses over 100 stations distributed throughout the state's Rocky Mountain region (Henn et al. 2017; Doesken & Schaefer 1987). Similarly, in the state of Utah, the SNOTEL network includes approximately 60 stations positioned across the Wasatch Range and other mountainous areas (Henn et al. 2017). These stations provide valuable real-time data on snowpack conditions, helping researchers, water managers, and avalanche forecasters make informed decisions. These measurements are essential for accurately assessing snow depth, snow water equivalent, and other critical parameters. The data collected by these ground-based networks, combined with satellite imagery, enhances the accuracy and reliability of snow monitoring efforts.

In addition to satellite remote sensing and ground-based observations, scientists also employ technology such as LiDAR to monitor snow. LiDAR technology measures the distance between the sensor and the snow surface, providing highly accurate and detailed information about snow depth, surface roughness, and snowpack structure. By conducting airborne LiDAR surveys across snow-covered areas, scientists can create high-resolution maps that depict variations in snow distribution and properties. These maps are invaluable for understanding snow dynamics, identifying avalanche-prone areas, and supporting land management decisions. The integration of satellite remote sensing, ground-based observations, and technologies like LiDAR has made it possible to monitor snow more accurately, enabling scientists to obtain a comprehensive and detailed understanding of snowpack conditions, including its spatial variability, temporal changes, and underlying processes. This information enables more accurate avalanche forecasting, better water resource management, and improved decision-making.

2.2 *What Does a Day of a Water Manager Look Like?*

Snow melt is a significant source of water for downstream regions, particularly during the spring and early summer when snowpacks melt (Barnett et al. 2005). A water manager is responsible for ensuring a reliable water supply for various uses, including agriculture, industry, and domestic consumption (Seeger 1960). They monitor the snowpack, track snow melt rates, and allocate water resources to meet the demands of different stakeholders. A water manager's day typically begins with a thorough review of various data and reports related to water management, like analyzing weather forecasts, streamflow measurements, reservoir levels, and snowpack data (Seeger 1960). They closely work with meteorologists, hydrologists, engineers, and other experts to understand the complex factors influencing water availability (Seeger 1960). Regular meetings and discussions occur to exchange insights, share data, and develop strategies for effective water management. They also engage with stakeholders, including government agencies, local communities, and industry partners, to gather input and incorporate diverse perspectives into decision-making processes.

They often conduct site visits to reservoirs, dams, and water treatment plants to oversee their operation and ensure optimal performance (Gomes et al. 2020). Rigorous inspections, maintenance activities, and adherence to safety protocols are part of their duties. For instance, they may coordinate the controlled release of water from reservoirs to strike a balance between storage capacity, flood control, and downstream water requirements (Seeger 1960). They also monitor snow accumulation and snowpack conditions in mountainous regions prone to spring or rapid snowmelt floods, taking proactive measures to manage water levels in reservoirs and control water releases. By analyzing historical data, real-time observations, and climate projections, they evaluate the reliability and sustainability of water resources, helping develop strategies for water allocation, managing water rights, and planning for potential water shortages or drought conditions.

2.3 *How Does the Snow Sport and Tourism Industry Operate?*

Snow sports and tourism are very popular in recreation and are a major component in the tourism industry, especially in regions with favorable winter climates and mountainous landscapes. This industry encompasses various activities, including skiing (Fig. 9.2), snowboarding, snowmobiling, ice climbing, snowshoeing, and more. The operation of snow sports and tourism involves several key aspects, including snow condition monitoring, ski resort management, safety measures, environmental considerations, and marketing strategies. These elements work together to create enjoyable and sustainable experiences for visitors while minimizing environmental impacts.

Snow condition monitoring is required by the operation of snow sports and tourism (Ross et al. 2020). Ski resorts and snow sports facilities have to closely monitor snowfall, snow depth, snow quality, and weather conditions to ensure optimal skiing and snowboarding conditions. Automated weather stations provide real-time information on snowpack, helping ski resort operators make decisions regarding trail maintenance, snow grooming, and avalanche control. Snow surveys and field observations are also conducted to assess snow conditions, including snowpack stability and avalanche risk. Trained personnel perform snowpack tests, analyze snow layers, and monitor weather patterns to ensure visitor safety (MacGillivray et al. 2006).

Safety is a top priority in snow sports and tourism (Bentley et al. 2004). Ski resorts normally implement measures to minimize risks and ensure the well-being of visitors. Trained ski patrol teams monitor the slopes, respond to accidents, and provide first aid when necessary. They also conduct avalanche control measures to reduce the risk of avalanches and maintain safe skiing conditions. For example, Whistler Blackcomb is one of the largest and most popular ski resorts in North America. It offers over 8171 acres of skiable terrain, world-class facilities, and hosted the alpine skiing events during the 2010 Winter Olympics. Ski resorts operate through effective management systems encompassing various aspects, including lift operations, slope maintenance, guest services, and accommodations. Modern ski resorts feature well-designed lift systems, including chairlifts, gondolas, and aerial tramways, to transport visitors efficiently across the mountain. Lift operators ensure the safe operation of lifts, monitor capacity, and maintain equipment. Meanwhile, slope maintenance involves snow grooming, trail marking, and signage to provide a safe and enjoyable skiing experience.

Environmental sustainability is an essential consideration in snow sports and tourism operations nowadays (Flagestad and Hope 2001). Ski resorts recognize the importance of preserving the natural environment and strive to minimize their ecological footprint. They implement practices such as water conservation, energy-efficient infrastructure, waste management, and environmental education programs. Some resorts have even adopted renewable energy sources, such as solar or wind energy, to reduce their reliance on non-renewable resources and lower greenhouse gas emissions. By prioritizing sustainability, ski resorts aim to protect the surrounding ecosystems and ensure the longevity of snow sports activities.

Research and innovation can definitely help in advancing snow sports and tourism, enhancing safety within the snow sports and tourism industry (Unbehaun et al. 2008). Scientists continuously study snow hydrology, climate change impacts, slope stability, and snowpack dynamics to understand better and manage snow resources (Steiger et al. 2019). This knowledge contributes to improved snow management techniques, safety protocols, and sustainable practices within the industry (Blackburn 2004). Collaboration between scientists, industry stakeholders, and policymakers fosters the development of effective strategies for snow sports and tourism that balance environmental conservation, visitor safety, and economic growth.

2.4 How Does the Government Make Policies and Decisions About Snow?

Governments formulate policies and make decisions related to snow management. The government recognizes the need to address issues related to snow, such as snow removal, road maintenance, and safety concerns. This may arise from public demand, expert recommendations, or an analysis of past experiences and challenges faced during snowy seasons and then research and collect relevant data to understand the specific snow-related issues comprehensively. This involves studying snowfall patterns, analyzing road conditions, assessing the impact on transportation, and considering the economic and social implications. They engage with various stakeholders, including transportation departments, meteorological agencies, snow removal service providers, local communities, and experts in relevant fields

Fig. 9.2 Ski Lift of Winter Park Resort in Colorado. (Image courtesy: Dr. Ziheng Sun)

(Engelhard et al. 2007). This collaboration ensures the policy development process incorporates diverse perspectives, expertise, and practical considerations. The government establishes clear goals and objectives for snow-related policies based on research and stakeholder input. These goals include ensuring safe and efficient transportation, minimizing disruptions, reducing accidents, promoting environmental sustainability, and supporting economic activities like tourism and trade (Kuemmel and Kuemmel 1994). The government designs the policy framework that outlines the specific measures, guidelines, and regulations to address the identified issues. This may include provisions for snow removal strategies, road maintenance standards, allocation of resources, coordination mechanisms, and safety protocols (Miller 2020). The formulated policies are shared with relevant stakeholders and the public for feedback and consultation. This allows for further refinement, consideration of alternative perspectives, and identification of potential challenges or unintended consequences. Once the policies are finalized, the government begins implementing them. This involves allocating resources, establishing coordination mechanisms, and disseminating guidelines and regulations to the relevant departments, agencies, and service providers responsible for snow removal, road maintenance, and safety management. Governments continuously monitor the implementation of snow-related policies to assess their effectiveness and make necessary adjustments. They collect data on key performance indicators, such as accident rates, road conditions, transportation efficiency, and environmental impact, to evaluate the policy outcomes and identify areas for improvement.

For example, the Snow and Ice Control Policy implemented by the City of Toronto in Ontario, Canada, was created to address the challenges posed by snow and ice on city roads during winter and ensure safe and efficient transportation for residents (Audit of Winter Road Maintenance Program – Toronto 2023). The policy is developed and implemented by the Transportation Services division of the city government. The government took a comprehensive approach to snow removal, road salting, and maintenance activities. The policy is based on extensive research, analysis of weather patterns, and consultation with various stakeholders, including transportation experts, city officials, and community representatives. The policy outlines specific procedures and guidelines for snow removal and ice control operations. It includes criteria for prioritizing roads based on traffic volume, transit routes, emergency access, and school zones. The policy also establishes performance standards for snow clearing, specifying the maximum time allowed for snow removal after a snowfall event. To implement the policy effectively, the City of Toronto has a dedicated fleet of snowplows, salt spreaders, and other equipment. When a snow event occurs, the Transportation Services division activates its snow response teams and deploys resources strategically across the city. Snow Plows are first dispatched to clear major arterial roads, bus routes, and priority areas, followed by residential streets. The policy also emphasizes communication and public awareness. The city utilizes various channels, such as its website, social media platforms, and traditional media, to provide timely updates on snow removal operations, road conditions, and any necessary travel advisories. Residents are encouraged to stay informed and cooperate by following parking regulations during snow-clearing operations. In addition, the policy includes provisions for continuous monitoring and evaluation. The Transportation Services division regularly assesses the effectiveness of snow and ice control operations, gathers feedback from residents, and identifies areas for improvement. This feedback loop helps to refine the policy and make necessary adjustments to enhance snow management practices.

The government's decision-making process involves several steps. Usually the initial thing done by meteorological agencies is to closely monitor weather patterns and collect data from various sources such as satellites and weather stations, analyzing this data to forecast snowstorms and their potential impacts (Mitchell et al. 1998). Based on the forecast, relevant government agencies, including transportation departments, public works departments, and emergency services, collaborate to assess the risks and impacts of snow on critical infrastructure and public safety. They exchange information, develop contingency plans, and coordinate resources. Once the assessment is complete, the government activates snow response operations, mobilizing snow removal teams, equipment, and resources to clear roads and public spaces. Snowplows, salt spreaders, and personnel are deployed based on predetermined criteria such as snow accumulation thresholds and visibility conditions (Campbell and Langevin 1995). Throughout the snow event, government agencies continuously monitor weather conditions and snowfall accumulation, updating their strategies and operations accordingly. After the snowstorm subsides, post-storm assessments are conducted to evaluate the effectiveness of the response, gathering feedback, assessing performance, and identifying areas for improvement. An example of effective government decision-making about snow can be seen in the case of the Massachusetts Department of Transportation (MassDOT). They have a comprehensive snow and ice control program in place, which includes pre-treating roads, deploying a large fleet of snowplows, and utilizing innovative technologies such as real-time weather and road condition monitoring. MassDOT's decision-making process ensures a proactive and efficient response to snow events, facilitating safe and reliable transportation for residents and visitors.

2.5 How Does the Snow Community Work and Live? What Science Knowledge Do They Use Every Day?

The snow community encompasses diverse examples of organizations dedicated to studying and living in snow-dominated environments. Take the Sami people for example. They are the indigenous inhabitants of Sápmi, a region that spans across northern parts of Norway, Sweden, Finland, and Russia (Kiniry 2020). The Sami have a deep connection with the snow and ice, relying on traditional knowledge and scientific understanding to navigate and adapt to their snow-covered surroundings. Their expertise in reindeer herding is vital for sustaining their way of life in harsh winter conditions. The Sami community's knowledge of snow and its ecological significance, passed down through generations, showcases the integration of traditional knowledge and scientific insights in snow-related practices. IceCube Neutrino Observatory located at the South Pole is an example of a state-of-art research facility that uses a cubic kilometer of Antarctic ice as a detector to capture high-energy neutrinos from outer space (Deiana et al. 2022). By studying these neutrinos, scientists gain insights into astrophysical phenomena such as supernovae, and gamma-ray bursts. The observatory utilizes an array of optical sensors buried deep within the ice, taking advantage of the exceptional clarity and stability of the polar ice sheet. This project represents the intersection of snow and ice science with astrophysics, showcasing the interdisciplinary nature of the snow community.

In snowy mountainous regions, local communities have developed specific skills and knowledge related to snow and its management. Take, for instance, the mountain farmers in the Swiss Alps who practice traditional snow farming techniques. They understand the importance of water availability during drier months and employ ingenious methods to ensure a controlled supply. By managing snow accumulation, they can control the timing and rate of snowmelt, allowing for a steady release of water when it is most needed for agricultural purposes. This traditional practice demonstrates their deep understanding of snowpack properties and the local climate. Moreover, the architectural techniques employed by snow communities reflect their intimate knowledge of snow. Houses are constructed with steep roofs, often layered with insulating materials such as thatch or snow blankets, which prevent excessive snow buildup and minimize the risk of structural damage. Houses are strategically positioned to leverage natural slopes and wind patterns, allowing snow to slide off more easily (Valle 2023). These traditional building methods showcase the integration of local wisdom and scientific principles in creating resilient structures that withstand the harsh snowy conditions. Beyond these practical aspects, local snow communities rely on their observational skills and traditional indicators to navigate their surroundings safely. Through generations of experience, they have learned to interpret visual cues provided by the snow and the environment. By observing snow surface patterns, wind direction, and temperature gradients, they can discern variations in snowpack stability and potential avalanche risk (Riseth et al. 2011). This knowledge, combined with their understanding of local weather patterns and the behavior of different snow layers, enables them to make informed decisions about travel, hunting, and other outdoor activities. While they may not employ sophisticated scientific instruments, their day-to-day reliance on observation and interpretation exemplifies the practical science embedded in their way of life.

Across different snow-covered regions, diverse communities integrate scientific knowledge with their traditional practices to thrive in snow-dependent environments. For instance, Andes Mountains, communities like the Quechua and Aymara have devised sophisticated irrigation systems called "qanats" or "acequias" (Neuwirth 2021). These systems harness the power of snowmelt, precisely managing water resources through canals to sustain agriculture in arid high-altitude regions. By incorporating scientific insights on snow accumulation and the timing of meltwater release, these communities optimize water usage for crop cultivation, showcasing the synergistic relationship between traditional knowledge and scientific principles in snow management (Eira et al. 2018).

In general, local snow mountainous communities face a range of hardships that require them to apply scientific knowledge and innovative techniques to overcome them. One significant challenge is the risk of avalanches, which is a constant concern for communities residing in avalanche-prone areas (Jamieson and Stethem 2002). In Alaska's Arctic communities, extreme cold temperatures, heavy snowfall, and frozen ground pose difficulties in maintaining essential services and infrastructure (Pastick et al. 2015). The Alaska Department of Transportation and Public Facilities (DOT&PF) constructs ice bridges on frozen rivers to create temporary roadways, allowing transportation of essential goods and services despite freezing temperatures. The impact of climate change on snow availability and quality presents significant challenges for snow communities. Indigenous communities like the Inuit in northern Canada utilize scientific principles in their traditional snow house, the igloo (Tremblay et al. 2006). The domed shape and insulating properties of igloos enable efficient heat retention, offering warmth in extremely cold conditions. By combining traditional wisdom with scientific approaches, they ensure their safety, adapt to changing environmental conditions, and develop innovative solutions for transportation, energy supply, and climate resilience. Through their resilience and resourcefulness, these communities showcase the successful integration of science and local knowledge in navigating the unique challenges of living in snowy environments.

3 Current Research in Snow Monitoring and Prediction

Current research in snow monitoring and prediction encompasses a range of cutting-edge topics that contribute to our understanding of snow processes, improve forecasting capabilities, and enhance snow-related decision-making. Scientists and researchers are continually pushing the boundaries of knowledge in this field, utilizing advanced technologies, innovative methodologies, and interdisciplinary approaches. In this section, we will explore some of the latest hot research topics in snow monitoring and prediction.

3.1 Rocky Mountains

This mountain range spans across western North America, including the United States and Canada. The Rocky Mountains provide a diverse range of snow conditions and are an important source of freshwater for downstream regions (Fig. 9.3). Pelto et al. (2019) focused on evaluating the feasibility of using remotely sensed methods to assess seasonal glacier mass balance. Tandem airborne laser scanning (ALS) surveys and field-based measurements were conducted over a 4-year period on six alpine glaciers in British Columbia, Canada. Geodetic balance, winter balance, and summer balance were calculated using digital elevation models (DEMs) and density estimates based on surface classification. The study found that geodetic estimates of winter, summer, and annual balance were slightly different from glaciological measurements, with surface classification playing a significant role in geodetic annual mass change. The results demonstrate the potential of ALS and geodetic methods for accurately assessing seasonal mass change in multiple glaciers. Another study that highlights the challenges in accurately estimating snow water storage is conducted by Wrzesien et al. (2019), due to limited observational networks and satellite sensing limitations. Four global data sets showed moderate agreement in estimating global SWS, but when compared to high-resolution regional models, they differed significantly, with biases up to 66% and potential underestimation of SWS by 1500 km^3 (Wrzesien et al. 2019) globally. This emphasizes the need for further research to improve water resource characterization in snow-dominated regions, particularly in mountains. Machine-learning techniques have been widely used in monitoring and forecasting snow in the Rockies (Sun et al. 2022b). Cannistra et al. (2021) highlight the importance of optical remote sensing techniques in measuring snow-covered area (SCA) and their impact on physical, ecological, and societal systems. The development of a convolutional neural network-based method using high-resolution satellite imagery shows promise in accurately observing SCA at fine-scale spatial and temporal resolutions, offering potential applications in ecological and water resource research. Sun et al. (2022a) used Geoweaver to construct the full-stack SWE forecasting workflow to make the snow AI more FAIRable (findable, accessible, interoperable, and reusable) (Alnaim and Sun 2022; da Silva et al. 2023) (Fig. 9.3).

3.2 *Himalayas*

In the realm of monitoring and forecasting, Sood et al. (2020) examined snow cover variability over the North Indian Himalayas, including the Western Himalayas and Karakoram mountain ranges, using MODIS data (Jain et al. 2008). The findings reveal changes in snow cover area, with a one-month shift in snow accumulation and melt over the past decade, providing valuable information for climatology, hydrology, cryosphere, and glaciology studies. This research contributes to our understanding of snow processes and enhances snow monitoring and prediction capabilities in the North Indian Himalayas. Kirkham et al. (2019) present the use of a passive gamma ray sensor and complementary meteorological instruments at high altitude in the Nepal Himalayas to measure snow water equivalent (SWE) and snow depth. The findings highlight the accuracy and limitations of the instrument setup, emphasizing the importance of continuous monitoring for developing snow models and improving understanding of the high-altitude water cycle in data scarce regions. This information enhances our understanding of snow processes and contributes to improved climatology, hydrology, cryosphere, and glaciology studies in the region.

Fig. 9.3 Snow cover on Rocky Mountain. (Image courtesy: Dr. Ziheng Sun)

Using MODIS Terr and Aqua snow cover products, Sahu and Gupta (2020) focus on estimating the annual and seasonal snow cover area (SCA) in the Chandra basin, Western Himalayas. The study reveals that the average SCA observed during the study period was 84.94% of the basin area, with the highest and lowest values in 2009 and 2016, respectively. A strong correlation is observed between SCA and temperature, indicating the high sensitivity of SCA variability to temperature changes.

3.3 *Alps*

Alps have extensive snow cover and have significant impact on various aspects of life in the region, they stretch across several countries including France, Switzerland, Italy, Austria, Germany, Slovenia, and Liechtenstein. The mountain range runs approximately 1200 km (750 miles) in a crescent shape, from the Mediterranean Sea in the south to the Danube River in the north. Alps are known for their landscapes, high peaks, and diverse ecosystems, making them a popular destination for outdoor activities, tourism, and winter sports.

Poussin et al. (2019) focused on monitoring snow cover in the Gran Paradiso National Park in the western Italian Alps using Landsat satellite observations stored in an Open Data Cube. The methodology includes temporal filtering and multi-sensor analysis to improve the estimation of snow cover area. However, the challenge of cloud cover still affects the accuracy of snow mapping from satellite data. Another study conducted by Lievens et al. (2022) explores the potential of the Sentinel-1 mission to monitor snow depth in the European Alps at sub-kilometer resolutions. The Sentinel-1 backscatter observations, particularly in cross-polarization, exhibit a strong correlation with regional model simulations of snow depth. An empirical change detection algorithm using radar backscatter changes is employed to retrieve snow depth, showing promising results with a spatiotemporal correlation of 0.89 compared to in situ measurements. The findings highlight the capability of Sentinel-1 to provide valuable snow estimates in mountainous regions, aiding applications such as water resource management, flood forecasting, and numerical weather prediction. Further research is needed to understand the underlying physical mechanisms of Sentinel-1's sensitivity to snow accumulation.

3.4 *Other Regions and Global Study*

Notarnicola (2020) used MODIS products from 2000 to 2018 revealing that approximately 78% of global mountain areas are experiencing a decline in snow cover, characterized by shorter snow cover duration and smaller snow cover area. Changes in snow cover are influenced by air temperature and precipitation, with implications for water resources, ecosystems, tourism, and energy production. However, the analysis is limited by the relatively short time period and uncertainties in snow cover estimates, particularly in complex terrain. In the domain of remote sensing, Tsai et al. (2019) use spaceborne remote sensing, particularly synthetic aperture radar (SAR), offer a valuable means of monitoring snow cover continuously, overcoming limitations such as cloud cover and polar darkness. This review discusses the advancements in SAR-based techniques, including interferometric SAR and polarimetric SAR, along with the integration of auxiliary data to improve snow cover analysis. The summary highlights the importance of monitoring snow cover and the potential of SAR for providing valuable insights into the cryosphere. Furthermore, studies conducted by Aves et al. (2022) in the Ross Island region of Antarctica collected snow samples and found microplastics present in all samples, with fibers and polyethylene terephthalate (PET) being the most common types. The research suggests that microplastics can be transported over long distances and may originate from both distant and local sources, including nearby research stations, highlighting the widespread presence of airborne microplastic pollution in Antarctica. Another interesting study by Zhang et al. (2019), based on long-term observational data, has found that the previously observed inverse relationship between central Eurasian spring snow cover and Indian summer monsoon rainfall has disappeared since 1990. This change is attributed to the altered regulation of mid-tropospheric temperature and land-ocean thermal contrast due to declining spring snow cover, indicating that Eurasian spring snow cover may no longer serve as a reliable predictor of Indian summer monsoon rainfall in a changing climate. In the domain of monitoring and forecasting, Largeron et al. (2020) reviewed the challenges in monitoring and forecasting snow cover characteristics in mountainous regions and discussed the use of data assimilation techniques to improve estimates of snow water equivalent (SWE). It highlights the suitability of different data assimilation methods based on snow model complexity and available observations, providing recommendations for enhancing snow hydrology monitoring and prediction systems in mountainous areas.

4 Why Are These Studies Difficult to Be Used in Reality?

Not all the mentioned studies achieve this desired level of actionability. This inability to bridge the gap between knowledge and action poses a significant challenge, as it hinders the effective utilization of research findings and diminishes their potential impact. By unraveling the reasons behind non-actionability, we can gain a deeper understanding of how to enhance the relevance and practicality of scientific research, ultimately striving for research outcomes that make a tangible difference in addressing societal challenges.

Most snow cover variation studies currently emphasizes the scientific endeavors rather than providing practical guidance on addressing the observed concerning changes. Without clear and specific recommendations, stakeholders such as policymakers, resource managers, and local communities face difficulties in translating the Himalaya study's results into actionable measures and interventions tailored to their unique circumstances. First of all, the technology has limitations considering snow conditions can vary significantly within relatively small regions. The data generated by these snow studies might not offer the local specificity that snow communities require. Actionable decisions often need data at finer spatial resolutions to address localized challenges accurately (Yang et al. 2013). For instance, datasets from the MODIS or Synthetic Aperture Radar (SAR) sensors are typically collected on predefined schedules, and their availability might not align with the real-time or near-real-time decision-making needs of snow communities. For managing issues like avalanche risk, water resource planning, or emergency response during snow events, stakeholders require timely and up-to-date information. In addition, remote sensing data alone may not provide the contextual insights needed for actionable decision-making. While these data can indicate the presence or absence of snow cover, they may not offer information about factors like snowpack density, water content, or local terrain conditions, which are critical for various applications, including water resource management and avalanche forecasting (Vickers et al. 2020; Tsai et al. 2019; Hedrick et al. 2018). Meanwhile, the adoption of snow research outcomes for actionable decision-making can be operationally challenging as it usually requires integrating these result datasets into existing decision-support systems, ensuring data continuity, and maintaining technical infrastructure, all of which demand resources and expertise that may not be readily available to snow communities. Last but not least, many snow covered regions, e.g., Himalayas, span multiple countries or jurisdictions. Geopolitical tensions, disputes over boundaries, resources, and differing national priorities can hinder collaborative efforts to address shared snow-related challenges. Lack of cooperation among countries and local communities with various culture can prevent the development of regionally coordinated strategies for collective snow management and effective climate adaptation.

Therefore, the absence of stakeholder involvement significantly hampers the study's actionability (Greenwood 2007). Failure to mention the engagement of relevant stakeholders and end-users limits the study's potential impact. Engaging stakeholders, including local communities residing in the snow-impacted region, policymakers, environmental agencies, scientific organizations, and non-governmental organizations (NGOs) working on climate change and natural resource management, is crucial to enhancing the study's actionability. Collaborating with these stakeholders allows researchers to gain valuable insights, validate findings, and develop specific actions and policies based on the study's outcomes. By incorporating diverse perspectives, knowledge, and expertise, the research can be aligned with the real-world needs and priorities of the stakeholders, increasing the likelihood of its findings being effectively utilized.

On the other hand, researchers should offer practical and tailored recommendations that go beyond presenting the analysis and findings. These recommendations may involve implementing improved snow monitoring systems given local available resources to enhance data collection and analysis capabilities. Additionally, suggesting community-based adaptation strategies can empower local communities to respond to changes in snow cover by utilizing their traditional knowledge and practices. Policy interventions, such as the development of sustainable land and water resource management policies, can also be proposed to mitigate the impacts of snow cover changes.

For example, Poussin et al. (2019) discussed the use of remotely sensed data to monitor and understand climate and environmental changes in mountain regions, with a focus on snow cover as a key variable. While the study provides valuable insights into the methodology for snow cover detection and its application to a case study in the western Italian Alps, it may have low actionableness in terms of public adaptability and scientific acceptance. The paper's highly technical nature is evident through the utilization of advanced remote sensing techniques, statistical analyses, and complex algorithms for snow cover detection. These methods may involve intricate image processing algorithms, spatiotemporal techniques, probability of snow approaches, and multi-sensor data fusion. Furthermore, scientists may employ specialized terminology related to remote sensing, such as Landsat data, Worldwide Reference System-2 paths and rows notation system, Collection 1 Level 1 data, green and shortwave Infrared (SWIR1) bands, and the snow observation from space (SOfS) algorithm (Frau et al. 2018). These technical aspects and specialized terminology can create barriers in understanding the intricacies of the study and its implications for climate change research and environmental management. Another factor that may limit the actionableness of the study is the lack of clear recommendations or actionable steps for policymakers, resource managers, or other stakeholders. While the study highlights the importance of monitoring and understanding snow cover changes in mountain regions, it does not provide specific guidance on how to use this information to inform decision-making or take action to mitigate the impacts of climate change.

To increase the actionableness of the mentioned studies above, the researchers could take one step further to provide more specific recommendations. For instance, they could identify key areas or regions within mountainous environments where the methodology for snow cover detection could be applied effectively. By highlighting these specific areas, policymakers, resource managers, and stakeholders can better understand the relevance of the study's findings and how they can be translated into actionable steps. They could suggest prioritizing conservation efforts or implementing targeted adaptation measures in areas where snow cover changes pose significant ecological or socio-economic risks. Collaboration with local resource managers and policymakers is essential to translate the study's findings into practical management strategies. They can work closely with these stakeholders to identify specific areas where the methodology for snow cover detection can be integrated into decision-making processes. By demonstrating how remotely sensed data can complement sparse in situ observations and provide valuable insights into snow cover dynamics, the researchers can help inform land-use planning, conservation efforts, and climate change adaptation strategies in mountainous regions. They could also provide more specific guidance on how to use the information generated by the study to inform decision-making or take action to mitigate the impacts of climate change. For instance, they could suggest specific measures that could be taken to reduce the vulnerability of mountain communities to the impacts of snow cover changes, such as investing in water storage infrastructure or promoting alternative livelihoods that are less dependent on snow cover. By providing clear and actionable recommendations, the authors can help ensure that their research has a tangible impact on the ground.

5 Playbook to Advance Actionable Science in Snow

To maximize the impact of snow research, it is essential to bridge the gap between knowledge generation and practical action. This section provides detailed suggestions for snow scientists and stakeholders, targeting those already familiar with the field. By focusing on specific areas such as methodology advancements, stakeholder engagement, and actionable recommendations, we aim to enhance the relevance and application of snow research findings.

5.1 *Suggestions for Snow Scientists*

Understanding the needs, concerns, and constraints of stakeholders such as policymakers, land managers, and local communities is vital for aligning research objectives with real-world problems. In addition to considering end-users, snow scientists must also take into account the actual costs associated with their research. While advanced technologies like Unmanned Aerial Vehicles (UAVs) with Light Detection and Ranging (LiDAR) may offer exciting possibilities, their financial feasibility should be carefully evaluated. Instead of relying solely on expensive and complex technologies, scientists should explore cost-effective alternatives that can still yield reliable and actionable results. Realistic barriers faced by people in snow-covered areas should also be acknowledged and factored into research design. Accessibility, equipment availability, and manpower limitations are all practical constraints that need to be considered. By developing research methodologies that can be implemented within these constraints, scientists can ensure that their findings are feasible and can be readily applied in real-world scenarios.

Time constraints are an important consideration for snow scientists conducting research in snow-covered areas (Bates 2021). Efficiently managing research activities within limited time frames is essential to ensure the collection of high-quality data and maximize the impact of the research (Durre et al. 2010). Scientists need to carefully plan research activities, taking into account the seasonal variations and weather patterns specific to the study area. Understanding the optimal timing for data collection, such as peak snowfall periods or stable snow conditions, enables scientists to capitalize on favorable weather and snow conditions, ultimately maximizing the opportunities for data acquisition. Streamlining fieldwork logistics can maximize efficiency, such as careful planning of transportation routes, minimizing travel time between sampling sites, and optimizing data collection workflows. Using efficient data collection techniques and equipment that require minimal setup time can also help maximize the productivity of fieldwork. To achieve these objectives, scientists can explore partnerships with equipment manufacturers, research institutions, or governmental agencies that may provide access to necessary equipment (Etzkowitz 1983). Adapting or developing low-cost alternatives or open-source solutions for measuring snow properties can help mitigate equipment availability challenges. Finding qualified personnel or local expertise in snow research can be difficult, particularly in remote areas. Scientists should invest in capacity building and knowledge transfer by organizing training programs or workshops for local communities, students, or researchers. By building local capacity, scientists can create a network of individuals with the necessary skills and expertise to assist in data collection and analysis. To manage costs effectively, scientists can seek funding opportunities specifically designed for snow-related research, such as grants or fellowships. National and international research funding agencies often provide grants, fellowships, and research funding opportunities for various scientific disciplines, including snow science. Examples of such organizations include the National Science Foundation (NSF), European Commission's Horizon Europe program, Natural Sciences and Engineering Research Council of Canada (NSERC), and Japan Society for the Promotion of Science (JSPS). Collaborative research projects between different countries or institutions can be a valuable avenue for sharing financial burdens and accessing resources. International programs and initiatives like World Climate Research Programme (WCRP) may offer funding opportunities or facilitate collaborations among snow scientists (Schiffer and Rossow 1983).

5.2 Suggestions for Decision Makers

For decision makers involved in funding allocation and policy making for snow monitoring, establishing a robust connection between scientific research and the decision making process is essential. Adopting advanced snow monitoring drones and leveraging their accurate and real-time data enables decision-makers to make informed assessments of risks, allocate resources effectively, and make timely decisions.

Decision-makers should prioritize key monitoring needs based on the significance and urgency of monitoring specific variables. For example, allocating resources for monitoring snowpack stability in high-risk avalanche-prone areas helps identify hazards and mitigate risks. Neglecting to prioritize these needs can lead to inadequate data collection, increased vulnerability to disasters, and potential loss of life and property. Collaborating with funding agencies is instrumental in securing financial support for snow monitoring programs. By effectively communicating the value of actionable science in snow monitoring, decision-makers can increase the likelihood of receiving the necessary funding. This collaboration empowers decision-makers to leverage additional resources, enabling them to enhance data collection capabilities, upgrade equipment, and improve response capabilities. Adequate financial support ensures accurate assessment of snow conditions, facilitates accurate avalanche forecasting, and empowers decision-makers to implement timely and effective mitigation measures to safeguard communities and critical infrastructure. Allocating a portion of the budget to technological advancements is essential for elevating the accuracy and efficiency of snow monitoring. By investing in cutting-edge technologies, sensor systems, and data processing tools, decision-makers can enhance their capabilities in assessing snow conditions. Utilizing advanced equipment enables decision-makers to accurately determine snow cover extent, predict snowmelt runoff, and effectively manage water resources. This strategic allocation of funds supports informed decision-making, aids in resource planning, and improves overall snow monitoring capabilities.

5.3 *Suggestions for Snow-Impacted Communities*

Snow-impacted communities face unique challenges when it comes to water availability and resource management for both high- and low-altitude areas (Jerez 2021). High-altitude regions experience heavy snowfall and prolonged snow cover characteristic of these areas necessitate careful consideration when implementing snow harvesting and storage practices. The design and construction of storage facilities should account for the potentially larger snow accumulation and the need for increased storage capacity to accommodate the substantial snow loads (Meløysund et al. 2006). Science-based techniques, such as snow density profiling and water content measurements, can aid in accurately estimating the amount of water stored in the snowpack and optimizing the capacity calculations (Proksch et al. 2016). Given the significant temperature fluctuations in high-altitude regions, which can lead to freeze-thaw cycles, insulation materials and storage designs must be tailored to minimize snowmelt and preserve the water resource effectively (Hamada et al. 2007). Meanwhile, the challenging accessibility of high-altitude regions during winter necessitates careful planning of infrastructure development and transportation logistics to ensure efficient snow collection, storage, and distribution.

To ensure the quality and safety of the harvested water, common water treatment methods such as filtration, sedimentation, and disinfection are commonly employed in snow-impacted communities (Harper et al. 2011). High-altitude regions can benefit from advanced water treatment technologies like UV disinfection or reverse osmosis, providing additional layers of protection if necessary. It is equally important to focus on the design and optimization of water distribution systems and infrastructure to efficiently deliver the stored water to the community. Considerations such as pressure, flow rate, and water demand should be taken into account to ensure a reliable and sufficient water supply that meets the needs of the community.

In contrast, some arid regions face distinct considerations in snow harvesting and storage due to lower snowfall variability and shorter snow cover periods. These areas often experience less snowfall, making it important to optimize snow collection efficiency during periods of intense snowfall. Techniques such as snow fences or catchment structures can be employed to enhance the effectiveness of snow collection. In low-altitude regions where snowmelt is relied upon for drinking water supply, careful attention must be given to water quality and the risks of contamination. Robust water treatment processes, including filtration and disinfection, should be implemented to address potential pollutants or impurities associated with low-altitude snowmelt. Moreover, the higher evaporation and sublimation rates in low-altitude regions require careful monitoring and modeling to estimate water losses and optimize storage capacities accordingly. In addition to these standard methods, low-altitude regions may also need to address specific contamination risks associated with pollutants commonly found in urban or agricultural environments. Water treatment technologies, such as activated carbon filters or ion exchange systems, can be employed to address these specific challenges.

In regions with varying altitudes, the incorporation of snow-sensitive designs into buildings and infrastructure can bring significant advantages. In high-altitude snow regions, where heavy snowfall and prolonged snow cover are common, it is essential to prioritize building designs that can safely withstand snow loads. Sloped roofs with adequate load-bearing capacity are crucial in preventing excessive snow accumulation and reducing the risk of roof collapse. By providing a steeper angle for snow to slide off, sloped roofs help alleviate the weight burden on structures (O'Rourke et al. 1982). Furthermore, incorporating materials such as steel or reinforced timber trusses enhances the structural integrity and resilience of buildings in the face of heavy snow loads. High-altitude communities can also consider utilizing snow sheds or snow fences strategically placed to create windbreaks and minimize snow accumulation in critical areas, such as access roads or pedestrian pathways. These measures, combined with regular monitoring and snow removal efforts, contribute to safer and more resilient built environments in high-altitude snow regions. During periods of abundant snowfall, communities can strategically stockpile excess snow in designated storage areas. Insulating materials or cooling systems can then be employed to preserve the stored snow, protecting it from external factors such as sunlight, temperature fluctuations, and wind erosion. By carefully managing the distribution of the stored snow, ski resorts and communities can ensure a consistent and high-quality snow base, even during periods of low snowfall. In both high- and low-altitude regions stockpiling excess snow during periods of abundant snowfall, communities can ensure a longer and more reliable skiing season. Preserving the stored snow using insulating materials or cooling systems helps maintain optimal snow conditions, allowing ski resorts to operate even during periods of low natural snowfall. This extension of the skiing season attracts more visitors and boosts the local economy.

Residents in regions prone to snowstorms should establish a dependable and regularly maintained snow removal strategy to swiftly clear pathways, streets, and entrances for safe movement. Stay informed about weather forecasts and road conditions, and if necessary, limit travel during severe snowstorms. Implement proactive measures such as installing snow-retention structures on rooftops and strategically trimming trees to prevent sudden snow slides or damage. Strengthen home's insulation and weatherproofing to conserve heat and reduce energy costs during extreme cold.

The level of implementation and adoption of these practices can vary from community to community. Some regions may have embraced these approaches extensively, while others may still be in the early stages of exploration and implementation. The decision to implement these practices rests with the communities themselves, considering their specific needs, resources, and long-term sustainability goals.

5.4 Suggestions for Downstream Water Users

Snow melt water users should leverage as many of the available datasets as possible, like remote sensing data from satellites like MODIS and AMSR-E, combined with ground-based measurements and citizen science initiatives like the Community Snow Observations project, to gain a comprehensive understanding of snow conditions. This enables informed decision-making regarding water management. Fostering collaboration and data sharing among stakeholders is vital. Snow melt water users should actively engage with local communities, scientists, and water management agencies to exchange knowledge and data. The SNODAS system, operated by the NSIDC (Barrett 2003), integrates various data sources to provide accurate snow cover and snowmelt information. By participating in such platforms, users can access a broader range of data and expertise, leading to improved water management strategies. Diversifying water sources is essential for reducing dependence solely on snow melt water. Users can explore alternative sources, including rainwater harvesting and groundwater recharge, to supplement their water supply. Engaging stakeholders and local communities is essential for effective snow melt water management. Community-based organizations, such as the CSAS in Colorado (Landry et al. 2014), actively involve local communities in snow monitoring, data collection, and water management efforts. By providing training and resources, these organizations empower communities to contribute their traditional knowledge and observations, enhancing the overall understanding of snow dynamics and improving water management strategies.

Addressing the challenges posed by climate change is of paramount importance. Snow melt water users should consider the long-term implications of climate change on snow accumulation and melt patterns. Collaborative research projects, such as the "Climate Impacts on Snow in the Pacific Northwest" initiative, integrate climate models with snow observations to assess future water availability and inform adaptation strategies. This knowledge enables users to adjust water allocation plans, implement water-saving technologies, and develop resilience measures to mitigate the impacts of climate change.

5.5 Suggestions for Snow Sport and Tourism Industry

One area for improvement is the management of snow resources. Snow sport and tourism industry stakeholders should focus on sustainable snowmaking practices to mitigate the impacts of climate change and ensure optimal snow conditions for skiing and other winter activities. This involves using efficient snowmaking systems, such as automated snow guns that produce high-quality snow with minimal water consumption. An exemplary case is to enhance snowpack monitoring and avalanche forecasting capabilities. This can be achieved through the integration of advanced technologies such as ground-based remote sensing instruments and satellite observations. For instance, the International Centre for Integrated Mountain Development (ICIMOD) (Schild 2008) and the Government of Nepal have utilized remote sensing data from satellites like Sentinel-1 to monitor snow cover and assess avalanche risks in the Himalayas. Implementing similar technologies and collaborations can improve safety measures and inform decision-making processes for snow sport and tourism activities. Furthermore, the industry can benefit from promoting climate change adaptation strategies. This involves diversifying winter tourism offerings beyond traditional snow-dependent activities. For instance, in regions experiencing reduced snowfall, developing alternative attractions like snowshoeing, cross-country skiing, or indoor winter sports facilities can help maintain tourism revenue and sustain local economies. The Swiss Alps provide an example where ski resorts have successfully expanded their offerings to include adventure parks, wellness centers, and cultural events, attracting tourists year-round.

The industry should also prioritize the preservation of natural habitats and ecosystems. Ski resorts can adopt environmentally friendly practices, such as minimizing the expansion of ski slopes into ecologically sensitive areas and implementing habitat restoration initiatives. The "Sustainable Slopes" program by the National Ski Areas Association (NSAA) (Rivera et al. 2006) in the United States serves as a benchmark for ski resorts committed to environmental stewardship, promoting initiatives like waste reduction, renewable energy use, and water conservation. Engaging in research partnerships, such as the Mountain Research Initiative (MRI) (Adler et al. 2020), enables the industry to access scientific expertise and enhance their understanding of climate change impacts on mountain regions. This knowledge can inform decision-making processes and contribute to the development of adaptation strategies tailored to specific regions and resorts.

6 Successful Use Cases of Snow Science

6.1 The Canadian Avalanche Association

The CAA has been at the forefront of employing snow science practices to enhance public safety in avalanche-prone areas of British Columbia, Canada. The region's rugged terrain and heavy snowfall attract outdoor enthusiasts, but also pose significant risks. The CAA recognized the need for effective risk management strategies and implemented an integrated approach, including avalanche forecasting, snowpack monitoring and research, education and outreach, and collaboration with stakeholders. By providing actionable information and promoting widespread adoption, the CAA has successfully addressed the issue of avalanche hazards, ensuring public safety in the region.

To tackle the avalanche risks in British Columbia, the CAA established an extensive avalanche forecasting and warning system (Stethem et al. 2003). Highly trained avalanche forecasters collected and analyzed data on snowpack conditions, weather patterns, and historical avalanche activity. The integration of advanced snow science techniques enabled the generation of daily avalanche bulletins and advisories, which provided the public with real-time information on hazards and recommended safety measures. This actionable approach empowered individuals, including backcountry users and winter sports enthusiasts, to make informed decisions and take appropriate precautions, such as route planning and slope avoidance. In addition, the CAA conducted thorough snowpack monitoring and research programs to better understand the region's snow characteristics and their contribution to avalanche formation. Field observations, including snow stability tests and snow profile analysis, were performed regularly to assess the stability and structural integrity of the snowpack. These scientific insights not only improved the accuracy of avalanche forecasts but also informed the development of standardized safety protocols. Through collaboration with governmental agencies, ski resorts, search and rescue organizations, and local communities, the CAA ensured a coordinated response to avalanche risks. This collaboration included data sharing, joint research projects, and integrating avalanche information into regional emergency response plans. By fostering partnerships and sharing expertise, the CAA optimized resources and promoted a culture of safety among stakeholders.

The Canadian Avalanche Association's application of snow science practices in British Columbia exemplifies a proactive and actionable approach to addressing avalanche risks. By employing an integrated system of forecasting, monitoring, education, and collaboration, the CAA has successfully enhanced public safety. The CAA's dedication to advancing snow science, fostering partnerships, and raising public awareness has made a significant impact on minimizing accidents and fatalities in avalanche-prone areas of British Columbia.

6.2 The Swiss Alps, Switzerland: Swiss Data Cube Project

In the Swiss Alps, renowned for their snow-covered landscapes and thriving winter tourism industry, scientists, local communities, and policymakers have come together to address the challenges posed by a changing snow environment (Jonas et al. 2009). This collaborative effort led to the establishment of the Swiss Data Cube project, a groundbreaking initiative that took place from 2016 to 2019 (Chatenoux et al. 2021). Involving researchers, data scientists, government agencies, and local communities, the project aimed to leverage Earth Observation data to monitor and understand climate and environmental changes in mountain regions. As part of this endeavor, remote sensing data from the Landsat series, spanning from 1984 to the present (Sun et al. 2019), was compiled and integrated into one comprehensive dataset. By combining traditional knowledge with scientific data, the project gained a deeper understanding of snowpack variability, ice conditions, and associated risks. This collaboration led to the development of culturally appropriate early warning systems for hazards like thin ice and unstable snow cover, ensuring the safety of community members engaged in traditional activities. Land-use planners and policymakers also benefited from the Swiss Data Cube dataset. The comprehensive snow cover and snow depth information supported evidence-based decision-making in managing water resources, conserving biodiversity, and planning infrastructure projects in mountainous regions. The impact of the Swiss Data Cube project extended beyond Switzerland, serving as an inspiring example of successful collaboration and actionable research in snow-dependent regions worldwide. The comprehensive dataset and methodologies developed during the project have inspired similar initiatives in other countries facing similar challenges. By sharing their experiences and knowledge, the Swiss researchers and stakeholders involved have contributed to a global understanding of the importance of integrating traditional knowledge and scientific expertise to address the impacts of climate change on snow-dependent ecosystems and livelihoods.

Through the Swiss Data Cube project, several positive outcomes were achieved, such as improved monitoring and decision-making in various sectors (Chatenoux et al. 2021). However, there were also challenges and potential areas for improvement that could have been avoided with more careful consideration. One such aspect is the accessibility and usability of the data. While the project successfully compiled a vast amount of remote sensing data, there could have been more emphasis on making it easily accessible and user-friendly for a wider range of stakeholders. This could have involved developing intuitive interfaces, providing training and support for data interpretation, and ensuring that the data is readily available in formats compatible with various analysis tools. Another aspect that could have been better addressed is the engagement of local communities throughout the project's duration. Although the project acknowledged the importance of traditional knowledge and involved local communities in certain stages, there could have been more proactive efforts to ensure their meaningful participation from the inception to the implementation of the project. This could have included establishing regular dialogue, fostering partnerships, and integrating traditional knowledge more deeply into the research and decision-making processes. By doing so, the project could have benefited from a broader range of perspectives, increased community ownership, and improved the relevance and applicability of the research outcomes.

7 Conclusion

The integration of scientific knowledge into snow monitoring practices plays a vital role in understanding and effectively managing snow resources. By leveraging scientific advancements and embracing innovative approaches, stakeholders of snow monitoring can enhance their capacity to adapt to changing snow conditions and improve the sustainable management of snow. Moving forward, it is essential for communities to focus on certain key areas and further improve their snow monitoring efforts. Firstly, enhancing snow accumulation and melt forecasting accuracy can provide valuable information for water resource planning, agriculture, and infrastructure management. By investing in research and technological advancements, communities can improve the predictive capabilities and lead time for snow-related events, enabling better preparation and response. Communities should strive to deepen their understanding of the impacts of climate change on snow dynamics. By conducting research and monitoring, they can gain insights into the changing patterns of snow accumulation, melt rates, and overall snowpack characteristics. This knowledge will enable communities to develop adaptive strategies to mitigate risks and adapt their practices accordingly.

Community engagement and public awareness initiatives are also mandatory in advancing snow practices. By involving local communities in data collection, monitoring, and decision-making processes, communities can benefit from the wealth of traditional knowledge and observations. Furthermore, raising awareness about the importance of snow monitoring and its implications for various sectors, including water management, agriculture, and tourism, will foster a culture of informed decision-making and collective responsibility. By prioritizing research, collaboration, community engagement, and public awareness, communities can harness the power of scientific knowledge in snow-centered activities to improve the quality of life for individuals living in snow-dependent regions. Embracing these improvements will pave the way for more sustainable and adaptive practices, ensuring the long-term resilience and well-being of communities in the face of changing snow conditions and associated challenges.

References

Adler, Carolina, Jörg Balsiger, Adrienne Grêt-Regamey, Andreas Heinimann, Christian Huggel, Rolf Weingartner, Irasema Alcántara-Ayala, et al. 2020. Making connections for our changing mountains: Future directions for the Mountain Research Initiative (MRI). *Mountain Research and Development* 40 (3): P1.

Alldredge, Alice L., and Mary W. Silver. 1988. Characteristics, dynamics and significance of marine snow. *Progress in Oceanography* 20 (1): 41–82.

Alnaim, Ahmed, and Ziheng Sun. 2022. Using Geoweaver to make snow mapping workflow FAIR. In *2022 IEEE 18th international conference on e-science (e-science)*, 409–410. IEEE.

Audit of Winter Road Maintenance Program – Toronto. Accessed 30 May 2023. https://www.toronto.ca/legdocs/mmis/2020/au/bgrd/backgroundfile-157521.pdf.

Aves, Alex R., Laura E. Revell, Sally Gaw, Helena Ruffell, Alex Schuddeboom, Ngaire E. Wotherspoon, Michelle LaRue, and Adrian J. McDonald. 2022. First evidence of microplastics in Antarctic snow. *The Cryosphere* 16 (6): 2127–2145.

Barnett, Tim P., Jennifer C. Adam, and Dennis P. Lettenmaier. 2005. Potential impacts of a warming climate on water availability in snow-dominated regions. *Nature* 438 (7066): 303–309.

Barrett, Andrew P. 2003. *National operational hydrologic remote sensing center snow data assimilation system (SNODAS) products at NSIDC*. Boulder: National Snow and Ice Data Center, Cooperative Institute for Research in Environmental Sciences.

Bates, Sofie. 2021. NASA snow campaign wraps for 2021. NASA, June. https://www.nasa.gov/feature/goddard/2021/nasa-snow-campaign-wraps-for-2021.

Bentley, Tim A., Stephen Page, and Linda Walker. 2004. The safety experience of New Zealand adventure tourism operators. *Journal of Travel Medicine* 11 (5): 280–286.

Blackburn, Robert R. 2004. *Snow and ice control: Guidelines for materials and methods*. Vol. 526. Transportation Research Board.

Campbell, James F., and André Langevin. 1995. Operations management for urban snow removal and disposal. *Transportation Research Part A: Policy and Practice* 29 (5): 359–370.
Cannistra, Anthony F., David E. Shean, and Nicoleta C. Cristea. 2021. High-resolution CubeSat imagery and machine learning for detailed snow-covered area. *Remote Sensing of Environment* 258: 112399.
Chatenoux, Bruno, Jean-Philippe Richard, David Small, Claudia Roeoesli, Vladimir Wingate, Charlotte Poussin, Denisa Rodila, et al. 2021. The Swiss data cube, analysis ready data archive using earth observations of Switzerland. *Scientific Data* 8 (1): 295.
City of Toronto. 2023. Sidewalk clearing. City of Toronto, January 6. https://www.toronto.ca/services-payments/streets-parking-transportation/road-maintenance/winter-maintenance/clearing-snow-and-ice-from-your-property/.
da Silva, Rafael Ferreira, Rosa M. Badia, Venkat Bala, Debbie Bard, Peer-Timo Bremer, Ian Buckley, Silvina Caino-Lores et al. 2023. Workflows community summit 2022: A roadmap revolution. *arXiv preprint arXiv:2304.00019*.
Deems, Jeffrey S., Thomas H. Painter, and David C. Finnegan. 2013. Lidar measurement of snow depth: A review. *Journal of Glaciology* 59 (215): 467–479.
Deiana, Allison McCarn, Nhan Tran, Joshua Agar, Michaela Blott, Giuseppe Di Guglielmo, Javier Duarte, Philip Harris, et al. 2022. Applications and techniques for fast machine learning in science. *Frontiers in Big Data* 5: 787421.
Doesken, Nolan J., and G.L. Schaefer. 1987. The contribution of SNOTEL precipitation measurements to climate analysis, monitoring and research in Colorado. In *Proc. Western Snow Conf*, 20–30.
Durre, Imke, Matthew J. Menne, Byron E. Gleason, Tamara G. Houston, and Russell S. Vose. 2010. Comprehensive automated quality assurance of daily surface observations. *Journal of Applied Meteorology and Climatology* 49 (8): 1615–1633.
Eira, Inger Marie, Anders Oskal Gaup, Inger Hanssen-Bauer, and Svein Disch Mathiesen. 2018. Snow cover and the loss of traditional indigenous knowledge. *Nature Climate Change* 8 (11): 928–931.
Engelhard, C., I. Se De Toffol, W. Rauch Lek, and R. Dallinger. 2007. Environmental impacts of urban snow management—The alpine case study of Innsbruck. *Science of the Total Environment* 382 (2–3): 286–294.
Etzkowitz, Henry. 1983. Entrepreneurial scientists and entrepreneurial universities in American academic science. *Minerva*: 198–233.
Flagestad, Arvid, and Christine A. Hope. 2001. Strategic success in winter sports destinations: A sustainable value creation perspective. *Tourism Management* 22 (5): 445–461. https://doi.org/10.1016/s0261-5177(01)00010-3.
Frau, Lorenzo, Syed R. Rizvi, Bruno Chatenoux, Charlotte Poussin, Jean-Philippe Richard, and Gregory Giuliani. 2018. Snow observations from space: An approach to map snow cover from three decades of landsat imagery across Switzerland. In *IGARSS 2018–2018 IEEE international geoscience and remote sensing symposium*, 8663–8666. IEEE.
Fuchs, Sven, and Michael Bründl. 2005. Damage potential and losses resulting from snow avalanches in settlements of the canton of Grisons, Switzerland. *Natural Hazards* 34 (1): 53–69.
Gomes, Marcos Geraldo, Victor Hugo Carlquist, Luiz da Silva, Fernando Rodrigues Pinto, Plinio Centoamore, Salvatore Digiesi, Francesco Facchini, and Geraldo Cardoso de Oliveira Neto. 2020. Economic, environmental and social gains of the implementation of artificial intelligence at dam operations toward Industry 4.0 principles. *Sustainability* 12 (9): 3604.
Greenwood, Michelle. 2007. Stakeholder engagement: Beyond the myth of corporate responsibility. *Journal of Business Ethics* 74: 315–327.
Hamada, Yasuhiro, Makoto Nakamura, and Hideki Kubota. 2007. Field measurements and analyses for a hybrid system for snow storage/melting and air conditioning by using renewable energy. *Applied Energy* 84 (2): 117–134.
Harper, Sherilee L., Victoria L. Edge, Corinne J. Schuster-Wallace, Olaf Berke, and Scott A. McEwen. 2011. Weather, water quality and infectious gastrointestinal illness in two Inuit communities in Nunatsiavut, Canada: Potential implications for climate change. *EcoHealth* 8: 93–108.

Hedrick, Andrew R., Danny Marks, Scott Havens, Mark Robertson, Micah Johnson, Micah Sandusky, Hans-Peter Marshall, Patrick R. Kormos, Kat J. Bormann, and Thomas H. Painter. 2018. Direct insertion of NASA Airborne Snow Observatory-derived snow depth time series into the iSnobal energy balance snow model. *Water Resources Research* 54 (10): 8045–8063.

Henn, Brian, Andrew F. Newman, Ben Livneh, Christopher Daly, and Jessica D. Lundquist. 2017. An assessment of differences in gridded precipitation datasets in complex terrain. *Journal of Hydrology* 556 (December): 1205–1219. https://doi.org/10.1016/j.jhydrol.2017.03.008.

Jain, Sanjay K., Ajanta Goswami, and Arun Kumar Saraf. 2008. Accuracy assessment of MODIS, NOAA and IRS data in snow cover mapping under Himalayan conditions. *International Journal of Remote Sensing* 29 (20): 5863–5878.

Jamieson, Bruce, and Chris Stethem. 2002. Snow avalanche hazards and management in Canada: Challenges and progress. *Natural Hazards* 26: 35–53.

Jerez, Mirian Masaquiza. 2021. Challenges and opportunities for Indigenous Peoples' sustainability.

Jonas, Tobias, Christoph Marty, and Jan Magnusson. 2009. Estimating the snow water equivalent from snow depth measurements in the Swiss Alps. *Journal of Hydrology* 378 (1–2): 161–167.

Karabatić, Ana, Robert Weber, and Thomas Haiden. 2011. Near real-time estimation of tropospheric water vapour content from ground based GNSS data and its potential contribution to weather now-casting in Austria. *Advances in Space Research* 47 (10): 1691–1703.

Kiniry, Laura. 2020. What we can learn from the Sámi tradition of Reading snow. *Condé Nast Traveler*, December 2, 2020. https://www.cntraveler.com/story/what-we-can-learn-from-the-sami-tradition-of-reading-snow.

Kirkham, James D., Inka Koch, Tuomo M. Saloranta, Maxime Litt, Emmy E. Stigter, Knut Møen, Amrit Thapa, Kjetil Melvold, and Walter W. Immerzeel. 2019. Near real-time measurement of snow water equivalent in the Nepal Himalayas. *Frontiers in Earth Science* 7: 177.

Kuemmel, David E., and David A. Kuemmel. 1994. *Managing roadway snow and ice control operations*. Vol. 207. Transportation Research Board.

Landry, Christopher C., Kimberly A. Buck, Mark S. Raleigh, and Martyn P. Clark. 2014. Mountain system monitoring at Senator Beck Basin, San Juan Mountains, Colorado: A new integrative data source to develop and evaluate models of snow and hydrologic processes. *Water Resources Research* 50 (2): 1773–1788.

Largeron, Chloé, Marie Dumont, Samuel Morin, Aaron Boone, Matthieu Lafaysse, Sammy Metref, Emmanuel Cosme, Tobias Jonas, Adam Winstral, and Steven A. Margulis. 2020. Toward snow cover estimation in mountainous areas using modern data assimilation methods: A review. *Frontiers in Earth Science* 8: 325.

Lievens, Hans, Isis Brangers, Hans-Peter Marshall, Tobias Jonas, Marc Olefs, and Gabriëlle De Lannoy. 2022. Sentinel-1 snow depth retrieval at sub-kilometer resolution over the European Alps. *The Cryosphere* 16 (1): 159–177.

MacGillivray, Brian H., Paul D. Hamilton, J.E. Strutt, and Simon J.T. Pollard. 2006. Risk analysis strategies in the water utility sector: An inventory of applications for better and more credible decision making. *Critical Reviews in Environmental Science and Technology* 36 (2): 85–139.

Meløysund, Vivian, Kim Robert Lisø, Jan Siem, and Kristoffer Apeland. 2006. Increased snow loads and wind actions on existing buildings: Reliability of the Norwegian building stock. *Journal of Structural Engineering* 132 (11): 1813–1820.

Miller, Kaycee. 2020, September 28. The complete guide to snow removal laws by state. Rentec Direct Blog. https://www.rentecdirect.com/blog/the-complete-guide-to-snow-removal-laws-by-state/

Mitchell, E. De Wayne, Steven V. Vasiloff, Gregory J. Stumpf, Arthur Witt, Michael D. Eilts, J.T. Johnson, and Kevin W. Thomas. 1998. The national severe storms laboratory tornado detection algorithm. *Weather and Forecasting* 13 (2): 352–366.

Neuwirth, Robert. 2021. New Mexico farmers manage scarce resource with centuries-old irrigation system. *Environment*, May 4, 2021.

Notarnicola, Claudia. 2020. Hotspots of snow cover changes in global mountain regions over 2000–2018. *Remote Sensing of Environment* 243: 111781.

O'Rourke, Michael J., Robert Redfield, and Peter von Bradsky. 1982. Uniform snow loads on structures. *Journal of the Structural Division* 108 (12): 2781–2798.

Pastick, Neal J., M. Torre Jorgenson, Bruce K. Wylie, Shawn J. Nield, Kristofer D. Johnson, and Andrew O. Finley. 2015. Distribution of near-surface permafrost in Alaska: Estimates of present and future conditions. *Remote Sensing of Environment* 168: 301–315.

Pelto, Ben M., Brian Menounos, and Shawn J. Marshall. 2019. Multi-year evaluation of airborne geodetic surveys to estimate seasonal mass balance, Columbia and Rocky Mountains, Canada. *The Cryosphere* 13 (6): 1709–1727.

Poussin, Charlotte, Yaniss Guigoz, Elisa Palazzi, Silvia Terzago, Bruno Chatenoux, and Gregory Giuliani. 2019. Snow cover evolution in the Gran Paradiso National Park, Italian Alps, using the earth observation data cube. *Data* 4 (4): 138.

Proksch, Martin, Nick Rutter, Charles Fierz, and Martin Schneebeli. 2016. Intercomparison of snow density measurements: Bias, precision, and vertical resolution. *The Cryosphere* 10 (1): 371–384.

Rasmussen, Roy, Bruce D. Baker, John Kochendorfer, Tilden P. Meyers, Scott Landolt, Alexandre P. Fischer, Jenny Black, et al. 2012. How well are we measuring snow: The NOAA/FAA/NCAR winter precipitation test bed. *Bulletin of the American Meteorological Society* 93 (6): 811–829. https://doi.org/10.1175/bams-d-11-00052.1.

Riseth, Jan Åge, Hans Tømmervik, Elina Helander-Renvall, Niklas Labba, Cecilia Johansson, Eirik Malnes, Jarle W. Bjerke, et al. 2011. Sámi traditional ecological knowledge as a guide to science: Snow, ice and reindeer pasture facing climate change. *Polar Record* 47 (3): 202–217.

Rivera, Jorge, Peter De Leon, and Charles Koerber. 2006. Is greener whiter yet? The sustainable slopes program after five years. *Policy Studies Journal* 34 (2): 195–221.

Ross, Cameron, Peter Lambs, Paul McAlpine, Gavin Kennedy, and Chris Button. 2020. Validation of gyroscope sensors for snow sports performance monitoring. *International Journal of Computer Science in Sport* 19 (1): 51–59.

Sahu, Rakesh, and R.D. Gupta. 2020. Snow cover area analysis and its relation with climate variability in Chandra basin, Western Himalaya, during 2001–2017 using MODIS and ERA5 data. *Environmental Monitoring and Assessment* 192: 1–26.

Schiffer, Robert A., and William B. Rossow. 1983. The International Satellite Cloud Climatology Project (ISCCP): The first project of the world climate research programme. *Bulletin of the American Meteorological Society* 64 (7): 779–784.

Schild, Andreas. 2008. ICIMOD's position on climate change and mountain systems. *Mountain Research and Development* 28 (3): 328–331.

Seeger, William R. 1960. Duties and responsibilities of the water utility manager. *Journal-American Water Works Association* 52 (9): 1122–1128.

Serreze, Mark C., Martyn P. Clark, Richard L. Armstrong, David A. McGinnis, and Roger S. Pulwarty. 1999. Characteristics of the western United States snowpack from snowpack telemetry (SNOTEL) data. *Water Resources Research* 35 (7): 2145–2160.

Sood, Vishakha, Sartajvir Singh, Ajay Kumar Taloor, Shivendu Prashar, and Ravneet Kaur. 2020. Monitoring and mapping of snow cover variability using topographically derived NDSI model over north Indian Himalayas during the period 2008–19. *Applied Computing and Geosciences* 8: 100040.

Steiger, Robert, Daniel Scott, Bruno Abegg, Marc Pons, and Carlo Aall. 2019. A critical review of climate change risk for ski tourism. *Current Issues in Tourism* 22 (11): 1343–1379.

Stethem, Chris, Bruce Jamieson, Peter Schaerer, David Liverman, Daniel Germain, and Simon Walker. 2003. Snow avalanche Hazard in Canada – A review - natural hazards. SpringerLink. Accessed 29 May 2023. https://link.springer.com/article/10.1023/A:1022998512227.

Sun, Ziheng, Liping Di, and Hui Fang. 2019. Using long short-term memory recurrent neural network in land cover classification on Landsat and Cropland data layer time series. *International Journal of Remote Sensing* 40 (2): 593–614.

Sun, Ziheng, Nicoleta C. Cristea, Kehan Yang, Ahmed Alnuaim, Lakshmi Chetana Gomaram Bikshapathireddy, Aji John, Justin Pflug, et al. 2022a. Making machine learning-based snow

water equivalent forecasting research productive and reusable by Geoweaver. In *AGU fall meeting abstracts*, vol. 2022, IN23A-04.
Sun, Ziheng, Laura Sandoval, Robert Crystal-Ornelas, S. Mostafa Mousavi, Jinbo Wang, Cindy Lin, Nicoleta Cristea, et al. 2022b. A review of earth artificial intelligence. *Computers & Geosciences*: 105034.
Tremblay, Martin, Christopher Furgal, Violaine Lafortune, Caroline Larrivée, Jean-Pierre Savard, Michael Barrett, Tuumasi Annanack, Noat Enish, Peter Tookalook, and Betsy Etidloie. 2006. Communities and ice: Bringing together traditional and scientific knowledge. *Climate Change: Linking Traditional and Scientific Knowledge, 123–138.*
Tsai, Ya-Lun S., Andreas Dietz, Natascha Oppelt, and Claudia Kuenzer. 2019. Remote sensing of snow cover using spaceborne SAR: A review. *Remote Sensing* 11 (12): 1456.
Unbehaun, Wiebke, Ulrike Pröbstl, and Wolfgang Haider. 2008. Trends in winter sport tourism: Challenges for the future. *Tourism Review* 63 (1): 36–47.
Valle, Giovanni. 2023. The best building materials for cold climates: Complete guide. *Your Own Architect*, April. https://www.yourownarchitect.com/the-best-building-materials-for-cold-climates-complete-guide/.
Vickers, Hannah, Stein Rune Karlsen, and Eirik Malnes. 2020. A 20-year Modis-based snow cover dataset for Svalbard and its link to phenological timing and sea ice variability. MDPI, April 1. https://www.mdpi.com/2072-4292/12/7/1123.
Wrzesien, Melissa L., Tamlin M. Pavelsky, Michael T. Durand, Jeff Dozier, and Jessica D. Lundquist. 2019. Characterizing biases in mountain snow accumulation from global data sets. *Water Resources Research* 55 (11): 9873–9891.
Yang, Jun, Peng Gong, Fu Rong, Minghua Zhang, Jingming Chen, Shunlin Liang, Xu Bing, Jiancheng Shi, and Robert Dickinson. 2013. The role of satellite remote sensing in climate change studies. *Nature Climate Change* 3 (10): 875–883.
Yang, Kehan, Aji John, Ziheng Sun, and Nicoleta Cristea. 2023. Machine learning for snow cover mapping. In *Artificial intelligence in earth science*, 17–39. Elsevier. https://doi.org/10.1016/B978-0-323-91737-7.00004-9
Zhang, Taotao, Tao Wang, Gerhard Krinner, Xiaoyi Wang, Thomas Gasser, Shushi Peng, Shilong Piao, and Tandong Yao. 2019. The weakening relationship between Eurasian spring snow cover and Indian summer monsoon rainfall. *Science Advances* 5 (3): eaau8932.
Zrnic, D.S., J.F. Kimpel, D.E. Forsyth, A. Shapiro, G. Crain, R. Ferek, J. Heimmer, W. Benner, F.T.J. McNellis, and R.J. Vogt. 2007. Agile-beam phased array radar for weather observations. *Bulletin of the American Meteorological Society* 88 (11): 1753–1766.

Chapter 10
Toward More Actionable Vulnerability Indices for Global Environmental Change

Elia Axinia Machado

Contents

E. A. Machado (✉)
Earth, Environmental, and Geospatial Sciences, Lehman College, City University of New York, Bronx, NY, USA

Earth and Environmental Sciences, The Graduate Center, City University of New York, New York, NY, USA
e-mail: elia.machado@lehman.cuny.edu

Z. Sun (ed.), *Actionable Science of Global Environment Change*,
https://doi.org/10.1007/978-3-031-41758-0_10

1 Introduction

More than 95% of terrestrial ecosystems have been modified to some degree by human activities (Kennedy et al. 2019) resulting in the loss of natural ecosystems and an increasingly warmer planet through Global Environmental Change (GEC) processes. GEC includes systemic changes that operate globally through the major systems of the geosphere-biosphere (e.g., greenhouse emissions) and cumulative changes that represent the global accumulation of localized changes (e.g., land use change) (Turner et al. 1990).

As the global population surpasses eight billion, the negative impacts of GEC processes have proven a threat to the survival and functioning of ecological and social systems, underscoring the urgency to implement mitigation and adaptation strategies for GEC. It is known that the impacts of GEC, including climate change, are not and will not be equally distributed among human populations and ecosystems (Dow 1992; O'Brien and Leichenko 2000; Kasperson et al. 2001; Liverman 2001; Eriksen and Kelly 2007; Thomas et al. 2019). This is not only due to differences in exposure among and within human populations and ecosystems to GEC impacts, but also to differences in their sensitivity, coping, and adaptive capacity to those impacts, leading to differential vulnerability across the globe.

Global Environmental Change vulnerability assessments, hereafter referred to as VAs, are used to identify the geographical areas and/or populations where the impacts of GEC will be the highest. Often, VAs are index-based[1] where vulnerability processes are characterized using indicators and aggregated into a vulnerability index (VI) representing the overall vulnerability of a unit of analysis (e.g., county, hydrological unit).

Effectively, VIs reduce the complexity of vulnerability into one value allowing the identification, ranking, visualization, and comparison of vulnerability levels in a study area across space and time (Cutter et al. 2003; Fekete et al. 2010; Abson et al. 2012; Tate 2012; Stafford and Abramowitz 2017; Fernandez et al. 2017; Reckien 2018; Anderson et al. 2019; de Sherbinin et al. 2019), which enables their use for targeting and vulnerability reduction tracking purposes. Over time, the use of VIs has become instrumental in global change research and policy arenas (Fekete et al. 2010; Preston et al. 2011). Identifying vulnerable populations is central to climate change National Adaptation Plans (NAPs), and to Nationally Determined Contributions (NDCs), and creating risk and vulnerability indicators to inform decision makers is an important component of the Hyogo Framework for Action 2005–2015 (Anderson et al. 2019).

Specific applications of VIs include prioritizing funding for GEC mitigation and adaption strategies, assessing their effectiveness, and guiding their development by comparing vulnerability levels under different GEC projections or scenarios (Moss et al. 2001; Adger et al. 2004; Rygel et al. 2006; Eriksen and Kelly 2007; Cutter and Finch 2008; Eakin and Bojórquez-Tapia; 2008; Hahn et al. 2009; Machado 2011; de

[1] These are also known as indicator-based vulnerability assessments.

Sherbinin 2014). Vulnerability indices in GEC have been developed for a wide range of geographical areas, scales, and foci, including natural hazards (e.g., Clark et al. 1998; Wu et al. 2002; Sullivan and Meigh 2005; Rygel et al. 2006; Aceves-Quesada et al. 2007; Fekete et al. 2010), climate change (e.g., Moss et al. 2001; de Bruin et al. 2009; Preston et al. 2009; Mainali and Pricope 2017; Mi et al. 2017; Menezes et al. 2018), water resources (e.g., Alessa et al. 2008; Döll 2009), rural livelihood vulnerability (Eakin and Borjorquez-Tapia 2008), and health impacts (Wolf and McGregor 2013; Conlon et al. 2020).

Despite the growth of VIs, a number of methodological, technical, and actionability challenges have been raised that limit their usefulness and that of VAs to advance GEC research and to inform policy (e.g., Kelly and Adger 2000; Luers et al. 2003; Cutter and Finch 2008; Weichselgartner and Kasperson 2010; Preston et al. 2011; Fekete 2012; Soares et al. 2012; Anderson et al. 2019; de Sherbinin et al. 2019; Spielman et al. 2020). This chapter seeks to enhance the usefulness of VIs and index-based VAs for research and policy and draws from the geography, sustainability science, and GEC literatures to offer best-practice guidelines for VI construction while highlighting limitations and challenges in the process. Relevant contributions in this regard, among others, have proposed methodological guidelines and criteria for VAs (e.g., Turner et al. 2003; Schröter et al. 2005; Polsky et al. 2007; Patt et al. 2008) as well as indicators and indices (e.g., Eriken and Kelly 2007; Beccari 2016; Spielman et al. 2020); examined methodological choices in VI construction to understand their effect in index results (e.g., Tate 2012; Yoon 2012; Reckien 2018; Anderson et al. 2019; Bucherie et al. 2022), and their potential implications for decision-making processes (e.g., Machado and Ratick 2018); analyzed specific aspects of VI construction and VAs such as scale (e.g., Stephen and Dowing 2001; Vicent 2007; Fekete et al. 2010) and validation (e.g., Bakkensen et al. 2017; Rufat et al. 2019; Brinkmann et al. 2022); identified, examined, and proposed vulnerability mapping guidelines (Preston et al. 2011; de Sherbinin et al. 2019); and provided technical guidelines for the construction of composite indices (Saisana and Tarantola 2002; OECD-JRC 2008). Other important studies have focused on bridging science and policy, particularly from the sustainability literature, and examined the attributes of scientific work and strategies that promote their actionability. Of particular interest are the work on boundary organizations and objects[2] (Guston 1999, 2001; Cash et al. 2003; Clark et al. 2011) and related work liking science and policy in the context of GEC, vulnerability, and adaptation assessments (Kasperson and Berberian 2011; DeCrappeo et al. 2018; Weichselgartner and Kasperson 2010; Giorgi 2020). While there is no commonly agreed upon

[2]The concept of boundary work in its broad sense is used to refer to "activities of those seeking to mediate between knowledge and action" (Clark et al. 2011, p. 1). In this context, boundary objects are defined as the "collaborative products such as reports, models, maps, or standards" (Clark et al. 2011:1) that can result from it. Therefore, it is argued here that VAs and VIs can be considered boundary objects when their objective is to inform policy and/or decision making. Preston et al. (2011, p. 2) considers vulnerability maps as boundary objects "that facilitate communication and learning among stakeholders and researchers."

methodology for VAs and VIs, collectively these works highlight the importance of: (i) the saliency, credibility, and legitimacy of the information produced (e.g., Cash et al. 2003; Turner et al. 2003; Patt et al. 2008; Weichselgartner and Kasperson 2010; Clark et al. 2011)[3]; (ii) the robustness and rigor of the methodology used and the transparency in the presentation of the results (e.g., Eriksen and Kelly 2007; Vicent 2007; Tate 2012); and (iii) applying an interdisciplinary approach that engages stakeholders and decision makers (e.g., Schröter et al. 2005; Preston et al. 2011; DeCrappeo et al. 2018).

This chapter is guided by these principles and continues with an overview of the concept of vulnerability in the GEC literature followed by stepwise guidelines to construct a VI, a discussion of final considerations, and proposed assignment activities.

2 Vulnerability in the Global Environmental Change Literature

While it is generally understood that vulnerability is the susceptibility to damage or harm, there is not a unifying concept of vulnerability in GEC research and practice (Dow 1992; Cutter 1996; Eriksen and Kelly 2007; Vogel and O'Brien 2004; Adger 2006; Birkmann 2006; Gallopín 2006; Liverman 2001; Eakin and Luers 2006; Füssel and Klein 2006; Soares et al. 2012; Paul 2013; Giupponi and Biscaro 2015; Langill et al. 2022).

Numerous definitions of vulnerability, theoretical frameworks, and assessments methodologies have emerged over time. These have resulted from different disciplinary contributions to this field as it has evolved from the risk-hazards tradition focusing on the biophysical aspects of vulnerability, to more integrated approaches seeking to assess the biophysical and social aspects of vulnerability as a coupled socio-ecological system. Additionally, a more complex understanding of vulnerability and earth systems has led to the recognition of vulnerability as a multifaceted, dynamic, multi-scalar, and place-based concept (O'Brien and Leichenko 2000; Liverman 2001; Stephen and Downing 2001; Cutter et al. 2003; Turner et al. 2003; Vogel and O'Brien 2004; Adger 2006; Eriksen and Kelly 2007; Preston et al. 2011; Soares et al. 2012), that results from the interaction of three dimensions: exposure, sensitivity, and adaptive/coping capacity/resilience (McCarthy et al. 2001; Turner et al. 2003; O'Brien et al. 2004; Schröter et al. 2005; Gallopín 2006; Cutter and Finch 2008). The literature differs in the definitions of these dimensions and how they interact and contribute to the overall vulnerability of an exposed unit or

[3] In similar lines, Giorgi (2020) proposed robustness, reliability, and relevance actionable climate change information.

system.[4] In short, these dimensions refer to the exposure of the unit of analysis or coupled socio-ecological system (CSES) to the hazard or stressors of interest, its sensitivity to the hazard(s) or stressor(s), and its capacity to cope, adapt, and/or recover from harm.

Soares et al. (2012) identifies three main GEC vulnerability conceptualizations and associated assessment approaches based on their focus: biophysical, social, and integrated. Other authors have described similar but slightly distinct categorizations such as risk-hazard, political ecology/economy, and ecological resilience approaches (Eakin and Luers 2006).[5] Additional references on vulnerability conceptualizations and frameworks are included at the end of the section.

The biophysical approach (also known as risk-hazard or top-down) developed from the natural hazards literature (Liverman 2001; Turner et al. 2003; Eakin and Luers 2006; Soares et al. 2012), with seminal contributions from White (1973, 1974), White and Haas (1975), and Burton et al. (1978). Here, the hazard event (or stressor) and exposure are the main foci and determinants of vulnerability (Liverman 2001; Turner et al. 2003; Cutter et al. 2009; Soares et al. 2012). Consequently, the emphasis is placed on assessing the hazards' impacts (e.g., assets loss and populations at risk), and the characteristics of the hazard such as its probability, impact, and spatial distribution (Turner et al. 2003; Eakin and Luers 2006; Cutter et al. 2009; Soares et al. 2012) as external sources of vulnerability. Risk hazards frameworks and early climate impacts assessments (e.g., Burton et al. 1978; Kates et al. 1985) are representative examples of this approach (Turner et al. 2003; Soares et al. 2012).

In contrast, the social approach (also known as bottom-up or political economy/ecology), focuses on social structures and vulnerability as a social construct. Political economy, political ecology, human ecology, and the food security/endowments literatures are major contributors to this approach (Eakin and Luers 2006; Soares et al. 2012) stemming from the work of O'Keefe (1976), Chambers (1989), Sen (1990), and Bohle et al. (1994), among others. In this approach, political, economic, and historical structures and processes are considered the main determinants of vulnerability through the generation of differential exposures and impacts, and particularly differential coping and adaptive capacity (Dow 1992; Liverman 2001; Eakin and Luers 2006; Füssel and Klein 2006; Soares et al. 2012). Importantly, here vulnerability is not conceptualized as an outcome resulting from exposure to hazard(s) or stressor(s), but rather embedded into the social system through political, socio-economic, and historical processes (Dow 1992; Cutter 1996; Eakin and Luers 2006). Later socially focused conceptualizations of

[4] For instance, Adger (2006, p. 270) defines them as: "Exposure is the nature and degree to which a system experiences environmental or socio-political stress [...] Sensitivity is the degree to which a system is modified or affected by perturbations. Adaptive capacity is the ability of a system to evolve in order to accommodate environmental hazards or policy change and to expand the range of variability with which it can cope.

[5] Others categorize vulnerability conceptualizations as end point, focal point, and starting point (Kelly and Adger 2000).

vulnerability depart from classical political economy and ecology conceptualizations and were pioneered by the work *At Risk* of Blaikie et al. (1994).[6] This book is also widely cited in the Disaster Risk Reduction literature (Giupponi and Biscaro 2015) and introduces the pressure and release model where risk is the product of hazard and vulnerability (Risk = Hazard × Vulnerability). Collectively, these works have driven the recognition of social vulnerability (i.e., "the characteristics of a person or group in terms of their capacity to anticipate, cope with, resist and recover from impacts of a hazard") (Wisner et al. 2003, p. 11) as a key component of vulnerability and adaptation assessments.

The need for more interdisciplinary and transdisciplinary approaches to VAs has led to the development of coupled human-environmental systems or CSES VAs approaches. CSES vulnerability analysis aims to integrate both, the social and biophysical aspects of vulnerability as a coupled socio-ecological system[7] that incorporates both systems and the complexity of their interactions across spatiotemporal scales (O'Brien and Leichenko 2000; Turner et al. 2003; Schröter et al. 2005; Adger 2006; Gallopín 2006; Füssel and Klein 2006; Soares et al. 2012).

The concept of scale, and especially the recognition that social and bio-physical conditions (hence vulnerability) vary across temporal and spatial domains, is central to this approach and embedded in the notion of place-based vulnerability analysis (Cutter 1996; Turner et al. 2003; Mainali and Pricope 2017). Rooted in the idea of hazard of place, place-based vulnerability analysis seeks to integrate the biophysical and social domains within a geographical space (e.g., the area occupied by populations of concern), while recognizing the differential and dynamic nature of vulnerability across time and space (Cutter 1996; Turner et al. 2003; Cutter et al. 2009). Bounding the CSES system facilitates focusing on its specific characteristics and the decision-making context, but should not impede considering relevant linkages outside the study area (Turner et al. 2003; Schröter et al. 2005; Eriksen and Kelly 2007).

Additionally, CSES research has been greatly influenced by ecology through the concepts of resilience and adaptive capacity as well as related theories on how systems respond to disturbances (e.g., hurricanes) and adapt to changing conditions (Turner et al. 2003; Eakin and Luers 2006; Gallopín 2006). Over time, the concept of resilience has evolved from its ecological definition, i.e., "a measure of the persistence of systems and of their ability to absorb change and disturbance and still maintain the same relationships between populations or state variables" (Holling 1973, p. 14). However, it remains central to much CSES and vulnerability research in GEC seeking to characterize the ability of socio-ecological systems to bounce back to a desired state after disturbance(s) and their adaptive capacity in response to change (e.g., Eriksen and Kelly 2007; Frazier 2012; Singh-Peterson and Underhill 2017).

[6] See also the second edition of At Risk, Wisner et al. (2003).

[7] Gallopín (2006, p. 294) defines a socio-ecological system as "a system that includes societal (human) and ecological (biophysical) subsystems in mutual interaction."

Not surprisingly, different conceptualizations of vulnerability have resulted in different methodological approaches to GEC VAs (see, e.g., Birkman 2006; Füssel and Klein 2006; Soares et al. 2012 for a characterization of assessments). At the very least, VAs identify the areas and/or populations that are more vulnerable to the impacts of GEC and environmental hazards. However, in many cases these assessments also seek to inform decision making to reduce vulnerability and/or adapt to GEC impacts (Schröter et al. 2005; Eriksen and Kelly 2007; Patt et al. 2008; Preston et al 2009; Preston et al. 2011). In the latter case, the term adaptation assessment is often used. Much discussion in GEC vulnerability research, especially within the CSES literature, has focused on identifying the characteristics and desired elements of VAs such as profiling differential vulnerability and adaptive capacity (e.g., Turner et al. 2003; Schröter et al. 2005; Soares et al. 2012) and identifying place-based characteristics of vulnerability. Additional characteristics of CSES VAs and indicators in GEC are discussed in RASSP (2001), Turner et al. (2003), Vogel and Obrien (2004), Schröter et al. (2005), Eriksen and Kelly (2007), Polsky et al. (2007), Polsky and Eakin (2011), and Soares et al. (2012), among others.

Summary definitions of vulnerability in the context of natural hazards and GEC can be found in Cutter (1996), Adger et al. (2004), Paul (2013), and Bharwani (2022).. Brooks (2003) also discusses and provides definitions of risk. For detailed reviews on vulnerability conceptualizations and frameworks see Cutter (1996), Adger (1999, 2006), Birkmann (2006), Eakin and Luers (2006), Füssel and Klein (2006), Cutter et al. (2009), Soares et al. (2012), Paul (2013), and Gumel (2022). References with a dedicated focus on VA and mapping in the context of climate change include Olmos (2001), Brooks (2003), Füssel and Klein, (2006), Preston et al. (2011), Nguyen et al. (2016), and de Sherbinin et al. (2019). See Giupponi and Biscaro (2015) for a review of the concept of vulnerability in the context of climate change adaptation and disaster risk reduction.

3 Guided Steps to Construct a Composite Vulnerability Index

Vulnerability indices aim at capturing quantitatively the overall vulnerability of one or more units of analysis in a study area through the combination of vulnerability indicators into a vulnerability value.

The main steps to construct a VI involve selecting or developing a vulnerability theoretical framework, defining an analytical design consistent with the framework, generating vulnerability data indicators, scaling them to a consistent measurement scale, and combining them or a subset of them into a composite index with or without prior weighting. Additional important steps include sensitivity and uncertainty analysis, the evaluation and validation of index results and indicators, and the graphical and spatial representation of the indicators and index results (Fig. 10.1).

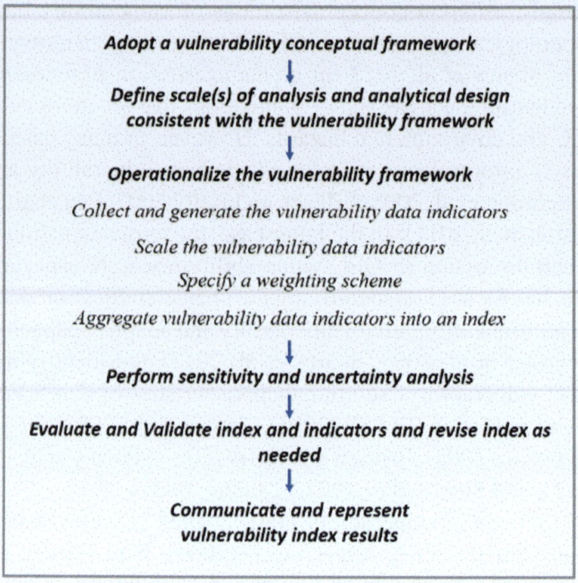

Fig. 10.1 Methodological steps to construct and communicate the results of a vulnerability index

The latter is obviously not necessary to construct the index, but it is key to communicate the results and to inform decision-making processes.

There are several methodological options available at each step (Beccari 2016; Runfola et al. 2017; Machado and Ratick 2018; Anderson et al. 2019), which loosely correspond to two index-construction approaches (deductive and inductive) according to how the indicators are selected.

The next section describes these approaches and steps to construct a composite vulnerability index. In practice, these steps are interdependent and should be guided by an overching vulnerability conceptual framework.

3.1 Adopt a Vulnerability Conceptual Framework

Loosely defined, a vulnerability framework describes how vulnerability is conceptualized in terms of the hazards(s) or stressor(s) of concern, vulnerability dimensions and indicators, and how these elements relate to each other and the overall vulnerability of a unit of analysis or CSES. In addition, the framework should also specify and justify the spatiotemporal scale(s) at which the processes that shape vulnerability operate (Eriksen and Kelly 2007; Frazier 2012).

This step is critical because the vulnerability framework is the foundation to construct a VI since it guides the selection of the scale of analysis, the input data variables representing the vulnerability indicators, and the index construction method (Fekete et al. 2010; Preston et al. 2011). As such, the framework should be grounded on a thorough understanding of the goals and purpose of the VA or VI, the hazard(s) or stressor(s) of concern, and the relevant causal processes driving vulnerability (Turner et al. 2003; Schröter et al. 2005; Eriksen and Kelly 2007; Preston et al. 2011). Preston et al. (2011) discusses two main (non-exclusive) objectives of VAs and underscores the need to align VAs with them.[8] The first objective (problem orientation) is to gain understanding about the system of interest and the nature of vulnerability and includes vulnerability mapping and developing vulnerability analytical methods. The second objective (decision support) is to identify vulnerability management strategies and to specify their implementation. Involving relevant actors and stakeholders at the beginning of the VA or VI process and their participation in designing a framework that is consistent with their needs and decision scale is essential for supporting decision making processes (Stephen and Downing 2001; Schröter et al. 2005; Preston et al. 2011; Singh-Peterson and Underhill 2017). See Schröter et al. (2005) and Preston et al. (2011) for more guidelines about engaging stakeholders in VAs and Singh-Peterson and Underhill (2017) for an illustration of involving local actors in a multi-scalar resilience assessment to flooding.

Lastly, as shown in section two, there are many ways to conceptualize vulnerability. For transparency[9] and credibility, it is instrumental to define and justify each of the elements of the vulnerably framework, but also to explain any assumptions or issues embedded in the chosen scale of analysis and indicator selection (Eriksen and Kelly 2007; Preston et al. 2011).

3.2 *Define a Scale of Analysis and Analytical Design Consistent with the Vulnerability Framework*

Defining the scale(s) of analysis is key in place-based VAs and has implications through the index construction process. Although this step is presented separately, it is closely intertwined with the vulnerability framework and indicator selection processes since both are scale-specific (Stephen and Downing 2001; Turner et al. 2003; Schröter et al. 2005; Eriksen and Kelly 2007; Vicent 2007; Tate 2012; McLaughlin and Cooper 2010; Preston et al. 2011; Mainali and Pricope 2017).

Scale determines "the spatial, temporal, quantitative or analytical dimensions used to describe a phenomenon" (Gibson et al. 2000, p. 218). In practice, this step involves defining the geographical scale(s)[10] of analysis in terms of the study area

[8] Patt et al. (2008) identify four goals of a VA: improve adaptation, frame the climate change mitigation problem, address social injustice, and conduct scientific research.

[9] Eriksen and Kelly (2007, p. 504) define transparency as: "the practice of presenting a methodological account or conceptual framework that is clear and precise free from ambiguity and easy to comprehend and contains a full account of assumptions and potential strengths and weakness."

[10] This book chapter uses the geographical definition scale where large scale (i.e., finer scale) is associated with more detail (e.g., a street map), and small scale is associated with less detail (e.g.,

extent and unit of analysis, but also the temporal period(s) of interest. The spatial unit of analysis is usually administrative (e.g., county) or a bio-physical (e.g., hydrological unit), but can also consist of one or more cells in a grid system (e.g., Sullivan and Meigh 2005; Hegglin and Huggel 2008; Preston et al. 2009; Abson et al. 2012; Mainali and Pricope 2017).

The temporal scale can refer to current, past, and/or future times and its appropriate identification requires understanding how the processes that determine vulnerability change over time and updating the indicators accordingly (Eriksen and Kelly 2007; Preston et al. 2011). The temporal scale reflects the dynamic nature of vulnerability and "reliance on single snapshot in time could be seriously misleading" (Eriksen and Kelly 2007, p. 518). To capture vulnerability dynamics, authors have projected or hypothesized future conditions affecting vulnerability using scenarios or climatic models (e.g., Moss et al. 2001; Metzger and Schröter 2006; Döll 2009; Bjarnadottir et al. 2011) and examined historical changes and future vulnerability changes using census data (e.g., Cutter and Finch 2008). Other studies have found substantial variations in social vulnerability drivers during different stages of a disaster (i.e., preparedness, response, and recovery) and have suggested a phase-oriented approach to the selection, weighting, and aggregation of projected vulnerability drivers to make VAs more salient for emergency managers (Rufat et al. 2015).

As with the theoretical framework, the study extent, spatial and temporal units of analysis chosen should correspond with the VA or VI objective and the spatiotemporal scale necessary to inform the actions or decisions of interest (Stephen and Downing, 2001; Schröter et al. 2005; Fekete et al. 2010; Preston et al. 2011). Importantly, the scale of analysis should reflect the geographic area(s) and time period(s) associated with the vulnerability processes and stressors considered (Stephen and Downing 2001; Schröter et al. 2005; Fekete et al. 2010; McLaughlin and Cooper 2010; Preston et al. 2011; Tate 2012; Mainali and Pricope 2017).

In addition, it is also important to consider the budget and time constraints of the VA (Schröter et al. 2005). For instance, local level analysis has been proven to be more useful than at smaller scales in a variety of ways. These include capturing vulnerability causes and characterizing adaptive capacity (O'Brien et al. 2004; Eriksen and Kelly 2007; Tate 2012; Singh-Peterson and Underhill 2017; Langill et al. 2022), facilitating participatory methods and qualitative data collection (O'Brien et al. 2004; Fekete et al. 2010; Singh-Peterson and Underhill 2017; Langill et al. 2022) and dealing with multivariate and complex problems (Stephen and Downing 2001). For transparency and technical transferability to other scales, Fekete et al. (2010) propose to label vulnerability indicators as scale dependent, scale independent (i.e., can be used at other scales), or non-scalable.

Regardless of how spatial and temporal bounds are defined, it is important to note that each unit of analysis will be associated with one index value only, and thus will be homogeneous analytically even though the levels of vulnerability are likely to vary within it and over time. Additionally, changes in the unit of analysis (e.g.,

---

a word map).

census track vs county) can affect the VI results. This issue, known as the Modifiable Unit of Analysis Problem (MAUP, Openshaw 1983) is a well-studied phenomenon associated with areal aggregated data and refers to the fact that results of spatial data analysis can change depending on the delineation of the unit of analysis boundaries and level of aggregation (Xie and Ziegler 2018). Changes in geographical extent can also affect the contribution of variables for an index as shown by Spielman et al. (2020) in the case of the well-known social vulnerability index (SoVI, Cutter et al. 2003). It is therefore important to explore how changing the geographical extent, and spatial (but also temporal) unit of analysis affects the spatial patterns of vulnerability in the study area, index values, and ranking of the units of analysis vulnerability values through sensitivity analysis. See Table 10.1 in Sect. (3.6) for examples of sensitivity analyses and Clark and Avery (1976), Schmidtlein et al. (2008), Preston et al. (2011), Frazier (2012), and Tate (2012) for more details on this issue.

Needless to say, getting the scale right is of outmost importance, but also challenging due to the cross-scalar nature of vulnerability in space and time (Turner et al. 2003; Ericksen and Kelly 2007; Fekete et al. 2010). Socio-ecological processes operate at a wide range of spatiotemporal scales (Kasperson et al. 2001; Turner et al. 2003; Schröter et al. 2005; Fekete et al. 2010; Preston et al. 2011; Frazier 2012) and their cross-scalar interactions amplify or reduce vulnerability, producing vulnerability outcomes at a particular geographical scale and time (Kasperson et al. 1988; Turner et al. 2003; Adger et al. (2004); O'Brien et al. 2004; Vincent 2007; Fekete et al. 2010). Vulnerability processes that appear homogenous at smaller (i.e., aggregated) scales are likely to appear more heterogenous at larger (finer) scales (Stephen and Downing 2001; Fekete, et al. 2010), leading to cross-scalar variations in vulnerability values as well. Consequently, several scholars have emphasized the need to include the nesting scales that shape vulnerability in index-based VAs (O'Brien et al. 2004; Sullivan and Meigh 2005) and to consider how vulnerability drivers interact across geographical and temporal scales (Rufat et al. 2015). Achieving this involves representing the socio-ecological characteristics of the study area, but also relevant spatial dependencies with other places that may be outside the main scale(s) of focus (Schröter et al. 2005; Eriksen and Kelly 2007; Preston et al. 2011; Frazier 2012).

In practice, capturing these dynamics in VAs remains a challenge. Most index-based assessments do not operationalize interacting vulnerability indicators and stressors at multiple scales. A notable exception is O'Brien et al.'s (2004) seminal paper assessing the vulnerability of Indian agriculture to two stressors (climate change and globalization) at district level with ground truthing using local case studies. Mindful of scalar issues, other assessments have conducted and compared results at different spatial scales (e.g. Sullivan and Meigh 2005; Vicent 2007; Balica et al. 2009; Fekete et al. 2010; McLaughlin and Cooper 2010; Singh-Peterson and Underhill 2017). Singh-Peterson and Underhill (2017), for example, propose and demonstrate a multi-scalar framework co-produced with stakeholders to determine resilience levels at household level, town, and local government scale integrating quantitative and qualitative methods. Mainali and Pricope (2017) conduct their assessment at 1 km resolution but identify multi-scalar drivers of vulnerability

Table 10.1 Summary of methodological choices tested by reference. Note: Input indicator data set refers to the type and number of data indicators used to construct the index

Reference	Methodological choice examined
Anderson et al. (2019)	Indicator dataset, aggregation method
Chakraborty et al. (2005)	Indicator dataset
Brooks et al. (2005)	Weighting
Jones and Andrey (2007)	Indicator dataset, scaling, weighting
Schmidtlein, et al. (2008)	Indicator dataset, study area, unit of analysis, index contraction (PCA: component selection, PCA rotation, weighting of the components)
Marulanda et al. (2009)	Indicator dataset, scaling, weighting
Bjarnadottir et al. (2011)	Exposure indicator values
Yoon (2012)	Index construction approach (inductive PCA, deductive), scaling
Cutter et al. (2013)	Indicator dataset, study area, unit of analysis, scaling, weighting, directionality of data indicator
Tate (2012)	Indicator dataset, unit of analysis, measurement error, scaling, indicator data transformation (counts, proportion, density), factor retention method, weighting
Tate (2013)	Indicator dataset, unit of analysis, measurement error, indicator data transformation (counts, percentage, density), scaling, weighting
Cutter et al. (2013)	Indicator dataset (number of indicator data), study area, unit of analysis, scaling, weighting, directionality of data indicator
Fernandez et al. (2017)	Weighting
Mainali and Pricope (2017)	Indicator dataset, index construction approach (inductive PCA, deductive WLC)
Runfola et al. (2017)	Importance weights in WOWA
Stafford and Abramowitz (2017)	Indicator dataset, study area, scaling
Tapia et al. (2017)	Aggregation method, weighting
Machado and Ratick (2018)	Aggregation method (WOWA, WLC, DEA, compromise programming), weighting, decision risk
Raghavan Sathyan et al. (2018)	Indicator dataset
Reckein (2018)	Index construction approach (inductive PCA, deductive WLC), indicator data transformation (percentage, density)
Spielman et al. (2020)	Geographic extent
Bucherie et al. (2022)	Indicator dataset and grouping, index construction approach (inductive PCA, deductive WLC), weighting

through a sensitivity analysis conducted at two different spatial scales (national and physiographic region). Another approach exemplified by Preston et al. (2009) is to build on a smaller a scale VA by engaging stakeholders to gain a better understanding of the local vulnerability context and to validate the assessment to promote its use. Lastly, Langill et al. (2022) use Ordinary Least Squares (OLS) regression modelling to assess the contribution of vulnerability indicators at different scales to overall vulnerability. Although these types of studies may not produce VI scores, they quantify vulnerability in many cases and their results can be very useful to select reliable indicators for index construction.

See Stephen and Downing (2001), Fekete et al. (2010), and Frazier (2012) for a more thorough discussion on the implications of spatial and temporal scale in VAs. Eriksen and Kelly (2007) and Vicent (2007) are also very relevant in the context adaptation and climate change. For a review of the concept of scale and associated terms see Gibson et al. (2000) and Fekete (2010). Preston et al. (2011) discusses scale in the context of mapping vulnerability to climate change.

3.3 Operationalize the Vulnerability Framework

This step starts with data collection to generate vulnerability data indicators. Subsequently, the data indicators are scaled and aggregated into an index according to the chosen weighting scheme.

Collect and/or Generate the Vulnerability Data Indicators

This step focuses on the generation of the *data indicators* representing the conceptual vulnerability indicators specified in the vulnerability framework. In essence, this involves collecting and processing the data needed to select and/or derive the data indicators.

Eriksen and Kelly (2007) address three desirable but challenging aspects of the indicator selection process (robustness, transparency, and verification) in the context of climate change adaptation policy. Like others (e.g., Cutter 1996; Clark et al. 1998; Schröter et al. 2005; Preston et al. 2011; Rufat et al. 2015) they emphasize the importance of the spatiotemporal context of vulnerability, stressing that indicators need to be reliable and representative of the processes that determine vulnerability in the context of the analysis, but also that verifying the indicators and the conceptual framework is key for the credibility of the indicators and index, as well as for a better understanding of vulnerability. In addition, Schröter et al. (2005) also stress that indicators should be scientifically sound, understood by the stakeholders, and spatially explicit so they can be mapped. Lastly, the comprehensiveness of the indictor set needs to be assessed and documented as well as correlations among data indicators. See Clark et al. (1998), Cutter et al. (2003), Yoon (2012), Rufat et al.

(2015), Beccari (2016), and Fatemi et al. (2017) for a thorough review of indicators used in disaster risk, vulnerability, and resilience literatures.

In practice, the selection of vulnerability indicators has been guided by the literature, data availability, statistical analysis, stakeholders, and local actors through surveys and workshops (Beccari 2016). These strategies correspond loosely to two approaches of indicator section known as deductive and inductive[11] (Adger et al. 2004; Eriksen and Kelly 2007; Vicent 2007[12]; Füssel 2010; Tate 2012).[13] Although both approaches should be grounded on a solid theoretical framework, they differ in the degree to which it guides the selection of vulnerability data indicators.

Deductive approaches typically use less indicators than inductive approaches. The indicators are selected and combined based on pre-existing (e.g., theoretical and/or literature-based) conceptualizations of the relationship between indicators and how they contribute to the overall vulnerability, for example, Aceves-Quesada et al. (2007), Hegglin and Huggel (2008), Mainali and Pricope (2017). Inductive approaches rely on statistical methods the most and data characteristics to select the final set of data indicators. Data reduction methods such as principal components (PCA) or factor analysis are often used, which are applied to extract uncorrelated components from a large set of potential data indicators and have been widely used to construct social vulnerability indices following Cutter et al. (2003). The final set of data indicators (components in this case) are selected from the uncorrelated set and aggregated to construct the index, for example, Clark et al. (1998), Rygel et al. (2006), Cutter and Finch (2008), Tate et al. (2010), Mainali and Pricope (2017), Conlon et al. (2020), and Bucherie et al. (2022). Other inductive approach studies have used past disaster outcome data from the EM-DAT database to construct a VI (e.g., Brooks and Adger 2003) or in combination with a correlation and Monte Carlo randomization procedure (Brooks et al. 2005) or used past flood event data to refine the selection of indicators (Bucherie et al. 2022). While inductive approaches using past event data can provide a detailed hazard and place understanding of vulnerability processes and their interactions, generalizing the findings may be unreliable (Rufat et al. 2015). On the other hand, deductive-based approaches should make a better effort at contextualizing the indicators and validating them (Rufat et al. 2015). In practice, both approaches can be used jointly, and some studies may not belong clearly to one of these categories (Eriksen and Kelly 2007).

Ideally, the desired set of data indicators would be available for the study area, time frame(s), and spatial unit(s) of analysis selected. Commonly used data sources include secondary data from national statistical agencies, and censuses, and primary data collected through population surveys, workshops, or focus groups with stakeholders and/or experts (Beccari 2016). However, data availability is a challenge

---

[11] Inductive and deductive approaches have also been referred to as data-driven and theory-driven, respectively (Eriksen and Kelly 2007).

[12] Vicent (2007, p. 15) also differentiates between aggregate indices "where constituent parts are not recognizable" and composite indices "where they are."

[13] Tate (2012) describes three types of approaches including inductive, deductive, and hierarchical. The latter can be considered a special case of the inductive approach.

(Eriksen and Kelly 2007; Vincent 2007; Fekete et al. 2010; Abson et al. 2012; Preston et al. 2011; Frazier 2012; Beccari 2016; Bucherie et al. 2022). Most long-term demographic data for the United States are only available every 10 (decennial census) or 5 years (American Community Survey) with varying availability at different spatial scales. Additionally, the data representing the conceptualized indicators may not exist at all, and/or the selected indicators may be very difficult or costly to capture quantitatively. To address this, studies have selected their data indicators based on data availability (e.g., Abson et al. 2012), changed the scale of analysis and indicators, collected primary data (Fekete et al. 2010), and/or disaggregated data to finer scales using dasymetric techniques (e.g., Maantay et al. 2007; Mennis 2009).

For credibility and transparency of the VI and VAs in general it is key to use data from reputable sources, and to cite, and to describe them appropriately. When primary data are used, the data collection methodology should be explained and justified thoroughly. Limitations in data collection should also be documented and, to the extent possible, their effect on vulnerability values should be assessed as potential sources of error and uncertainly. Once the data are collected, data quality screening and preprocessing processes such as evaluating data completeness and outliers need to be undertaken to prepare the data for use in the next steps and documented as part of the VI construction methodology.

Scale the Data Indicators

It is likely that data indicators are measured in different units and have different range of values. Before being combined, they need to be transformed to the same measurement scale and units (if needed) through a scaling process. This process is also referred to as standardization or normalization, although the meaning of these terms is not interchangeable in all contexts.[14] Scaling can also be done after the aggregation process, particularly when using PCA (e.g., Abson et al. 2012; Monterroso et al. 2014).

Numerous scaling methods have been used to generate VIs including: linear scaling (e.g., Willis and Fitton 2003; Sharma and Patwardhan 2008; Collins et al. 2009; Yoon 2012; Ahsan and Warmer 2014, ratios (e.g., Cutter et al. 2000; Wu et al. 2002; Yoon 2012), conversion to z-scores (e.g., Monterroso et al. (2014); Yoon 2012; Baeck et al. 2014; Bucherie et al. 2022), thresholds (e.g., Hurd et al. 1999; Moss et al. 2001; Aceves-Quesada et al. 2007; Hegglin and Huggel 2008), and fuzzy set membership functions (e.g., Eakin and Bojórquez-Tapia 2008). Other studies have ranked data indicator to a common ordinal scale according to their contribution to overall vulnerability (e.g., Hegglin and Huggel 2008; McLaughin and Cooper 2010) or based on percentiles (e.g., Menezes et al. 2018). Studies working with demographic data have transformed demographic variables to population density or percentages in the construction of social vulnerability indices (e.g., Tate 2012; Reckien

[14] Some studies use the term standardization to refer to the conversion to *z*-scores only.

2018) prior to the PCA. See Yoon (2012) for a detailed explanation of *z*-score, ratio, and min-max rescaling methods.

Min-max linear scaling is the most common scaling method followed by z-scores standardization (Tate 2012), using of categorical scales, and ranking (Beccari 2016). Sensitivity studies of scaling suggest that the impact of scaling methods on the outcome vulnerability is minimal (Jones and Andrey 2007; Yoon 2012; Tate 2012). However, Tate (2012) and Reckein (2018) found that the choice of transformation method for demographic data (i.e., density vs percentages) is highly influential in the VI results, although density input data seem to produce more robust results across aggregation methods (Reckein 2018). As in other methodological options, the choice of scaling method(s) should correspond to the best representation of vulnerability and be guided by the previous steps in index generation.

Specify the Weighting Scheme for the Data Indicators

Data indicators are weighted to reflect their differential contribution to the overall vulnerability of a unit of analysis through importance weights (Machado and Ratick 2018). However, weights also reflect the degree of trade-off or compensation (i.e., substitutability) among the data indicators being aggregated (OECD-JRC 2008; Fernandez et al. 2017) and can be used to explore different decision risk strategies (i.e., from risk adverse to risk taking) in decision making processes[15] using order weights (see Runfola et al. 2017 and Machado and Ratick 2018 for more details on this).

Commonly used weighting strategies include using equal weights (e.g. Clark et al. 1998; Cutter et al. 2000; Moss et al. 2001; Abson et al. 2012; Tate 2012; Yoon 2012), consulting literature (Cutter et al. 2003), consulting experts, stakeholders, or the wider population using an analytical hierarchy process (AHP) (e.g., Eakin and Bojorquez-Tapia 2008; Roy and Blaschke 2015), Delphi methods (e.g., ESPON Programme 2012), focus groups, interviews, or surveys (e.g., Brooks et al. 2005; Hegglin and Huggel 2008; Tate 2012; Ahsan and Warmer 2014; Oulahen et al. 2015; Runfola et al. 2017; Bucherie et al. 2022), and using the percent variance explained in PCA (Cox et al. 2006; Stafford and Abramowitz 2017). Less commonly used weighting strategies include optimization processes such as in data envelopment analysis (DEA), and reviewing planning documents or assessment reports (e.g., Reckien 2018).

Using equal weights is the most common weighting method (Tate 2012; Beccari 2016). This is mostly due to the lack of other alternatives, understanding, or knowledge of the importance of each indicator, and perception of "objectivity" in the

[15] Rinner and Malczewski (2002) differentiate between risk averse and risk-taking decision strategies. Applied to vulnerability analysis, a risk averse strategy prioritizes avoiding risk actions by weighting data indicators with higher values more (i.e., weighting negative outcomes more). A risk-averse strategy weighs data indicators with lower values more, resulting in a more "optimistic" (i.e., less vulnerable values).

resulting index (Tate 2012), but also to the absence of an accepted weighting method (Oulahen et al. 2015). However, using equal weights (or not specifying weights) assumes that all data indicators contribute equally to vulnerability, which is not likely (Rufat et al. 2015). At the sub-index level, studies have recommended and used equal weights (Villa and McLeod 2002; Tate 2012), though. See also Stafford and Abramowitz (2017) for a rationale on using equal and differential weights in PCA analysis and Oulahen et al. (2015) for a discussion on weighing and a case study on how to engage practitioners in the process.

Since using different sets of weights has proven to change index results (Fernandez et al. 2017; Machado and Ratick 2018; Reckien 2018) and the effect of different weighting schemes on VI results can vary depending on the VI structure (Tate 2012), it is recommended to perform a sensitivity analysis in this regard (see Sect. 3.6 for more details).

Aggregate Data Indicators

In this step the data indicators are aggregated into a VI. Effectively, this step represents the operationalization of the relationship between vulnerability indicators and how they contribute to the overall vulnerability of each unit of analysis.

Weighted Linear Combination (WLC)

This is the most widely used method to generate a composite index (Clark et al. 1998; Tate 2012; de Sheribin et al. 2019) and used in deductive and inductive approaches. It involves adding the scaled data indicators (e.g., Hegglin and Huggel 2008; Reckien 2018) or PCA components/factors (e.g., Clark et al. 1998, Cutter et al. 2003) previously multiplied (or not) by a weight. The result is a weighted average of the data indicators. If no weight is applied explicitly, the weights are considered equal. The aggregation could be done into one value at once or undergo sequential aggregation steps into sub-indices following a hierarchical process where indicators are aggregated successively into one or more vulnerability components, dimensions, and ultimately index values (Tate 2012; Machado and Ratick 2018).

Ordered Weighted Average (OWA), Weighted OWA (WOWA), and Order Rated Effectiveness (ORE)

These methods are used in a variety of fields, but to a lesser extent in GEC. Examples of applications include Feizizadeh and Blaschke (2013), Runfola et al. (2017), Machado and Ratick (2018), and Klimberg and Ratick (2020a). OWA offers the unique advantage of allowing the exploration of different decision risk strategies through the output of different VIs using different order weights, while WOWA incorporates both, importance and order weights. ORE combines the rank scaled

OWA results from different order weights into one value through Data Envelopment Analysis (DEA) optimization (Klimberg and Ratick 2020a). For more details on these see references above and Yager (1988), Malczewski (1999), Jiang and Eastman (2000), and Klimberg and Ratick (2020b).

Data Envelopment Analysis (DEA), Compromise Programing (CP), and Pareto Ranking

Known as frontier methods, these methods assign weights to the input data indicators and calculate VI values in one step. In general, these methods calculate index values in a multivariate space defined by the different data indicators based on the distance of each unit of analysis to the unit of analysis with the highest indicator data values. Examples of DEA applications in GEC include Clark et al. (1998), Yuan et al. (2015), and Machado and Ratick (2018). Rygel et al. (2006) apply Pareto ranking, and Machado and Ratick (2018) illustrate the use of compromise programing.

The methods described are the most used to generate VIs. Other studies have combined an index approach with other techniques to identify vulnerability hotspots (e.g., Sharma and Patwardhan 2008; Tapia et al. 2017), or used a ratio approach of vulnerability dimensions (e.g., Balica et al. 2009). Also, it is worth noting three composite risk indices focused on risk that include vulnerability as a component. The first two are global indicators of risk at country level and include the WorldRiskIndex (WRI) and the INFORM indices, recently validated by Birkmann et al. (2022).

The WRI focuses on risk as a product of exposure and several vulnerability components[16] aggregated through a WLC approach. INFORM[17] includes several quantitative indices aimed at supporting decision-making on humanitarian crises and disasters and result from the collaboration of the Inter-Agency Standing Committee Reference Group on Risk, Early Warning and Preparedness, and the European Commission. Of particular interest here are the INFORM risk index that adopts the UNISDR definition of risk as the product of hazard, exposure, and vulnerability[18] and the INFORM Climate Change, an upgrade of INFORM Risk Index that includes climate and socio-economic projections. The third composite index is the National Risk Index (NRI),[19] a mapping application developed by the Federal Emergency Management Agency (FEMA) of the United States to visualize risk to 18 natural hazards at county and census tract level. The generalized equation for the NRI

[16] *WorldRiskIndex = Exposure × ((1/3) × (Susceptivity + Lack of Coping Capacity + Lack of Adaptive Capacity). Source: https://www.ireus.uni-stuttgart.de/en/International/WorldRiskIndex/*

[17] https://drmkc.jrc.ec.europa.eu/inform-index

[18] *Risk = Hazard&Exposure* $^{1/3}$ × *Vulnerability* $^{1/3}$ × *Lack of coping capacity* $^{1/3}$ *source:* https://drmkc.jrc.ec.europa.eu/inform-index/INFORM-Risk/Methodology

[19] https://hazards.fema.gov/nri/determining-risk

calculates risk as the product of expected annual loss and social vulnerability divided by community resilience.

There is a wide variety of methodological options to generate a VI and it is out of the scope of this chapter to provide an exhaustive thorough guide to all of them. Interested readers are encouraged to review the references provided here and Saisana and Tarantola (2002) and OECD-JRC (2008) for technical guidance.

Each methodological choice along the index construction process is critical since it can impact the results of the VI significantly (Abson et al. 2012; Yoon 2012; Tate 2012; García de Jalón et al. 2016; Runfola et al. 2017; Machado and Ratick 2018; Reckien 2018; Bucherie et al. 2022) and all have their pros and cons.

Deductive approaches using WLC are widely used due to their simplicity, which facilitates their implementation and understanding by wide audiences (Malczewski 2006; Füssel 2010; Ratick and Osleeb 2013; Rufat et al. 2015). For instance, importance weights are explicit in a simple equation, so they are easy to identify. However, finding and justifying the weights used can be a challenge (Clark et al. 1998; Malczewski 2000), which is an issue with other methods as well. Additionally, the averaging effect of the WLC could be considered advantageous since it decreases sensitivity to outliers, but this can be undesirable since low data indicators can average out larger values (Klimberg and Ratick 2020a) leading to masking out highly vulnerable units. It is also important to note that WLC can result in redundancies or overemphasis in certain vulnerability aspects or dimensions if the data indicators used are not independent. Therefore, it is crucial to examine correlations among data indicators before generating the index.

Data reduction techniques allow the reduction of a large set of potential data indicators into uncorrelated components to select from without redundancies and have become a preferred method to generate social vulnerability indices to environmental hazards (Stafford and Abramowitz 2017). However, the PCA methodology is not easily understood by most of its users (Stafford and Abramowitz 2017) and selecting which components to retain is largely subjective (Abson et al. 2012; Stafford and Abramowitz 2017). Furthermore, interpreting the components and how they relate to real-word decision making can be challenging as well (Reckien 2018; Spielman et al. 2020). In this regard, Spielman et al. (2020) demonstrate how the overall contribution of a variable to a PCA-based index can be measured, which can improve the interpretability of the index. Abson et al. (2012) recommend using PCA when there are correlating and complex interactions among the initial pool of indicator data that cannot be captured arithmetically and does not recommend it when there are differences among the data indicators regarding their contribution to the overall vulnerability. This is because PCA approach is typically used with equal weights or weights based on the percent variance explained of each component (Reckien 2018), which is not a measure of their relationship with vulnerability (Barankin et al. 2021). It is also advisable to limit data indicators to those that are strongly associated with vulnerability based on the literature and/or relevant community (Stafford and Abramowitz 2017). Statistical approaches such as correlation (Brooks et al. 2005) and regression analysis (Barankin et al. 2021) can be very informative to select data indicators and their weights based on the relationship between data indicators and outcome vulnerability measures. However, finding appropriate outcome data and communicating the results can be difficult, among other methodological issues (see Barakin et al. 2021 for details). Further considerations and discussion on PCA in the GEC context can be found in Abson (2012), Cutter et al. (2013), Monterroso et al. (2014), Reckien (2018), Stafford and Abramowitz (2017), and Barankin et al. (2021). Spielman et al. (2020) describe the application of PCA in detail in the context of SoVI and provide a thorough evaluation of SoVI's limitations.

OWA, WOWA, and frontier methods are also more difficult to understand and to implement than WLC but offer several advantages too. These include producing a full range of VIs associated with different levels of decision risk (OWA) and importance weights (WOWA), while both incorporate WLC as a "variant" in the analysis (Machado and Ratick 2018).

Frontier methods such as DEA are computationally more complex, although the pareto approach to ranking units of analysis is fairly intuitive. DEA does not require scaling of data indicators (Klimberg and Ratick 2020a) and can provide a systematic weighting method in the absence of solid weighting guidelines or when there are high discrepancies in this regard. This method optimizes data indicator weights and results in the largest index value possible for each unit of analysis (Machado and Ratick 2018). The latter makes DEA well suited for risk-averse decision-making contexts since it reduces the possibility of missing vulnerable units of analysis, but can result in very high VI values in the study region, impeding the prioritization of most vulnerable areas (Machado and Ratick 2018; Klimberg and Ratick 2020a).

Ultimately, the best methodological choice is the one that best represents the vulnerability processes at play in a given context and satisfies the goals of the VA (Machado and Ratick 2018). It is key therefore to have a thorough understanding of each method, how it adheres to the vulnerability framework chosen, and its implications to meet the needs of the VI users. Lastly, applying and comparing different methods, including sensitivity analysis and validation (e.g., Mainali and Pricope 2017; Reckien 2018; Stafford and Abramowitz 2017; Bucherie et al. 2022) is needed for the reasons stated earlier and to make more informed methodological decisions. Regardless of the approach taken, the literature seems to agree that using vulnerability outcomes to inform the selection of data indicators adds robustness to the composite index (e.g., Barankin et al. 2021; García de Jalón et al. 2016; Bucherie et al. 2022). It is also recommended to describe all the method(s) used thoroughly for transparency and to ensure their repeatability (Beccari 2016) as well as the reproducibility of the VI results.

3.4 Perform Sensitivity and Uncertainty Analysis

Performing sensitivity and uncertainly analysis[20] is essential to understand which index construction decision(s) influence(s) index values the most (Tate 2012) and the variability in index ranges resulting from different methodological choices. Without this information, VIs can give a false sense of objectivity and reliability (Beccari 2016), in addition to compromising the credibility of the information provided in support of decision making (Schröter et al. 2005). Of particular interest are changes in the stability of index values over space and time and the relative rankings of the unit of analysis (Tate 2012). It is also proposed in this chapter to broaden the sensitivity analysis to include methodological decisions that can affect the interpretation of VIs (hence decision-making processes), especially when maps and other visual representations are used to represent the values.

Sensitivity analysis can be conducted at local or global level (Saitelli et al. 2004). Local sensitivity analysis assesses changes in the output (i.e., index) by changing one methodological option (e.g., weighting) at a time while keeping the others constant (Tate 2012). The resulting alternative indices have been compared using statistical methods such as parametric and non-parametric correlation analysis (e.g., among index ranks), analysis of the variance (Tate 2012), bootsraping (Mainali and Pricope 2017; Raghavan et al. 2018), Constant Elasticity Substitution Functions (Fernandez et al. 2017), randomizing exposure values (Bjarnadottir et al. 2011), and Monte Carlo simulation (Runfola et al. 2017).

Methodological choices known to affect index values and their interpretation include the choice of the unit of analysis, geographical extent, indicator selection, scaling, weighting, aggregation, and spatial visualization, although the latter has not received sufficient interest in the context of VI construction. Vicent (2007) describes three sources of uncertainly in creating deductive indices of adaptive capacity that can also apply to VIs: i) the selection of the driving forces of adaptive capacity, ii) the indicator(s) chosen to represent the driving force(s), and iii) the direction of the relationship (i.e., whether increasing a data indicator results in a positive or negative effect on adaptive capacity). Mainali and Pricope (2017) add a scalar component to the methodological considerations and identify multi-scalar vulnerability driving factors using bootstrapping at two aggregated levels (national and physiographic region). Their analysis is expanded by a statistical descriptive analysis of the different components to better understand changes in the distribution of different index configurations.

Two shortcomings of local sensitivity analysis are its inefficiency to test a large set of methods and its inability to detect interactions at different index construction stages (Tate 2012). In contrast to local sensitivity analysis, global sensitivity

[20] Uncertainty analysis "focuses on how uncertainty in the input factors propagates through the structure of the composite indicator and affects the values of the composite indicator" while sensitivity analysis "studies how much each individual source of uncertainty contributes to the output variance" (Saisana et al. 2005, p. 308).

analysis varies all input factors (e.g., index construction parameters) simultaneously (Zhou et al. 2008) and examines "how the uncertainty in the output of a model (numerical or otherwise) can be apportioned to different sources of uncertainty in model input" (Saltelli et al. 2004, p. 45). In the context of VIs, global sensitivity analysis quantifies the relative importance of each index construction step in the composite index and the interaction among steps (Tate 2012).

The implications of some of the methodological choices mentioned above are addressed in other sections of this chapter. Table 10.1 provides examples of the methodological options examined in several studies. See Saltelli et al. (2004) for more details on implementing sensitivity analysis in general, and Saisana et al. (2005), García de Jalón et al. (2016), and others in Table 10.1 for applications to composite indices.

3.5 *Evaluate and Validate Index and Indicators*

Assessing the validity of a VI or vulnerability indicator involves determining to what degree it represents the real-world phenomenon it represents. Rufat et al. (2019, p. 2) define model validity as "the degree to which a model adequately represents its underlying construct" and identify three relevant aspects of vulnerability (its multidimensionality, interactivity, and causal processes) in the validation of a social vulnerability index. While the index cannot be validated until it is constructed, the validation or verification of the indicators can be performed to inform indicator selection and weighting prior to generating the index (e.g., Barankin et al. 2021) or to refine it (e.g., Bucherie et al. 2022).

The issue of validation has been approached as an internal and external process (e.g., Rufat et al. 2019; Birkmann et al. 2022). Internal validation approaches include robustness and convergence methods. The first evaluates the correctness of a model from the analytical and statistical perspective through sensitivity and uncertainty analysis (e.g., Tate 2012), see Sect. 3.6 for more details. The latter assesses agreement between different models representing the same phenomenon (Rufat et al. 2019), which has been examined using different methods. Examples include calculating the standard deviation of index values in each unit of analysis resulting from different VIs (e.g., Preston and Jones 2008); calculating spearman's rank (Yoon 2012; Spielman et al. 2020) or Pearson's (Anderson et al. 2019) correlation of different index configurations; calculating the percentage of study units undergoing class change of vulnerability (Anderson et al. 2019); mapping the range of index values ranks across index configurations (Spielman et al. 2020); computing descriptive statistics of different vulnerability dimensions and/or scatter plots (Mainali and Pricope 2017; Klimberg and Ratick 2020a); and using various strategies at once including mapping multiple index results highlighting the units of analysis with the highest and lowest scores, mapping vulnerability indicators and dimension ranges, and graphs comparing the ranked VI values using different methods (Machado and Ratick 2018).

This section focuses on external validation methods, which often use post-event data (e.g., disaster mortality data) as a proxy of vulnerability to determine how well an index or vulnerability indicator(s) explain vulnerability outcomes. Applications of this approach have used multivariate regression analysis along with property damages, fatalities, and frequency of disaster declarations to validate three resilience and two VIs (Bakkensen et al. 2017); flood-affected household data and logistic regression to validate social vulnerability indicators in Germany (Fekete 2009); total property damages per capita in a county and OLS regression (Yoon 2012); spatial regression and outcome data from Hurricane Sandy on housing assistance applicants, affected renters, and housing damage and property loss to validate four social models of vulnerability controlling for flood exposure (Rufat et al. 2019); three forms of regression analysis with insurance claim payouts and assistance grants data from Hurricane Sandy in New Jersey along with social and exposure indicators (Barankin et al. 2021); and applied internal and external validation methods to validate two global indices at national level (Birkmann et al. 2022). Less commonly used approaches include expert elicitation (Brooks et al. 2005); consulting stakeholders (Oulahen et al. 2015) and/or in addition to hazard planning schemes, and satellite imagery of hazard events (Preston et al. 2009); and collecting data at finer scales to validate vulnerability measures at coarser scales (e.g., O'Brien et al. 2004). The latter is only appropriate if the hazard context, spatial, and cultural context of the validation is the same as the coarser scale analysis (Eriksen and Kelly 2007; Fekete 2009). See Rufat et al. (2019) for other validation examples and Birkmann et al. (2022) for a detailed internal and external validation analysis.

Several studies have emphasized the need to increase validation efforts of VIs and indicators (Preston et al. 2011; Tate 2012; de Sherbinin et al. 2019; Rufat et al. 2019; Anderson et al. 2019; Birkmann et al. 2022). However, validating index results using external data remains a challenge because vulnerability is a complex, multidimensional concept, not directly observable (Eriksen and Kelly 2007; Vincent 2007; Tate 2012; de Sherbinin et al. 2019; Spielman et al. 2020). Outcome data often represent one aspect of vulnerability and therefore may be more appropriate to validate data indicators or vulnerability dimensions rather than the VI as a measure of overall vulnerability. When a VI is validated (e.g., vulnerability to flooding) using outcome data (e.g., flood property damage), it is good practice to control for confounding factors or vulnerability aspects not reflected in the outcome data (Anderson et al. 2019) (e.g., flood exposure). However, even when this is done the issue remains of attempting to validate current vulnerability with data from the past (Vicent 2007). Additionally, some VIs are generic and do not correspond to specific stressors or hazards, which makes it difficult to identify adequate validation data (de Sherbinin et al. 2019). Using survey data or experts for validation to assess or increase the validity of VIs helps increase the validity of results, but also has limitations such as targeting appropriate experts (Anderson et al. 2019).

Lastly, some VIs are produced to inform future policies using climatic or demographic projections, for which no outcome data exist (Schröter et al. 2005; de Sherbinin et al. 2019).

In addition to validation, Preston et al. (2011) recommends performing an evaluation of VAs, which is also applicable to VIs. Such evaluation involves determining if the VI objectives have been met and helps identify the most appropriate VA or VI method(s) in a particular context. Along these lines Anderson et al. (2019) and Spielman et al. (2020) extend this further by including theoretical frameworks. Importantly, Spielman et al. (2020) develop a set of criteria for complex social indicators and apply them to evaluate the internal and theoretical consistency of SoVI, proposing to validate SoVI against generic criteria rather than an external outcome.

3.6 Communicate and Represent Vulnerability Index Results

Communicating and publishing VI results in different formats (especially digital and interactive dashboards) facilitates their comparison with other studies and their use (Beccari 2016). Maps and dashboards are one of the most common ways to communicate and represent VI values (Preston et al. 2011; Beccari 2016; de Sherbinin et al. 2019) and the focus of this section. Mapping vulnerability values increases viewers' relatability, helps interpret and internalize information, and reveal geographic and temporal vulnerability patterns (Preston et al. 2009; de Sheribin et al. 2019). Additionally, mapping vulnerability can also be used for planning purposes (Preston et al. 2011) since it can stimulate discussions on vulnerability drivers and potential strategies to reduce it through different scenarios and/or projections.

Typically, choropleth or grid-based maps are used to represent index values which are often classified into several categories and less commonly represented using a continuous color scale (e.g., Bucherie et al. 2022). The former can be achieved using different classification methods including standard deviations (e.g., Cutter et al. 2013; Anderson et al. 2019), equal intervals (e.g., Abson et al. 2012; Mainali and Pricope 2017; Machado and Ratick 2018; Menezes et al. 2018; Rufat et al. 2019), natural breaks (e.g. Tate et al. 2010), and quantiles or percentages (e.g., Cutter et al. 2000; Wu et al. 2002; Burton 2010; Anderson et al. 2019). The same map can be interpreted in different ways by different audiences (Preston et al. 2011) and hence lead to different decisions. This is particularly the case when different color schemes (Schloss et al. 2018), number of classes, and data classification methods are used such as in choropleth (Monmonier 1991) and some grid-based renderings. Figure 10.2 illustrates how applying different classification methods can affect the spatial patterns of a simplified VI to flooding in Northeastern United States calculated using a WLC method. The data and methods used are described in Machado and Ratick (2018).

Different data classification methods and class numbers are suited for different data distributions and decision-making methods. For instance, the natural-breaks method maximizes differences between classes of data (Tate et al. 2010), while using standard deviation is useful to highlight differences with respect to the mean index value but is not appropriate to identify which units of analysis comprise a certain percentage of the highest values (e.g., top 10%) in the study area for funding purposes. Also, Bucherie et al. (2022) recommend using quantile-based rather than

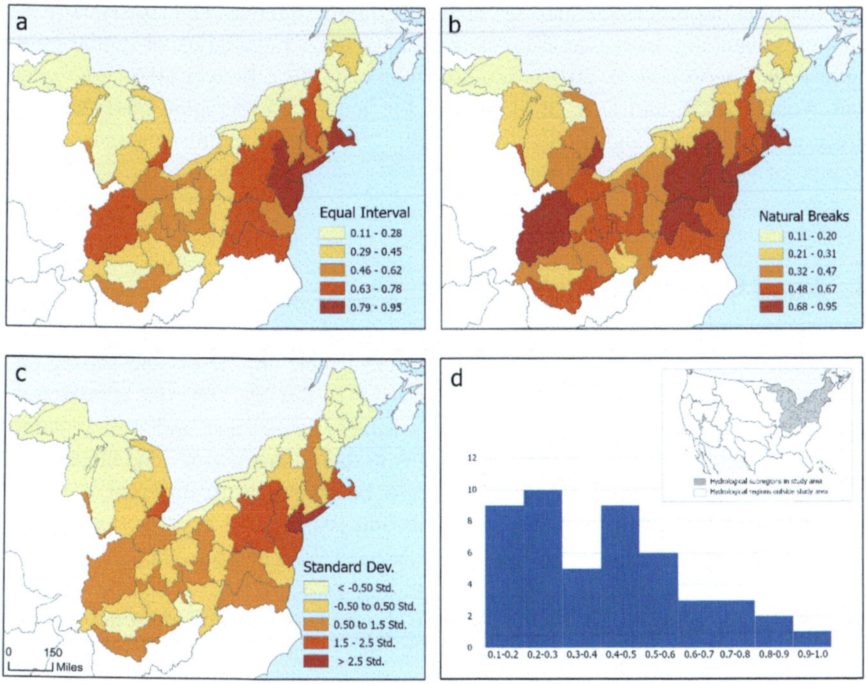

Fig. 10.2 Effect of different data classification methods on the spatial distribution of VI values, (**a**) Equal interval, (**b**) Natural breaks, (**c**) Standard deviations, (**d**) Study area map inset and VI values distribution. The unit of analysis is hydrological subregions (HUC-4)

value-based thresholds to account for different skewness of index value distributions.

A vulnerability map should be tailored to its intended audience and efforts should be made to ensure that the values will be presented and interpreted as intended (Preston et al. 2011). More specifically, it is recommended to: (i) explore different data classification methods and how intended users interpret the resulting maps; (ii) choose the method that best fits the data distribution, usage, and decision-making process needs (e.g., prioritization processes); (iii) be explicit about the data classification method used and indicate the numerical thresholds that correspond to each class when using ordinal labels in the legend (e.g., low, moderate, high); (iv) use a colorblind safe palette where hues saturate at higher VI values and restrict the use of divergent color to when there is a meaningful midpoint threshold, such as *z*-scores (Brewer 1994). Colorbrewer[21] has great color palette suggestions; (v) map the data indicators independently (Schröter et al. 2005) and at different aggregation stages (e.g., sub-indices, vulnerability dimensions) in addition to the index values; (v) explore using interactive portals (Beccari 2016; de Sherbinin et al. 2019) to support

[21] https://colorbrewer2.org/#type=sequential&scheme=BuGn&n=3

decision-making processes; and (vi) map uncertainty in the data and analysis (e.g., with crosshatching, increased color saturation, or fuzzy boundaries, or spatial averaging) (de Sheribin et al. 2019). See Monmonier (1991), Brewer (1994), Krygier and Wood (2005), and Dent et al. (2008) for general cartographic guidance and Preston et al. (2011) and de Sherbinin et al. (2019) for mapping guidance in the context of vulnerability mapping.

4 Final Considerations

Effective VIs need to be scientifically valid, meet their objectives, and satisfy the needs of their intended users. This chapter provides a flexible, stepwise approach to generate VIs aimed at enhancing their usefulness in science and policy while acknowledging the challenges and limitations in their construction. The intent is to provide sufficient flexibility and guidance to be adaptable to different contexts rather than an exhaustive review of methods and issues in VI construction. While there is no commonly agreed upon methodology for VAs and VIs, there has been a convergence toward best practices and guiding principles for index-based VAs which form the basis of this chapter, such as those mentioned the introduction section (i.e., ensuring the saliency, credibility, and legitimacy of the information produced).

Vulnerability indices have become increasingly used in research, policy, and outreach efforts for two main interrelated reasons. First, like other composite indices, they facilitate the interpretation of multi-dimensional issues by reducing their complexity (Saisana and Tarantola 2002). This helps reveal spatial and temporal vulnerability patterns that are difficult to identify through the examination of vulnerability indicators separately. Second, VIs support the prioritization, targeting, and assessment of vulnerability reduction efforts by allowing ranking and comparing VI values across space and time. Additionally, composite indices and vulnerability mapping have been successful at educating the public and increasing awareness about issues (Saisana and Tarantola 2002) such as climate change (Preston et al. 2011).

Paradoxically, the main reasons that promote the use of VIs are their source of criticism and concern.

Vulnerability is a dynamic and multifaceted concept. Reducing its complexity to a composite value is challenging, can be misleading (Eriksen and Kelly 2007; Vicent 2007; Fekete 2012), and lead to simplistic policy conclusions (Saisana and Tarantola 2002; Preston et al. 2011). Also, as shown here, constructing a VI involves several methodological choices. When these choices are not aligned with the objectives of VIs and the indices are not grounded theoretically and/or empirically, the validity and usefulness of VIs are reduced due to arbitrary methodological decisions. Furthermore, different methodological choices can generate diverging VI results, and hence potentially decision outcomes. Lastly, capturing qualitative aspects of vulnerability into VIs is also a challenge. Excluding these aspects or not integrating them appropriately will likely bias the results of VIs and result in indices that are not fully representative.

These and other issues and limitations of constructing VIs described in this chapter are embedded in the index construction process but can be addressed to various extents as shown here. To be successful, efforts to construct VI should acknowledge the limitations of their use and uncertainties in their construction process as is the responsibility of the VI author(s) to examine, document, and communicate them as thoroughly and transparently as possible.

5 Proposed Activities

1. Follow the process described in this chapter (except the sensitivity and validation steps) to generate a hypothetical but realistic VI. Before starting it is necessary to fully understand the guidelines provided here and have read several of the case studies cited demonstrating the application of different VI construction methods. The VI should include at least three data indicators. All the steps in the methodology should be thoroughly documented, including a map of the indicators and the results.
2. Conduct a sensitivity analysis of a VI (from activity 1 or other). The analysis should examine at least one methodological choice (e.g., weighting, indicator set). Justify and describe the method(s) used and interpret your results.
3. Explore different cartographic methods (e.g., color ramps, classification methods) to map VI values: generate at least two different maps, discuss how they compare, and how the different maps may influence decision-making processes.

References

Abson, D.J., A.J. Dougill, and L.C. Stringer. 2012. Using principal component analysis for information-rich socio-ecological vulnerability mapping in Southern Africa. *Applied Geography* 35 (1–2): 515–524.

Aceves-Quesada, J.F., J. Díaz-Salgado, and J.L. López-Blanco. 2007. Vulnerability assessment in a volcanic risk evaluation in Central Mexico through multi-criteria GIS approach. *Natural Hazards* 40 (2): 339–356. https://doi.org/10.1007/s11069-006-0018-6.

Adger, W.N. 1999. Social vulnerability to climate change and extremes in coastal Vietnam. *World Development* 27 (2): 249–269.

———. 2006. Vulnerability. *Global Environmental Change* 16 (3): 268–281.

Adger, W.N., N. Brooks, G. Bentham, M. Agnew, and S. Eriksen. 2004. *New indicators of vulnerability and adaptive capacity*. Vol. 122. Tyndall Centre for Climate Change Research: Norwich.

Ahsan, M.N., and J. Warner. 2014. The socioeconomic vulnerability index: A pragmatic approach for assessing climate change led risks – A case study in the south-western coastal Bangladesh. *International Journal of Disaster Risk Reduction* (8): 32–49.

Alessa, L., A. Kliskey, R. Lammers, C. Arp, D. White, L. Hinzman, and R. Busey. 2008. The arctic water resource vulnerability index: An integrated assessment tool for community resilience and vulnerability with respect to freshwater. *Environmental Management* 42 (3): 523–541.

Anderson, C.C., M. Hagenlocher, F.G. Renaud, Z. Sebesvari, S.L. Cutter, and C.T. Emrich. 2019. Comparing index-based vulnerability assessments in the Mississippi Delta: Implications of contrasting theories, indicators, and aggregation methodologies. *International Journal of Disaster Risk Reduction* 39: 101128.

Baeck, S.H., S.J. Choi, G.W. Choi, and D.R. Lee. 2014. A study of evaluating and forecasting watersheds using the flood vulnerability assessment index in Korea. *Geomatics, Natural Hazards and Risk* 5 (3): 208–231.

Bakkensen, L.A., C. Fox-Lent, L.K. Read, and I. Linkov. 2017. Validating resilience and vulnerability indices in the context of natural disasters. *Risk Analysis* 37 (5): 982–1004.

Balica, S.F., N. Douben, and N.G. Wright. 2009. Flood vulnerability indices at varying spatial scales. *Water Science and Technology* 60 (10): 2571–2580.

Barankin, R.A., M.E. Portman, P. Kirshen, and R.E. Bowen. 2021. Evidence-driven approach for assessing social vulnerability and equality during extreme climatic events. *Frontiers in Wate.* 2: 544141.

Beccari, B. 2016. A comparative analysis of disaster risk, vulnerability and resilience composite indicators. *PLoS Currents* 14: 8.

Bharwani S. 2022. *Vulnerability definitions*. https://www.weadapt.org/knowledge-base/vulnerability/vulnerability-definitions Accessed November 14, 2022.

Birkmann, J. 2006. Measuring vulnerability to promote disaster-resilient societies: Conceptual frameworks and definitions. In *Measuring Vulnerability to Natural Hazards: Towards Disaster Resilient Societies*, J. Birkmann (Ed.), 1: 9–54. UN University Press.

Birkmann, J., A. Jamshed, J.M. McMillan, D. Feldmeyer, E. Totin, W. Solecki, Z.Z. Ibrahim, D. Roberts, R.B. Kerr, H.O. Poertner, and M. Pelling. 2022. Understanding human vulnerability to climate change: A global perspective on index validation for adaptation planning. *Science of the Total Environment* 803: 150065.

Bjarnadottir, S., Y. Li, and M.G. Stewart. 2011. Social vulnerability index for coastal communities at risk to hurricane hazard and a changing climate. *Natural Hazards* 59 (2): 1055–1075.

Blaikie, P., T. Cannon, I. Davis, and B. Wisner (Eds.). 1994. *At risk: Natural hazards, people's vulnerability and disasters*. 1st ed. London: Routledge.

Bohle, H.G., T.E. Downing, and M.J. Watts. 1994. Climate-change and social vulnerability—Toward a sociology and geography of food insecurity *Global Environmental Change*. 4: 37–48.

Brewer, C.A. 1994. Color use guidelines for mapping. *Visualization in Modern Cartography* 1994 (123–148): 7.

Brooks, N. 2003. *Vulnerability, risk and adaptation: A conceptual framework, working paper 38*. Norwich: Tyndall Centre for Climate Change Research.

Brooks N., and N.M. Adger. 2003. Country level risk measures of climate-related natural disasters and implications for adaptation to climate change. Tyndall Centre for Climate Change Research, working paper 26.

Brooks N., W.N. Adger, and P.M. Kelly. 2005. The determinants of vulnerability and adaptive capacity at the national level and the implications for adaptation. *Global Environmental Change* 15 (2): 151–163.

Bucherie, A., C. Hultquist, S. Adamo, C. Neely, F. Ayala, J. Bazo, and A. Kruczkiewicz. 2022. A comparison of social vulnerability indices specific to flooding in Ecuador: Principal component analysis (PCA) and expert knowledge. *International Journal of Disaster Risk Reduction* 73: 102897.

Burton, C.G. 2010. Social vulnerability and hurricane impact modeling. *Natural Hazards Review* 11 (2): 58–68.

Burton, I., G. White, and R. Kates. 1978. *The Environment as Hazard*. New York: Oxford University Press.

Cash, D.W., W.C. Clark, F. Alcock, N.M. Dickson, N. Eckley, D.H. Guston, J. Jäger, and R.B. Mitchell. 2003. Knowledge systems for sustainable development. *Proceedings of the National Academy of Sciences* 100 (14): 8086–8091.

Chakraborty, J., G.A. Tobin, and B.E. Montz. 2005. Population evacuation: Assessing spatial variability in geophysical risk and social vulnerability to natural hazards. *Natural Hazards Review* 6 (1): 23–33.

Chambers, R. 1989. Editorial introduction: Vulnerability, coping and policy. *IDS Bulletin* 21: 1–7.

Clark, W.A., and K.L. Avery. 1976. The effects of data aggregation in statistical analysis. *Geographical Analysis* 8 (4): 428–438.

Clark, G.E., S.C. Moser, S.J. Ratick, K. Dow, W.B. Meyer, S. Emani, W. Jin, J.X. Kasperson, R.E. Kasperson, and H.E. Schwarz. 1998. Assessing the vulnerability of coastal communities to extreme storms: The case of Revere, MA., USA. *Mitigation and Adaptation Strategies for Global Change* 3 (1): 59–82.

Clark, W.C., T.P. Tomich, M. Van Noordwijk, D. Guston, D. Catacutan, N.M. Dickson, and E. McNie. 2011. Boundary work for sustainable development: Natural resource management at the Consultative Group on International Agricultural Research (CGIAR). *Proceedings of the National Academy of Sciences* 113 (17): 4615–4622.

Collins, T.W., S.E. Grineski, and M.D. Aguilar. 2009. Vulnerability to environmental hazards in the Ciudad Juárez (Mexico)–El Paso (USA) metropolis: A model for spatial risk assessment in transnational context. *Applied Geography* 29 (3): 448–461.

Conlon, K.C., E. Mallen, C.J. RGronlund, V.J. Berrocal, L. Larsen, and M.S. O'Neill. 2020. Mapping human vulnerability to extreme heat: A critical assessment of heat vulnerability indices created using principal components analysis. *Environmental Health Perspectives* 128 (9): 097001.

Cox, J.R., C. Rosenzweig, W.D. Solecki, R. Goldberg, and P.L. Kinney. 2006. Social vulnerability to climate change: A neighborhood analysis of the northeast US Megaregion. *Technical paper prepared for the Northeast Climate Impacts Assessment*. Cambridge.

Cutter, S.L. 1996. Vulnerability to environmental hazards. *Progress in Human Geography* 20 (4): 529–539.

Cutter, S., J.T. Mitchell, and M.S. Scott. 2000. Revealing the vulnerability of people and places: a case study of Georgetown County, South Carolina. *Annals of the Association of American Geographers* 90 (4): 713–737. https://doi.org/10.1111/0004-5608.00219

Cutter, S.L., and C. Finch. 2008. Temporal and spatial changes in social vulnerability to natural hazards. *Proceedings of the National Academy of Sciences* 105 (7): 2301–2306.

Cutter, S.L., B.J. Boruff, and W.L. Shirley. 2003. Social vulnerability to environmental hazards. *Social Science Quarterly* 84 (2): 242–261.

Cutter, S.L., C.T. Emrich, J.J. Webb, and D. Morath. 2009. Social vulnerability to climate variability hazards: A review of the literature. *Final Report to Oxfam America.* 5: 1–44.

Cutter, S.L., C.T. Emrich, D.P. Morath, and C.M. Dunning. 2013. Integrating social vulnerability into federal flood risk management planning. *Journal of Flood Risk Management* 6 (4): 332–344.

de Bruin, K., R.B. Dellink, A. Ruijs, L. Bolwidt, A. van Buuren, J. Graveland, R.S. De Groot, P.J. Kuikman, S. Reinhard, R.P. Roetter, and V.C. Tassone. 2009. Adapting to climate change in The Netherlands: An inventory of climate adaptation options and ranking of alternatives *Climatic Change* 95 (1): 23–45.

de Sherbinin, A. 2014. Climate change hotspots mapping: What have we learned? *Climatic Change* 123 (1): 23–37.

de Sherbinin, A., A. Bukvic, G. Rohat, M. Gall, B. McCusker, B. Preston, A. Apotsos, C. Fish, S. Kienberger, P. Muhonda, and O. Wilhelmi. 2019. Climate vulnerability mapping: A systematic review and future prospects. *Wiley Interdisciplinary Reviews: Climate Change.* 10 (5): e600.

DeCrappeo, N.M., G.A. Bisbal, and A.M. Meadow. 2018. A path to actionable climate science: Perspectives from the field. *Environmental Management* 61 (2): 181–187.

Dent, B., J. Torguson, and T. Hodler (Eds.). 2008. *Cartography: Thematic map design*. 6th ed. New York: McGraw-Hill Higher Education. ISBN 1283388022.

Döll, P. 2009. Vulnerability to the impact of climate change on renewable groundwater resources: A global-scale assessment. *Environmental Research Letters*. 4 (3): 035006.

Dow, K. 1992. Exploring differences in our common future (s): The meaning of vulnerability to global environmental change. *Geoforum* 23 (3): 417–436.

Eakin, H., and L.A. Bojórquez-Tapia. 2008. Insights into the composition of household vulnerability from multicriteria decision analysis. *Global Environmental Change*. 18 (1): 112–127. https://doi.org/10.1016/j.gloenvcha.2007.09.001.

Eakin, H., and A.L. Luers. 2006. Assessing the vulnerability of social-environmental systems. *Annual Review of Environment and Resources* 31 (1): 365–394.

Eriksen, S.H., and P.M. Kelly. 2007. Developing credible vulnerability indicators for climate adaptation policy assessment. *Mitigation and Adaptation Strategies for Global Change*. 12 (4): 495–524.

ESPON. 2012. Climate change may hamper territorial cohesion. ESPON 2013 Programme/Map of the Month. January 2012. *ESPON Programme, the European Observation Network for Territorial Development and Cohesion*. https://www.espon.eu/sites/default/files/attachments/ESPON_MoM_190112.pdf. Accessed 14 Nov 2022.

Fatemi, F., A. Ardalan, B. Aguirre, N. Mansouri, and I. Mohammadfam. 2017. Social vulnerability indicators in disasters: Findings from a systematic review. *International Journal of Disaster Risk Reduction* 22: 219–227.

Feizizadeh, B., and T. Blaschke. 2013. GIS-multicriteria decision analysis for landslide susceptibility mapping: Comparing three methods for the Urmia lake basin, Iran. *Natural Hazards* 65 (3): 2105–2128.

Fekete, A. 2009. Validation of a social vulnerability index in context to river-floods in Germany. *Natural Hazards and Earth System Sciences* 9 (2): 393–403.

———. 2012. Spatial disaster vulnerability and risk assessments: Challenges in their quality and acceptance. *Natural Hazards* 61 (3): 1161–1178.

Fekete, A., M. Damm, and J. Birkmann. 2010. Scales as a challenge for vulnerability assessment. *Natural Hazards* 55 (3): 729–747.

Fernandez, M.A., S. Bucaram, and W. Renteria. 2017. (Non-) robustness of vulnerability assessments to climate change: An application to New Zealand. *Journal of Environmental Management* 203: 400–412.

Frazier, T.G. 2012. Selection of scale in vulnerability and resilience assessments. *Journal of Geography and Natural Disasters* 2: e108.

Füssel, H.M. 2010. Review and quantitative analysis of indices of climate change, exposure, adaptive capacity, sensitivity and impacts. *Background note to the world development report*. Potsdam Institute for Climate Impact Research.

Füssel, H.M., and R.J. Klein. 2006. Climate change vulnerability assessments: An evolution of conceptual thinking. *Climatic Change* 75 (3): 301–329.

Gallopín, G.C. 2006. Linkages between vulnerability, resilience, and adaptive capacity. *Global Environmental Change* 16 (3): 293–303.

Garcia de Jalon, S., A. Iglesias, and A.P. Barnes. 2016. Drivers of farm-level adaptation to climate change in Africa: An evaluation by a composite index of potential adoption. *Mitigation and Adaptation Strategies for Global Change* 21 (5): 779–798.

Gibson, C.C., E. Ostrom, and T.K. Ahn. 2000. The concept of scale and the human dimensions of global change: A survey. *Ecological Economics* 32 (2): 217–239.

Giorgi, F. 2020. Producing actionable climate change information for regions: The distillation paradigm and the 3R framework. *The European Physical Journal Plus* 135 (5): 435.

Giupponi, C., and C. Biscaro. 2015. Vulnerabilities—Bibliometric analysis and literature review of evolving concepts. *Environmental Research Letters* 10 (12): 123002.

Gumel, D.Y. 2022. Assessing climate change vulnerability: A conceptual and theoretical review. *Journal of Sustainability and Environmental Management* 1 (1): 22–31.

Guston, D.H. 1999. Stabilizing the boundary between US politics and science: The role of the Office of Technology Transfer as a boundary organization. *Social Studies of Science* 29 (1): 87–111.
———. 2001. Boundary organizations in environmental policy and science: An introduction. *Science, Technology, & Human Values* 26 (4): 399–408.
Hahn, M.B., A.M. Riederer, and S.O. Foster. 2009. The livelihood vulnerability index: A pragmatic approach to assessing risks from climate variability and change—A case study in Mozambique. *Global Environmental Change.* 19 (1): 74–88.
Hegglin, E., and C. Huggel. 2008. An integrated assessment of vulnerability to glacial hazards. *Mountain Research and Development* 28 (3): 299–309.
Holling, C.S. 1973. Resilience and stability of ecological systems. *Annual Review of Ecology and Systematics* 1: 1–23.
Hurd, B., N. Leary, R. Jones, and J. Smith. 1999. Relative regional vulnerability of water resources to climate change. *JAWRA Journal of the American Water Resources Association* 35 (6): 1399–1409.
Jiang, H., and J.R. Eastman. 2000. Application of fuzzy measures in multi-criteria evaluation in GIS. *International Journal of Geographical Information Science* 14 (2): 173–184.
Jones, B., and J. Andrey. 2007. Vulnerability index construction: Methodological choices and their influence on identifying vulnerable neighborhoods. *International Journal of Emergency Management* 4 (2): 269–295.
Kasperson, R.E., and M. Berberian (Eds.). 2011. *Integrating science and policy: Vulnerability and resilience in global environmental change*. Hoboken: Routledge.
Kasperson, R.E., O. Renn, P. Slovic, H.S. Brown, J. Emel, R. Goble, J.X. Kasperson, and S. Ratick. 1988. The social amplification of risk: A conceptual framework. *Risk Analysis* 8 (2): 177–187.
Kasperson, R.E., J.X. Kasperson, and K. Dow. 2001. Vulnerability, Equity, and Global Environmental Change. In *Global environmental risk*, J.X. Kasperson and R.E. Kasperson (Eds.), 247–272. United Nations University Press.
Kates, R.W., J.H. Ausubel, and M. Berberian (Eds.). 1985. *Climate impact assessment.* New York: Wiley.
Kelly, P.M., and W.N. Adger. 2000. Theory and practice in assessing vulnerability to climate change and facilitating adaptation. *Climatic Change* 47 (4): 325–352.
Kennedy, C.M., J.R. Oakleaf, D.M. Theobald, S. Baruch-Mordo, and J. Kiesecker. 2019. Managing the middle: A shift in conservation priorities based on the global human modification gradient. *Global Change Biology* 25 (3): 811–826.
Klimberg, R., and S. Ratick. 2020a. Developing composite indices for assessing vulnerability to floods due to climate change: Applying the order rated effectiveness modeling method. In *Applications of management science* 20, 117–142. Bingley: Emerald Publishing Limited.
———. 2020b. Evaluating effective performance using ordered weighted averaging and data envelopment analysis optimization. In *Applications of management science*, 20, 207–232. Bingley: Emerald Publishing Limited.
Krygier, J., and D. Wood. 2005. *Making maps: A visual guide to map design for GIS*. New York: The Guilford Press.
Langill, J.C., C. Abizaid, Y. Takasaki, and O.T. Coomes. 2022. Integrated multi-scalar analysis of vulnerability to environmental hazards: Assessing extreme flooding in western Amazonia. *Global Environmental Change* 76: 102585.
Liverman, D. 2001. Vulnerability to global environmental change. In *Global environmental risk*, J.X. Kasperson and R.E. Kasperson (Eds.). 201–216. United Nations University Press.
Luers, A.L., D.B. Lobell, L.S. Sklar, C.L. Addams, and P.A. Matson. 2003. A method for quantifying vulnerability, applied to the agricultural system of the Yaqui Valley, Mexico. *Global Environmental Change* 13 (4): 255–267.
Maantay, J.A., A.R. Maroko, and C. Herrmann. 2007. Mapping population distribution in the urban environment: The cadastral-based expert dasymetric system (CEDS). *Cartography and Geographic Information Science* 34 (2): 77–102.
Machado, E.A. 2011. *Assessing vulnerability to dengue fever in Mexico under global change.* Dissertation, Clark University.

Machado, E.A., and S. Ratick. 2018. Implications of indicator aggregation methods for global change vulnerability reduction efforts. *Mitigation and Adaptation Strategies for Global Change* 23 (7): 1109–1141.

Mainali, J., and N.G. Pricope. 2017. High-resolution spatial assessment of population vulnerability to climate change in Nepal. *Applied Geography* 82: 66–82.

Malczewski, J. 1999. *GIS and multicriteria decision analysis*. New York: Wiley.

———. 2000. On the use of weighted linear combination method in raster GIS: Common and best practice approaches. *Transactions GIS*. 4 (1): 5–22. https://doi.org/10.1111/1467-9671.00035.

———. 2006. GIS-based multicriteria decision analysis: A survey of the literature. *International Journal of Geographic Information Science*. 20 (7): 703–726. https://doi.org/10.1080/13658810600661508.

Marulanda, M.C., O.D. Cardona, and A.H. Barbat. 2009. Robustness of the holistic seismic risk evaluation in urban centers using the USRi. *Natural Hazards* 49 (3): 501–516.

McCarthy, J.J., O.F. Canziani, N.A. Leary, D.J. Dokken, and K.S. White. 2001. Climate change 2001: Impacts, adaptation and vulnerability. *A Report of Working Group II of the Intergovernmental Panel on Climate. Summary for policy makers*. Cambridge: Cambridge University Press.

McLaughlin, S., and J.A. Cooper. 2010. A multi-scale coastal vulnerability index: A tool for coastal managers? *Environmental Hazards* 9 (3): 233–248.

Menezes, J.A., U. Confalonieri, A.P. Madureira, I.D. Duval, R.B. Santos, and C. Margonari. 2018. Mapping human vulnerability to climate change in the Brazilian Amazon: The construction of a municipal vulnerability index. *PLoS One* 13 (2): e0190808.

Mennis, J. 2009. Dasymetric mapping for estimating population in small areas. *Geography Compass* 3 (2): 727–745.

Metzger, M.J., and D. Schröter. 2006. Towards a spatially explicit and quantitative vulnerability assessment of environmental change in Europe. *Regional Environmental Change* 6 (4): 201–216.

Mi, Z.F., Y.M. Wei, C.Q. He, H.N. Li, X.C. Yuan, and H. Liao. 2017. Regional efforts to mitigate climate change in China: A multi-criteria assessment approach. *Mitigation Adaption Strategies for Global Change* 22 (1): 45–66. https://doi.org/10.1007/s11027-015-9660-1.

Monmonier, M. 1991. *How to lie with maps*. Chicago: University of Chicago Press. ISBN 10: 0226534154ISBN 13: 9780226534152.

Monterroso, A., C. Conde, C. Gay, D. Gómez, and J. López. 2014. Two methods to assess vulnerability to climate change in the Mexican agricultural sector. *Mitigation and Adaptation Strategies for Global Change* 19 (4): 445–461.

Moss R., A. Brenkert, and E.L. Malone. 2001. *Vulnerability to climate change: A quantitative approach*. Prepared for the US Department of Energy.

Nguyen, T.T., J. Bonetti, K. Rogers, and C.D. Woodroffe. 2016. Indicator-based assessment of climate-change impacts on coasts: A review of concepts, methodological approaches and vulnerability indices. *Ocean & Coastal Management* 123: 18–43.

O'Brien, K., R. Leichenko, U. Kelkar, H. Venema, G. Aandahl, H. Tompkins, A. Javed, S. Bhadwal, S. Barg, L. Nygaard, and J. West. 2004. Mapping vulnerability to multiple stressors: Climate change and globalization in India *Global Environmental Change*. 14 (4): 303–313.

O'Brien, K.L., and R.M. Leichenko. 2000. Double exposure: Assessing the impacts of climate change within the context of economic globalization. *Global Environmental Change* 10 (3): 221–232.

———. 2003. Winners and losers in the context of global change. *Annals of the Association of American Geographers* 93 (1): 89–103.

OECD – JRC. 2008. *Handbook on constructing composite indicators: Methodology and user guide*. Organization for Economic Co-operation and Development, ISBN 978-92-64-04345-9.

O'Keefe, P. 1976. Taking the "Naturalness" out of "Natural disaster". *Nature* 260: 566–567.

Olmos, S. 2001. *Vulnerability and adaptation to climate change: Concepts, issues, assessment methods*. Climate Change Knowledge Network. http://lib.riskreductionafrica.org/bitstream/handle/123456789/454/vulnerability%20and%20adaptation%20to%20climate%20 change.%20concepts,%20issues,%20assessment%20methods.%20Santiago%20olmos.pdf?sequence=1. Accessed 5 Oct 2022.

Openshaw, S. 1983. *The modifiable areal unit problem*. Norwich: Geo Books.

Oulahen, G., L. Mortsch, K. Tang, and D. Harford. 2015. Unequal vulnerability to flood hazards: "ground truthing" a social vulnerability index of five municipalities in Metro Vancouver, Canada. *Annals of the Association of American Geographers*. 105 (3): 473–495.

Patt, A.G., D. Schröter, A.C. de la Vega-Leinert, and R.J. Klein. 2008. Vulnerability research and assessment to support adaptation and mitigation: Common themes from the diversity of approaches. In *Assessing vulnerability to global environmental change*, 1–25. London: Routledge.

Paul, S.K. 2013. Vulnerability concepts and its application in various fields: A review on geographical perspective. *Journal of Life and Earth Science* 8: 63–81.

Polsky, C., and H. Eakin. 2011. Global change vulnerability assessments: Definitions, challenges and opportunities. In *The Oxford handbook of climate change and society*, 205–216. Oxford: Oxford University Press.

Polsky, C., R. Neff, and B. Yarnal. 2007. Building comparable global change vulnerability assessments: The vulnerability scoping diagram. *Global Environmental Change* 17 (3–4): 472–485.

Preston, B.L., and R.N. Jones. 2008. Screening climatic and non-climatic risks to Australian catchments. *Geographical Research* 46 (3): 258–274.

Preston, B.L., C. Brooke, T.G. Measham, T.F. Smith, and R. Gorddard. 2009. Igniting change in local government: Lessons learned from a bushfire vulnerability assessment. *Mitigation and Adaptation Strategies for Global Change* 14 (3): 251–283.

Preston, B.L., E.J. Yuen, and R.M. Westaway. 2011. Putting vulnerability to climate change on the map: A review of approaches, benefits, and risks. *Sustainability Science* 6 (2): 177–202.

Raghavan Sathyan, A., C. Funk, T. Aenis, P. Winker, and L. Breuer. 2018. Sensitivity analysis of a climate vulnerability index-a case study from Indian watershed development programmes. *Climate Change Responses* 5 (1): 1–4.

RASSP. 2001. Vulnerability and resilience for coupled human-environment systems: report of the Research and Assessment Systems for Sustainability Program 2001 Summer Study. *Research and Assessment Systems for Sustainability Program Discussion Paper 2001–17, Environment and Natural Resources Program, Belfer Center for Science and International Affairs, Kennedy School of Government*, Harvard University, Cambridge, MA, Airlie House, Warrenton, VA, 29 May-1 June.

Ratick S., and J.P. Osleeb. 2013. Measuring the vulnerability of populations susceptible to lead contamination in the Dominican Republic: evaluating composite index construction methods. GeoJournal 78 (2): 259–272. https://doi.org/10.1007/s10708-011-9414-x

Reckien, D. 2018. What is in an index? Construction method, data metric, and weighting scheme determine the outcome of composite social vulnerability indices in new York City. *Regional Environmental Change* 18 (5): 1439–1451.

Rinner, C., and J. Malczewski. 2002. Web-enabled spatial decision analysis using ordered weighted averaging (OWA). *Journal of Geographical Systems* 4 (4): 385–403. https://doi.org/10.1007/s101090300095.

Roy, D.C., and T. Blaschke. 2015. Spatial vulnerability assessment of floods in the coastal regions of Bangladesh. *Geomatics, Natural Hazards and Risk* 6 (1): 21–44.

Rufat, S., E. Tate, C.G. Burton, and A.S. Maroof. 2015. Social vulnerability to floods: Review of case studies and implications for measurement. *International Journal of Disaster Risk Reduction* 14: 470–486.

Rufat, S., E. Tate, C.T. Emrich, and F. Antolini. 2019. How valid are social vulnerability models? *Annals of the American Association of Geographers* 109 (4): 1131–1153.

Runfola, D.M., S. Ratick, J. Blue, E.A. Machado, N. Hiremath, N. Giner, K. White, and J. Arnold. 2017. A multi-criteria geographic information systems approach for the measurement of vulnerability to climate change. *Mitigation and Adaptation Strategies for Global Change* 22 (3): 349–368.

Rygel, L., D. O'sullivan, and B. Yarnal. 2006. A method for constructing a social vulnerability index: An application to hurricane storm surges in a developed country. *Mitigation and Adaptation Strategies for Global Change* 11 (3): 741–764.

Saisana, M., and S. Tarantola. 2002. Report EUR 20408 EN. *State-of-the-art report on current methodologies and practices for composite indicator development*. Ispra: European Commission–Joint Research Centre.

Saisana, M., A. Saltelli, and S. Tarantola. 2005. Uncertainty and sensitivity analysis techniques as tools for the quality assessment of composite indicators. *Journal of the Royal Statistical Society: Series A (Statistics in Society)* 168 (2): 307–323.

Saltelli, A., S. Tarantola, F. Campolongo, and M. Ratto. 2004. *Sensitivity analysis in practice: A guide to assessing scientific models*. Chichester: Wiley.

Schloss, K.B., C.C. Gramazio, A.T. Silverman, M.L. Parker, and A.S. Wang. 2018. Mapping color to meaning in colormap data visualizations. *IEEE Transactions on Visualization and Computer Graphics* 25 (1): 810–819.

Schmidtlein, M.C., R.C. Deutsch, W.W. Piegorsch, and S.L. Cutter. 2008. A sensitivity analysis of the social vulnerability index. *Risk Analysis: An International Journal* 28 (4): 1099–1114.

Schröter, D., C. Polsky, and A.G. Patt. 2005. Assessing vulnerabilities to the effects of global change: An eight step approach. *Mitigation and Adaptation Strategies for Global Change* 10 (4): 573–595.

Sen, A. 1990. Food, economics and entitlements. In *The political economy of hunger*, J. Dreze and A. Sen (Eds.). 50–67. Oxford: Clarendon.

Sharma, U., and A. Patwardhan. 2008. Methodology for identifying vulnerability hotspots to tropical cyclone hazard in India. *Mitigation and Adaptation Strategies for Global Change* 13 (7): 703–717.

Singh-Peterson, L., and S.J. Underhill. 2017. A multi-scalar, mixed methods framework for assessing rural communities' capacity for resilience, adaptation, and transformation. *Community Development* 48 (1): 124–140.

Soares, M.B., A.S. Gagnon, and R.M. Doherty. 2012. Conceptual elements of climate change vulnerability assessments: A review. *International Journal of Climate Change Strategies and Management* 4 (1): 6–35.

Spielman, S.E., J. Tuccillo, D.C. Folch, A. Schweikert, R. Davies, N. Wood, and E. Tate. 2020. Evaluating social vulnerability indicators: Criteria and their application to the social vulnerability index. *Natural Hazards* 100: 417–436.

Stafford, S., and J. Abramowitz. 2017. An analysis of methods for identifying social vulnerability to climate change and sea level rise: A case study of Hampton roads. *Virginia. Natural Hazards* 85 (2): 1089–1117.

Stephen, L., and T.E. Downing. 2001. Getting the scale right: A comparison of analytical methods for vulnerability assessment and household-level targeting. *Disasters* 25 (2): 113–135.

Sullivan, C., and J. Meigh. 2005. Targeting attention on local vulnerabilities using an integrated index approach: The example of the climate vulnerability index. *Water Science and Technology* 51 (5): 69–78.

Tapia, C., B. Abajo, E. Feliu, M. Mendizabal, J.A. Martinez, J.G. Fernández, T. Laburu, and A. Lejarazu. 2017. Profiling urban vulnerabilities to climate change: An indicator-based vulnerability assessment for European cities. *Ecological Indicators* 78: 142–155.

Tate, E. 2012. Social vulnerability indices: A comparative assessment using uncertainty and sensitivity analysis. *Natural Hazards* 63 (2): 325–347.

———. 2013. Uncertainty analysis for a social vulnerability index. *Annals of the Association of American Geographers* 103 (3): 526–543.

Tate, E., S.L. Cutter, and M. Berry. 2010. Integrated multihazard mapping. *Environment and Planning B: Planning and Design* 37 (4): 646–663.

Thomas, K., R.D. Hardy, H. Lazrus, M. Mendez, B. Orlove, I. Rivera-Collazo, J.T. Roberts, M. Rockman, B.P. Warner, and R. Winthrop. 2019. Explaining differential vulnerability to climate change: A social science review. *Wiley Interdisciplinary Reviews: Climate Change* 10 (2): e565.

Turner, B.L., II, R.E. Kasperson, W.B. Meyer, K.M. Dow, D. Golding, J.X. Kasperson, R.C. Mitchell, and S.J. Ratick. 1990. Two types of global environmental change: Definitional and spatial-scale issues in their human dimensions. *Global Environmental Change* 1 (1): 14–22.

Turner, B.L., R.E. Kasperson, P.A. Matson, J.J. McCarthy, R.W. Corell, L. Christensen, N. Eckley, J.X. Kasperson, A. Luers, M.L. Martello, and C. Polsky. 2003. A framework for vulnerability analysis in sustainability science. *Proceedings of the National Academy of Sciences* 100 (14): 8074–8079.

Villa, F., and H.E. McLeod. 2002. Environmental vulnerability indicators for environmental planning and decision-making: Guidelines and applications. *Environmental Management* 29 (3): 335–348.

Vincent, K. 2007. Uncertainty in adaptive capacity and the importance of scale. *Global Environmental Change* 17 (1): 12–24.

Vogel, C., and K. O'Brien. 2004. Vulnerability and global environmental change: rhetoric and reality. In *Aviso: an information bulletin on global environmental change and human security*, issue no. 13, March 2004.

Weichselgartner, J., and R. Kasperson. 2010. Barriers in the science-policy-practice interface: Toward a knowledge-action-system in global environmental change research. *Global Environmental Change* 20 (2): 266–277.

White, G. 1973. Natural hazards research. In *Directions in geography*, R.J. Chorley (Ed.), 193–216. London: Methuen.

———. 1974. *Natural Hazards*. New York: Oxford Press.

White, G.F., and J.E. Haas. 1975. *Assessment of research on natural hazards*. Cambridge, MA: MIT Press.

Willis, I., and J.M. Fitton. 2003. A review of multivariate social vulnerability methodologies; a case study of the river Parrett catchment, Somerset. *Natural Hazards and Earth System Sciences* 2016: 1–7.

Wisner, B., P. Blaikie, T. Cannon, and I. Davis (Eds.). 2003. *At risk: Natural hazards, people's vulnerability and disasters*. New York: Routledge.

Wolf, T., and G. McGregor. 2013. The development of a heat wave vulnerability index for London. *United Kingdom. Weather and Climate Extremes* 1: 59–68.

Wu, S.Y., B. Yarnal, and A. Fisher. 2002. Vulnerability of coastal communities to sea-level rise: A case study of Cape May County, New Jersey, USA. *Climate Research* 22 (3): 255–270.

Xie, Z. and H. Ziegler. 2018. Mapping urban population at housing unit level with integrated geospatial technology. In *Comprehensive Remote Sensing*, Liang S (Ed.). Oxford: Elsevier.

Yager, R.R. 1988. On ordered weighted averaging aggregation operators in multicriteria decision making. *IEEE Transactions on Systems, Man, and Cybernetics* 18 (1): 183–190.

Yoon, D.K. 2012. Assessment of social vulnerability to natural disasters: A comparative study. *Natural Hazards* 63 (2): 823–843.

Yuan, X.C., Q. Wang, K. Wang, B. Wang, J.L. Jin, and Y.M. Wei. 2015. China's regional vulnerability to drought and its mitigation strategies under climate change: Data envelopment analysis and analytic hierarchy process integrated approach. *Mitigation and Adaptation Strategies for Global Change* 20 (3): 341–359.

Zhou, X., H. Lin, and H. Lin. 2008. Global sensitivity analysis. In *Encyclopedia of GIS*, ed. S. Shekhar and H. Xiong. Boston: Springer. https://doi.org/10.1007/978-0-387-35973-1_538.

Chapter 11
Actionable Science in Environmental Health

Qian Huang, Diego F. Cuadros, and Ziheng Sun

Contents

Q. Huang (✉)
Eastern Tennessee State University, Johnson City, TN, USA
e-mail: huangq@etsu.edu

D. F. Cuadros
University of Cincinnati, Cincinnati, OH, USA
e-mail: diego.cuadros@uc.edu

Z. Sun
Center for Spatial Information Science and Systems, Department of Geography and Geoinformation Science, George Mason University, Fairfax, VA, USA
e-mail: zsun@gmu.edu

Z. Sun (ed.), *Actionable Science of Global Environment Change*,
https://doi.org/10.1007/978-3-031-41758-0_11

1 Introduction

Environmental health elucidates the intricate relationship between human health and the environment we live in (APHA 2022). It is a subfield of public health that aims to promote human health and well-being. Neglecting environmental health could produce severe medical conditions and escalate into deficient communities. For instance, exposure to air pollution from vehicles and industrial sources can lead to respiratory conditions such as asthma, chronic obstructive pulmonary disease (COPD), and respiratory infections such as pneumonia (Li et al. 2003; Jiang et al. 2016a). Studies have shown that high levels of air pollution in urban areas are associated with increased hospital admissions for respiratory illnesses (Kuerban et al. 2020). Inhalation of air pollutants, such as particulate matter (PM) and gases, can trigger an inflammatory response in the respiratory system. Fine particulate matter (PM2.5) can penetrate deep into the lungs and cause inflammation in the airways (Xing et al. 2016). This inflammation can lead to coughing, wheezing, and shortness of breath. Other air pollutants, like ozone (O_3)and nitrogen dioxide (NO_2), can induce oxidative stress in the respiratory system. Oxidative stress occurs when there is an imbalance between the production of reactive oxygen species (ROS) and the ability of the body's antioxidant defenses to neutralize them. Prolonged exposure to oxidative stress can damage lung tissue and impair respiratory function, contributing to the development of respiratory diseases (Kelly 2003). Air pollutants can also directly affect the structure and function of the lungs, like causing bronchoconstriction and increased mucus production, making breathing more difficult (Laumbach and Kipen 2012).

Environmental factors including biological hazards have a profound impact on human health, causing a range of diseases and posing significant threat to public health and community safety. For example, waterborne diseases, which could be linked to inadequate water treatment and sanitation, can cause severe consequences on individual and community health. Coming into contact with microbial or fecal matter contaminated water could result in diarrheal diseases with symptoms like cholera, typhoid fever, and cryptosporidiosis (World Health Organization 2011). For example, in 2010, Haiti was stricken by the most deadly cholera epidemic in recent history. The United Nations peacekeeping mission inadvertently brought cholera to Haiti after a massive earthquake, and the disease was introduced into the country's most extensive water system due to inadequate sanitation and water treatment. This led to nearly 10,000 deaths and hundreds of thousands falling ill, underlining the severe consequences of poor environmental health practices (Barzilay et al. 2013). Similarly, the outbreak of cryptosporidiosis in Milwaukee, Wisconsin, in 1993, caused 400,000 positive cases, which marked the largest waterborne disease outbreak in US history (Mac Kenzie et al. 1994).

These are just a few examples of the intricate interaction between environment and human health. Despite the well-established links between environmental factors and serious health threats, implementing scientific knowledge into actionable data-driven measures remains a challenge (Fraisl et al. 2022). Several factors contribute to the gap. Addressing environmental health concerns often requires significant resources, including funding, infrastructure development, and capacity building. Resource limitations can hinder the implementation of effective interventions and preventive measures. Meanwhile, environmental health issues involve complex interactions between multiple factors, such as pollutants, exposure routes, and individual susceptibility. Likewise, as every decision has to go through the political decision-making processes, environmental health issues may not receive adequate attention or prioritization in policy-making and resource allocation, leading to limited action on prevention and mitigation strategies (Russell and Gruber 1987). Understanding and addressing these complexities require interdisciplinary collaboration and comprehensive approaches. Furthermore, these environmental health challenges may be intertwined with social, economic, and political factors (Pohjola and Tuomisto 2011). Overcoming governance, regulation, and stakeholder engagement barriers is crucial for translating scientific knowledge into actionable policies and practices.

This chapter will make an overview of the current data-driven research and analyze the intricate relationship between human health and the environment. It highlights the detrimental health impacts of air pollution, such as respiratory conditions, and waterborne diseases resulting from inadequate water treatment and sanitation. Data visualization techniques will be discussed for the implementation of a dashboard monitoring the air quality in a given community to illustrate the use of data-driven analytics into actionable results. The chapter emphasizes the need for practical actions to address these issues and bridge the gap between scientific knowledge and programmatic implementation. Key challenges will be discussed, including awareness gaps, resource constraints, complex interactions, and sociopolitical factors, as barriers to transforming scientific discoveries into actionable measures. By addressing these challenges, it would be possible to promote a healthier environment and safeguard human well-being.

2 Biological Hazards and Associated Diseases

As we dive into the expansive realm of environmental health, it is important to understand the spectrum of biological hazards that pose threats to human health and the subsequent health outcomes they can induce. Herein, we will examine air quality-related diseases, waterborne diseases, vector-borne diseases, and chemical exposure-related diseases to illustrate the complex interactions between the environment and human health. Each category will encompass an examination of the sources of these threats, their pathways into the human body, the symptoms and damage they cause, and current methods of prevention, cure, and mitigation.

2.1 Air Quality-Related Diseases

A host of pollutants, predominantly of anthropogenic origin, are detrimental to public health. The key offenders include PM, NO_2, sulfur dioxide (SO_2), carbon monoxide (CO), O_3, and volatile organic compounds (VOCs) (WHO n.d.-a, -b). These pollutants primarily originate from human activities, including the combustion of fossil fuels, industrial operations, and the use of a myriad of chemical products (Perera 2018). Among these pollutants, PM – a composite of minuscule solid and liquid particles suspended in air – poses a significant health risk. The ability of these particles to infiltrate deep into lung tissue and permeate the bloodstream renders them particularly harmful (American Lung Association n.d.). PM is broadly categorized into PM2.5 and PM10, based on their respective diameters measured in micrometers. Both, being potential carriers of toxic compounds, are associated with a spectrum of health issues, including but not limited to respiratory diseases like asthma and bronchitis, cardiovascular conditions, and lung cancer (US EPA 2014). Gaseous pollutants such as NO_2 and SO_2, which primarily stem from the combustion of fossil fuels, have the potential to cause respiratory irritation (US EPA 2016). Symptoms caused by these environmental agents may include coughing and shortness of breath, and these gasses can exacerbate existing respiratory conditions. Chronic exposure can lead to the onset of respiratory diseases and compromise lung function. Another gas of concern is ground-level O_3, formed by the interaction of sunlight with pollutants such as nitrogen oxides and volatile organic compounds. Ozone can initiate a plethora of health problems, ranging from chest discomfort and coughing to an augmented risk of infection, and COPD (Kim et al. 2020). Diseases related to poor air quality also include those caused by high pollen counts or extreme humidity conditions, such as mold growth leading to respiratory issues. In addition,

indoor air quality merits as much attention as outdoor air quality. Household activities such as cooking, heating, smoking, and the usage of cleaning products can result in elevated levels of indoor pollutants (WHO n.d.-a, -b). Given the considerable time individuals spend within indoor environments, these pollutants can be particularly harmful.

The health ramifications of poor air quality are not equally distributed among all strata of society. Certain demographics – children, the elderly, individuals with preexisting health conditions, and those disadvantaged socioeconomically – tend to bear the brunt of these health implications (Hooper and Kaufman 2018). Multiple factors account for these disparities, including unequal levels of pollutant exposure and variable capacities to manage health threats. A growing body of evidence underpins the association of chronic exposure to high levels of air pollution with the development of asthma and other respiratory complications in children (Burbank and Peden 2018), diminished lung function growth, and neurodevelopmental issues. In adults, chronic exposure is associated with heightened risks of heart disease, stroke, lung cancer, and COPD (Alexeeff et al. 2021). Moreover, preliminary evidence hints at potential associations between air pollution and mental health, diabetes, and other health conditions (Alderete et al. 2018). Moreover, the consequences of poor air quality extend beyond the health of individuals, imposing substantial burdens on healthcare systems and economies. Health complications linked to poor air quality necessitate increased healthcare utilization, leading to elevated healthcare expenditures (Li et al. 2020). Additionally, they contribute to productivity losses due to absenteeism and diminished work capacity.

Treatment involves managing symptoms and primarily mitigating exposure. For example, asthma may be managed through the use of inhalers and other medication, while bronchitis might be treated with cough suppressants, rest, and hydration (Cleveland Clinic 2022). Both individual actions and broader public health measures are critical for prevention. Individuals can reduce their exposure to air pollution by checking daily air quality monitoring and forecasts and limiting outdoor activities during poor air quality time, and public health campaigns can raise risk awareness of air pollution (Brook et al. 2010). Likewise, scientific research has driven the development of interventions to reduce exposure to air pollution, such as advocating for stricter emission standards and implementing urban planning initiatives to improve air quality. Policies and regulations are also essential for improving air quality and preventing air quality-related diseases.

2.2 *Waterborne Diseases*

Waterborne diseases are predominantly triggered by pathogens such as bacteria, viruses, and protozoa, most commonly introduced into water sources through fecal contamination (World Health Organization 2019). Inadequate sanitation and lack of access to clean drinking water significantly increase the incidence of waterborne diseases, especially in developing countries and post-disaster situations where the water supply infrastructure is compromised (Barzilay et al. 2013).

Pathogens and other agents from contaminated water can enter the human body through several pathways. These include ingestion (e.g., drinking contaminated water or eating food washed with it), contact with the skin (e.g., bathing in contaminated water), or through vectors (e.g., mosquitoes) that have bred in such water. Once ingested, these pathogens can affect various human organs, especially impacting the gastrointestinal tract (Kotloff et al. 2013).

Symptoms of waterborne diseases can range from mild gastrointestinal discomfort to severe dehydration and death. Diarrheal diseases such as cholera and cryptosporidiosis present symptoms including severe diarrhea, vomiting, abdominal cramps, and fever (Troeger et al. 2017). In severe cases, these illnesses can lead to life-threatening dehydration or malnutrition, particularly among children.

Medical treatment for waterborne diseases typically involves rehydration therapy, antibiotics, and supportive care (Acheson 2009). For example, oral rehydration salts (ORS) are used to treat dehydration from cholera. Preventative measures primarily focus on improving water quality and sanitation. Water treatment methods, including filtration, chlorination, and the use of ultraviolet light, have been effective in eliminating pathogens. Public health campaigns for hand hygiene and safe food handling are also significant (WHO 2017). It is also crucial to preserve water quality and implement effective waste management systems. Stringent environmental regulations and regular monitoring of water sources help prevent waterborne disease outbreaks.

2.3 Vector-Borne Diseases

Vector-borne diseases are illnesses caused by pathogens transmitted to humans and animals by vectors such as mosquitoes, ticks, and fleas. Vector-borne pathogens usually enter the human body through the bites of infected vectors. Climate conditions like temperature, rainfall, and humidity can affect vector behavior and survival, and consequently the transmission of diseases (Ryan et al. 2019). Human activities like urbanization, deforestation, and poor waste management can create favorable breeding conditions for vectors. For example, the *Aedes aegypti*, a species of mosquito known for its preference to bite humans and commonly found in tropical and subtropical regions worldwide, thrives in stagnant water and disposed tires often found in urban areas with poor sanitation (Messina et al. 2014; Kweka et al. 2018). They become carriers of the Zika virus when they feed on the blood of an infected individual. Once inside the mosquito's body, the Zika virus replicates and multiplies, establishing an infection within the mosquito. The virus moves to the mosquito's salivary glands, where it can be transmitted to a new host during subsequent feedings. When an infected mosquito bites a human, it injects saliva containing the Zika virus into the skin. The virus can then enter the human bloodstream through the mosquito's saliva, initiating an infection. After entering the human body, the Zika virus replicates and spreads to various systems and disseminates through the bloodstream, allowing the virus to reach different organs and tissues throughout the body (Reynoso et al. 2023). One of the most significant impacts of Zika virus infection is its effect on the nervous system. The virus has been associated with neurological disorders, including microcephaly, a condition characterized by an underdeveloped brain and head size. Zika can also cause Guillain-Barré syndrome, a rare neurological disorder that leads to muscle weakness and paralysis (Esposito and Longo 2017). In addition to neurological complications, Zika virus infection can cause a range of other health problems. Common symptoms in adults include fever, rash, joint pain, and conjunctivitis (red eyes). Some individuals may experience long-term complications, such as autoimmune and neurological disorders. Another important thing worrying us is that *Aedes aegypti* mosquitoes can also transmit the Zika virus vertically, from an infected pregnant woman to her fetus. This mode of transmission is a significant concern as it can result in severe birth defects and developmental abnormalities (Agusto et al. 2017).

Vector-Borne diseases can cause a wide range of symptoms depending on the specific pathogen involved. Malaria, for instance, causes fever, headache, and chills, and if not treated promptly, can lead to severe illness and death (World Health Organization 2020). Lyme disease, transmitted by tick bites, can cause fever, headache, fatigue, and a skin rash. If left untreated, the infection can spread to joints, the heart, and the nervous system (CDC 2022a).

Medical treatment for vector-borne diseases typically involves managing symptoms and eliminating the pathogen. Antimalarial drugs and antibiotics are developed for malaria and Lyme disease (CDC 2022a; World Health Organization 2020). Preventative measures involve reducing contact with vectors. These include using insect repellents, wearing protective clothing, and using bed nets in malaria-endemic areas. Community-wide strategies include vector control programs like environmental management (removing vector breeding sites), biological control (introducing predators), and chemical control (insecticides) (World Health Organization 2020). In addition, promoting sustainable urban planning and sanitation, implementing climate-adaptation measures to manage changing vector habitats, and improving surveillance and response systems for early detection and control of outbreaks are significant as both short- and long-term environmental health mitigation strategies (Ryan et al. 2019).

2.4 *Chemical Exposure-Related Diseases*

Exposure to harmful chemicals from industrial processes, improper waste disposal, pesticides, and household chemicals in the air, water, soil, or even everyday products can lead to a range of health conditions (Landrigan et al. 2018). For example, contracting lead, which often exists in peeling lead-based paint in old houses or contaminated water, can result in lead poisoning. Lead can enter our bodies through various routes like ingestion, inhalation, and dermal exposure. Once absorbed, lead is transported through the bloodstream to various organs and tissues. It can bind to red blood cells, which act as carriers, allowing lead to spread throughout the body. Lead has the ability to accumulate in different tissues and organs, with a particular affinity for bones, teeth, the brain, and kidneys. It can cross the blood-brain barrier, enabling its entry into the central nervous system, where it can have profound neurological effects. One of the key mechanisms of lead toxicity is its interference with enzymes and cellular function. Lead can disrupt the activity of enzymes involved in heme synthesis, which can lead to anemia. It can also interfere with calcium metabolism, disrupting bone health and contributing to the demineralization of bones (Goyer 1997). Moreover, our central nervous system is highly vulnerable to lead toxicity. Lead can affect the development and functioning of the brain, particularly in children whose brains are still developing. It can interfere with neurotransmitters, disrupt neuronal signaling, and impair cognitive function, leading to learning difficulties, developmental delays, and behavioral problems. In adults, lead exposure can cause neurological symptoms such as memory loss, mood disorders, and impaired coordination (Tyler and Allan 2014). The more concerning thing is that lead is not the only poison around. Similar to lead, asbestos, historically used in construction for its fire-resistant properties, can cause mesothelioma, a form of cancer (National Cancer Institute 2017).

When chemicals can enter the human body through inhalation, ingestion, or skin contact, these substances can interact with tissues and organs and damage them directly or indirectly. For instance, inhaled asbestos fibers can lodge in lung tissues, causing inflammation and scarring, which may eventually lead to mesothelioma (National Cancer Institute 2017).

Symptoms of chemical exposure-related diseases can vary widely depending on the specific chemical and the level and duration of exposure. Lead poisoning, for example, can cause developmental delay in children, abdominal pain, neurological changes, and anemia. In severe cases, it can lead to seizures, coma, or even death (CDC 2022b; Wani et al. 2015). Mesothelioma can cause symptoms like chest pain, shortness of breath, and fatigue, and is often fatal (National Cancer Institute 2017).

Treatment for chemical exposure-related diseases often involves eliminating the source of exposure and managing symptoms. For lead poisoning, this might include chelation therapy to remove lead from the body and treating any symptoms or complications (CDC 2022b). For prevention, measures like using less hazardous alternatives, improving ventilation, and using personal protective equipment in occupational settings are usually suggested (Landrigan et al. 2018). Moreover, policy and regulations of limiting the release of harmful chemicals into the environment, promoting safer alternatives, and improving waste disposal practices are necessary. For example, the US Environment Protection Agency's Clean Air Act and Clean Water Act list regulatory measures to prevent chemical exposure-related diseases (US EPA 2013).

In summary, actionable implementation of environmental science can be applied to manage and mitigate the impacts of environmental health hazards previously described like air quality-related diseases, waterborne diseases, vector-borne diseases, and chemical exposure-related diseases. This methodological approach can be implemented to understand and address the complex interactions between humans and their environment. Environmental science provides the foundation for identifying harmful elements in the environment, while it can be actionable by translating it into strategies that can be applied to reduce exposure and enhance health outcomes. This includes the development of effective treatments and preventative measures, such as improving sanitation, reducing contact with vectors, and managing chemical exposures. Moreover, these sciences can inform policies and regulations that further help to safeguard our environment and health. Ultimately, the actionable use of environmental science allows us to devise more effective strategies for mitigating environmental health risks and enhancing public health.

3 State of the Art Research in Environmental Health

In the following sections, we will discuss key advancements and ongoing challenges in emerging domains that directly intersect with environmental health. It begins with the nascent field of precision environmental health, followed by the studies of environmental epigenetics, nanotechnology for water purification, and ends with bioelectrochemical systems for waste treatment.

3.1 Precision Environmental Health

Precision environmental health is an emerging field and most of the advancements are definitely in the early stages or have limited practical application. For example, the work from Steinle et al. (2015) focused on developing personal air pollution sensors to measure an individual's exposure to various pollutants. These wearable sensors provide real-time data on air quality and can be integrated with mobile devices. They offer personalized exposure information and enable individuals to make informed decisions about their daily activities. However, challenges such as sensor accuracy, calibration, and data interpretation need to be addressed before widespread adoption. Another well-known study by Wild (2012) introduces the concept of the exposome, which encompasses the totality of environmental exposures an individual experiences throughout their lifetime. It emphasizes the need to measure multiple environmental factors, including air pollution, water quality, chemical contaminants, and social determinants, to better understand their combined effects on individual health. The exposome approach requires advanced technologies, such as high-throughput omics techniques, to capture comprehensive exposure data. However, data integration and analysis remain significant challenges. Vandenberg et al. (2016) used advanced statistical and computational approaches to analyze the complex mixtures of environmental exposures as they think traditional single-chemical risk assessments may not adequately capture the health risks associated with exposure to multiple pollutants simultaneously. They proposed integrating high-dimensional exposure data, such as biomarkers and sensor measurements, with advanced statistical modeling techniques to identify exposure patterns and their health effects. However, the complexity of data analysis, data sharing, and standardization pose challenges to its practical implementation.

Certainly, the transition from academic research to real-world application in the field of precision environmental health faces numerous challenges. First, the development and validation of precise and reliable measurement tools, such as sensors and omics techniques, are still ongoing (Nieuwenhuijsen et al. 2014). Also, further improvements in accuracy, sensitivity, and affordability of these technologies are needed to facilitate widespread use in real-world settings. Besides, integrating and analyzing large and diverse datasets from multiple sources pose significant computational and analytical challenges. Regarding ethical concerns, precision environmental health involves the collection of sensitive personal data, which raises questions about privacy, data ownership, and ethical considerations (Vayena et al. 2018).

3.2 Environmental Epigenetics

Environmental epigenetics is another rapidly evolving data-driven field that explores how environmental factors can influence gene expression patterns through epigenetic modifications. Exciting research is thriving on this topic. For example, Feinberg (2018) studied how environmental exposures can modify epigenetic marks, such as DNA methylation and histone modifications, leading to changes in gene expression and disease susceptibility. Madrigano et al. (2012) investigated the epigenetic effects of air pollution on human health. Joubert et al. (2012) examined the relationship between environmental chemicals and DNA methylation patterns in adults. They discovered significant differential DNA methylation at 26 CpG sites (CpGs) across 10 genes, with replication of findings for CpGs in aryl-hydrocarbon receptor repressor (AHRR), cytochrome P450 family 1 subfamily A member 1 (CYP1A1), and growth factor-independent 1 transcriptional repressor (GFI1) at strict statistical significance. AHRR and CYP1A1 are involved in tobacco smoke detoxification, while GFI1, previously unassociated with tobacco smoke responses, is involved in various developmental processes, which implicates epigenetic mechanisms in the pathogenesis of the adverse health outcomes associated with this important in utero exposure. However, most of these studies are still in the experimental stage, as environmental exposures are multifaceted, and it is challenging to attribute specific epigenetic changes to individual exposures. Understanding the cumulative effects of multiple exposures and their interactions requires holistic exposure assessment and study designs. Also, epigenetic modifications can vary across individuals and populations, making it challenging to establish consistent associations between environmental exposures and epigenetic changes, especially in long-term studies.

3.3 Nanotechnology for Water Purification

This is another exciting application to look forward to in environmental sustainability with direct implications on human health. Focusing on solutions to the waterborne diseases and water contamination, scientists try to purify the water using innovative nanomaterials, such as graphene oxide membranes, carbon nanotubes, nanocomposite materials, and nanocatalysts (Iravani 2021). These materials offer enhanced adsorption, filtration, and catalytic capabilities, effectively removing contaminants like heavy metals, organic pollutants, and microorganisms from water sources. The basic flow is to first synthesize advanced nanomaterials using techniques like sol-gel, hydrothermal, or chemical vapor deposition methods (Yoon et al. 2015), allowing precise control over material properties, such as size, shape, and surface functionality. Then modify the synthesized nanomaterials to enhance their adsorption or catalytic properties via surface functionalization or coating processes. Subsequently, integrating them into water treatment systems through various processes, including membrane filtration, adsorption, photocatalysis, and disinfection. Nanocomposite membranes with nanoscale pores selectively separate contaminants based on size, charge, and molecular interactions, effectively removing pollutants while allowing the passage of clean water. Those nanomaterials with high surface area and affinity for contaminants will serve as adsorbents to capture and immobilize pollutants through adsorption or chemisorption processes (Lim et al. 2021). Nanocatalysts, such as titanium dioxide nanoparticles, can be activated by light and generate reactive oxygen species that can degrade organic pollutants and destroy microbial pathogens. Another layer of filtering are antimicrobial nanomaterials which include silver nanoparticles or nanocomposites, and can exhibit strong bactericidal properties and remove harmful microorganisms.

Despite the promising advancements in nanotechnology for water purification, several factors hinder its widespread implementation in the real world (Gehrke et al. 2015). The production and large-scale deployment of advanced nanomaterials can be expensive, limiting their affordability and accessibility, especially in resource-limited areas. The manufacturing processes, stability, and performance of nanomaterials are still immature and need further optimization for real-world water treatment systems. The potential environmental and health risks associated with nanomaterials necessitate careful evaluation and regulation before widespread implementation. Comprehensive studies on their long-term effects are required to ensure their safe use. Lastly, integrating nanotechnology-based purification systems into existing water infrastructure and treatment plants requires careful consideration of compatibility, retrofitting, and operational feasibility, and have to go through the careful review, budgeting and decision-making processes by the water companies and operators.

3.4 Bioelectrochemical Systems for Waste Treatment

Waste generation and improper disposal are other factors that significantly impact the environment, leading to soil, air, and water pollution, and adversely affecting the environment and ultimately human health. This waste, whether it is industrial, domestic, or agricultural, often contains harmful substances that can infiltrate ecosystems and pose risks to human health. For example, hazardous waste may leach heavy metals into groundwater, or waste decomposition in landfills may release methane, a potent greenhouse gas. Therefore, effective waste treatment is crucial in mitigating these environmental impacts. There are many studies focusing on developing advanced technologies for waste treatment and to avoid waste from being directly released to our environment and affecting people's health. Bioelectrochemical systems (BES), including microbial fuel cells (MFCs), microbial electrolysis cells (MECs), and electrochemical bioreactors, are one of the potential solutions for this task (Kato 2015). Here we briefly introduce how it works. First, BES harnesses the metabolic activities of diverse microbial communities, and these microorganisms, including bacteria and archaea, interact with the electrodes and catalyze various electrochemical reactions (Butti et al. 2016). Microbes at the anode oxidize organic compounds in the waste, releasing electrons and protons (Lovley 2006). This process is facilitated by the use of specific electrode materials and biofilm formation. The released electrons travel from the anode to the cathode through an external circuit. At the cathode, a reduction reaction occurs, typically involving oxygen or other electron acceptors. This reaction consumes electrons and protons, thereby completing the electrochemical circuit (He and Angenent 2006). As the organic compounds are oxidized at the anode, they are effectively broken down, leading to the degradation and removal of pollutants in the waste (Zhang et al. 2022).

Several challenges hinder their widespread implementation. Scaling up BES from lab-scale to practical applications poses challenges in terms of maintaining stable and efficient performance (Das et al. 2020). Ensuring long-term stability, preventing biofouling, and optimizing system design are critical for successful implementation. Another concern is that BES involves intricate microbial–electrode interactions and electrochemical processes. Similar to nanotechnology, this is very advanced technology and the side effects are not very clear. Understanding and controlling these complex dynamics require further research and advanced monitoring techniques. Meanwhile, the cost-effectiveness of BES, including electrode materials, membrane requirements, and system components, needs to be improved for practical implementation. Similarly, when it comes to practice, it will require modifications and adaptations to the current systems and need the decision makers, stakeholders, and users to go through the business deciding process together.

To conclude, the actionable component of environmental health is continuously expanding with the advent of novel fields such as precision environmental health, environmental epigenetics, nanotechnology for water purification, and bioelectrochemical systems for waste treatment. These cutting-edge disciplines enhance our ability to personalize exposure assessments, pinpoint epigenetic alterations induced by environmental factors, innovate sustainable water purification methods, and improve waste treatment practices. However, to transform these advancements into more actionable solutions, several hurdles need to be overcome, including technological maturity, data interpretation, privacy issues, scalability, and integration with existing infrastructure. Furthermore, careful consideration of cost-effectiveness, safety, and ethical implications is essential to ensure the broad implementation of these novel techniques. To navigate these challenges and to successfully translate scientific findings into actionable applications, a comprehensive approach, including cross-disciplinary collaboration, robust policy frameworks, and societal engagement, is needed. As environmental science continues to evolve and become more actionable, it holds the potential to provide insights that can drive interventions, promote environmental sustainability, and safeguard human health in an increasingly complex and changing world.

4 A Successful Use Case: Cheverly Community Air Quality Monitoring Dashboard

Limited access to data on environmental hazards and their impact on human health can hinder research progress or drag our capability of responding to environmental health issues behind. Dashboards and geographic information systems (GIS) are powerful tools that can be utilized to overcome challenges associated with data access, data quality, and data interpretation in the field of actionable science in environmental health (Few 2006). These tools provide valuable means to visualize, analyze, and communicate complex environmental data, enabling scientists to identify and prioritize environmental health problems, develop effective interventions, and communicate risks to the public. For example, dashboards offer a user-friendly interface that consolidates and presents data in an easily understandable format (Rojas et al. 2020). These visualizations can include real-time or historical data on environmental parameters such as air quality, water quality, or the presence of hazardous substances. By making such data accessible and readily available, dashboards enhance the ability of scientists and policymakers to identify and prioritize environmental health problems, facilitating evidence-based decision-making. Furthermore, environmental health problems are often complex and influenced by multiple factors, posing challenges in data interpretation. Dashboards can provide interactive features that allow users to explore and analyze data in different contexts, facilitating a deeper understanding of the relationships and interactions between environmental hazards and health outcomes (Batty 2015). For instance, users can filter data based on specific criteria, visualize spatial or temporal patterns, and identify correlations between different environmental variables and health indicators. These functionalities enhance the interpretability of data and aid in identifying potential causes and trends. In conjunction with dashboards, GIS plays an important role in environmental health science by providing spatial analysis capabilities. GIS allows scientists to map environmental hazards, overlay them with population data, and visualize their spatial distribution (Tim 1995). This enables the identification of populations that may be most vulnerable to certain hazards, such as communities living near industrial sites or areas prone to natural hazards. By integrating GIS data with demographic information, socioeconomic factors, and health data, scientists can gain insights into the spatial patterns of environmental health risks and develop targeted interventions to protect affected populations.

The use of dashboards and GIS in environmental health is a rapidly growing field. The integration of dashboards and GIS allows scientists, policymakers, and communities to collaborate more effectively in identifying, understanding, and addressing environmental health challenges. Air quality has a direct impact on their health and poor air quality resulting from pollution can lead to respiratory issues, heart disease, cancer, and premature death. Tools derived from actionable science, such as dashboards, can help address air quality-related health threats. By utilizing data to identify and prioritize problems, developing and implementing effective interventions, and communicating risks to the public, actionable science applied through the development of dashboards provides a pathway to mitigate the adverse health effects of air pollution. In the next sections we illustrate the implementation of dashboards for air quality data visualization.

4.1 Dashboard Purpose

The relationship between air quality and public health forms the foundation for actionable science in the realm of environmental health. Harnessing our understanding of how air quality profoundly impacts human well-being, we can forge pragmatic solutions aimed at mitigating the detrimental health consequences of poor air quality. An illustrative use case in this context is the implementation of an air quality monitoring dashboard, a tool designed to collect real-time data on air pollutants and present it in an accessible format. This dashboard empowers individuals, communities, and policymakers alike, enabling them to make informed decisions that safeguard public health. It furnishes actionable information that not only guides immediate interventions, such as the avoidance of outdoor activities during episodes of heightened pollution, but also informs the development of long-term strategies to curtail pollution sources and enhance air quality. This use case exemplifies how actionable science, firmly rooted in the intricate relationship between air quality and public health, holds the potential to drive tangible and impactful outcomes, fostering the well-being of communities at large.

Cheverly, Maryland, is a small town located in Prince George's County, known for its close-knit community and suburban feel. The town had a population of about 6500 (US Census Bureau 2022). Cheverly is just over a mile away from Washington, DC, making it a popular residential area for people working in the city. It is located inside the Capital Beltway and has its own Metrorail station, providing easy access to the wider DC metropolitan area.

In 2021, Cheverly developed a community air quality monitoring (AQM) PurpleAir sensor network in partnership with the University of Maryland (UMD) and the Maryland Department of Environment. In 2022, the UMD team is expanding the sensor network into two neighboring communities. These communities are adjacent to heavy diesel traffic and industrial pollution sources as noted in Environmental Justice (EJ) screening tools. The Cheverly Community Air Quality Monitoring dashboard (https://arcg.is/1T0Gvb0) (Fig. 11.1) was created to address environmentally burdened communities' need for Air Quality information to support policy and personal health decisions. This project was funded by the Earth Science Information Partners (ESIP) FUNding Friday Program.

Fig. 11.1 Overview of Cheverly Community Air Quality Monitoring Dashboard

4.2 Data and Methods

This comprehensive air quality monitoring dashboard encompasses an array of dynamic features designed to provide a holistic understanding of the environmental landscape. The interactive maps (e.g., AirNow Interactive Map and Air Quality Aware) offer a visual representation of air quality across different regions, facilitating localized assessments (Fig. 11.2). Graphs and charts elucidate trends and patterns, empowering users to discern fluctuations and correlations between air quality and various factors. Through guidance, blogs, and news updates, users gain access to expert insights and relevant information to navigate the intricacies of air quality management effectively. The dashboard goes beyond mere monitoring, extending to encompass air quality forecasts, weather data, demographic information, and a county zoning atlas. Each facet incorporates real-time or up-to-date data, ensuring that the community remains equipped with the most pertinent information to inform decision-making and prioritize public health concerns.

The monitoring and forecasting data come from a variety of sources, including monitoring stations, satellite data, radar, and low-cost sensors. Traditional air quality monitoring stations are typically managed by government or environmental agencies. These stations measure a variety of pollutants, such as PM2.5 and PM10, NO_2, SO_2, CO, O_3, and sometimes VOCs. Satellite-based remote sensing can

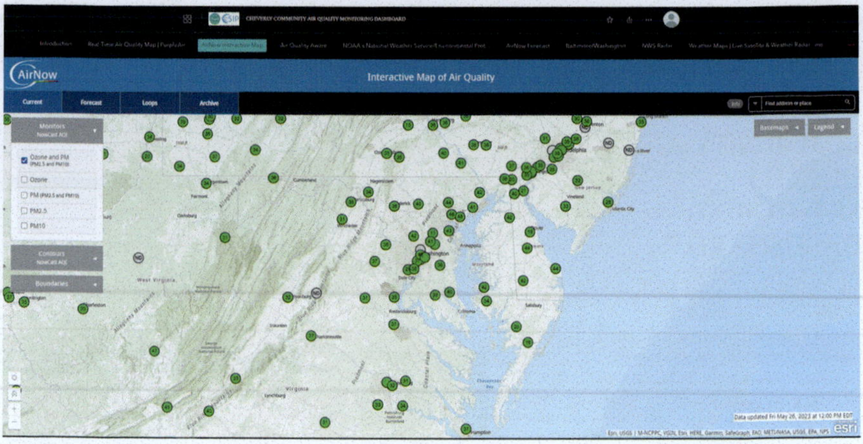

Fig. 11.2 Interactive Map of Air Quality on AirNow. (Screenshot taken on May 26, 2023)

provide data on air quality over broad geographic areas. Instruments on satellites can measure various aerosols and gases in the atmosphere (e.g., National Weather Service Radar Map) (Fig. 11.3). Online weather service, such as Meteoblue, utilizes various sources, including meteorological satellites, weather stations, weather prediction models, and radar, to capture weather information and generate weather maps (Fig. 11.4). In addition, low-cost sensors are increasingly being used to complement traditional monitoring stations. For example, PurpleAir makes sensors that empower Community Scientists who collect and share hyper-local air quality data with the public, public access to the PurpleAir.com map, and they offer a more cost-effective way to gather data, especially in areas where traditional monitoring is sparse. It displays current Air Quality Index (AQI) readings and a graph of 10 min PM2.5 AQI data for the previous 7 days currently exists (Fig. 11.5). The AQI was developed to monitor and communicate the level of air pollution, aiming to establish a standardized measure for assessing air pollution globally (AirNow n.d.). It quantifies the concentrations of different pollutants in the air – including the key offenders mentioned above – and converts these complex data into a simple scale ranging from 0 to 500. Lower AQI values indicate good air quality with minimal health concerns, while higher values signal potential health risks. The public, health officials, and environmental agencies use the AQI as a guide to understand air quality conditions and to take appropriate preventive or protective measures when required, such as limiting outdoor activities during high AQI readings (Kelly 2003).

4.3 *Results and Applications*

The Cheverly Community Air Quality Monitoring Dashboard yields impactful results, driving positive outcomes for public health and environmental management. As an example of actionable science in the environmental health field, it highlights

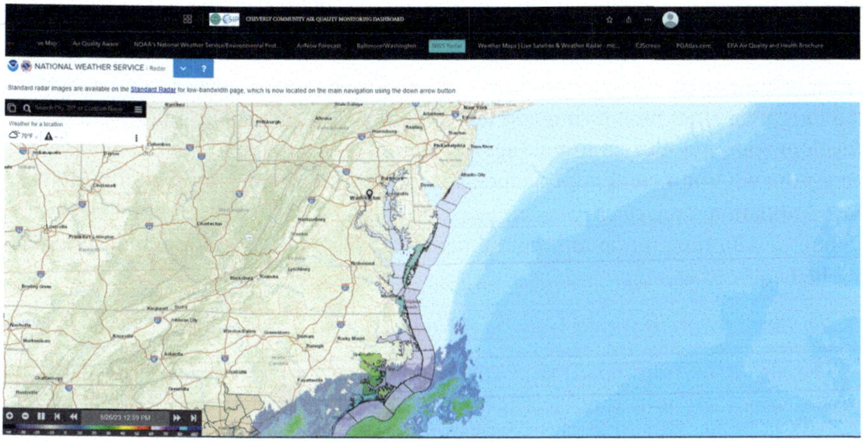

Fig. 11.3 National Weather Service Radar images. (Screenshot taken on May 26, 2023)

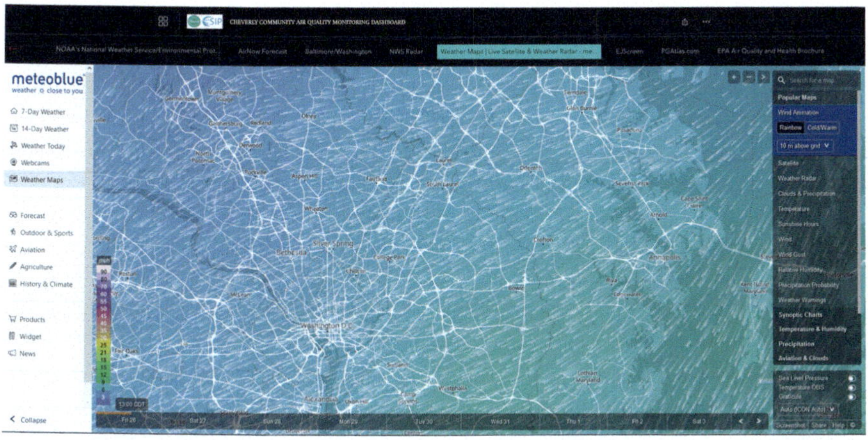

Fig. 11.4 MeteoBlue wind maps. (Screenshot taken on May 26, 2023)

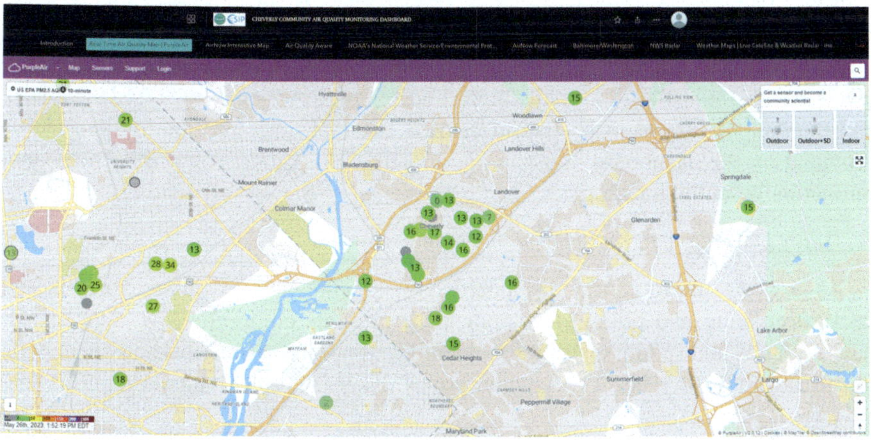

Fig. 11.5 PurpleAir Real-Time Air Quality Map. (Screenshot taken on May 26, 2023)

the transformative applications that emerge for improved awareness and understanding, real-time monitoring and alerts, guidance for personal protection, policy development and decision-making, community engagement and collaboration, long-term environmental improvements, support for research and scientific studies, data validation and quality assurance, early warning systems, and emergency response, data integration and public health surveillance, and communication and public health education.

1. *Improved Awareness and Understanding*

The air quality dashboard is instrumental in raising public awareness about air quality issues. By providing accessible and visually appealing information, it educates individuals and communities about the health risks associated with poor air quality in Cheverly town and surrounding areas. The dashboard empowers people to take proactive measures to protect their health by understanding the impacts of pollutants and making informed decisions.

2. *Real-Time Monitoring and Alerts*

The dashboard facilitates real-time monitoring of air quality by collecting and displaying up-to-date data from various sources such as Meteoblue and AirNow. This enables users to stay informed about the current air quality conditions in the town and receive timely alerts or notifications when pollution levels exceed safe thresholds. Such real-time information empowers individuals to modify outdoor activities or take preventive measures to minimize exposure, safeguarding their well-being.

3. *Guidance for Personal Protection and Emergency Response*

With its comprehensive information and recommendations, the dashboard offers guidance on how individuals can protect themselves during periods of poor air quality. It provides valuable insights into suitable precautions, such as wearing masks, adjusting outdoor activities, or seeking shelter in environments with better air filtration. By promoting informed actions, the dashboard supports personal health protection strategies. Moreover, the dashboard serves as an early warning system during environmental emergencies or rapidly deteriorating air quality. It promptly alerts public health personnel to sudden spikes or significant changes in pollutant levels, for example, the abnormally high value in PurpleAir, enabling rapid response, and implementation of protective measures. The real-time information and alerts provided by the dashboard assist in coordinating emergency response efforts and mitigating potential health risks.

4. *Policy Development and Decision-Making*

As a powerful tool, the dashboard assists policymakers in evidence-based decision-making and policy development. It presents comprehensive data on air quality trends and spatial distribution, identifying pollution hotspots and areas with the greatest h0ealth risks. This enables policymakers to prioritize interventions and implement targeted policies and regulations, ultimately improving air quality and protecting public health in Cheverly community and surrounding areas.

5. *Community Engagement and Collaboration*

The air quality dashboard fosters community engagement by actively involving citizens in environmental health initiatives. It encourages individuals to participate by reporting observations, contributing data through citizen science projects, and sharing insights or concerns related to air quality. Throughout the dashboard development process, active involvement and valuable input were received from the UMD, NASA, ESIP, U.S. Environmental Protection Agency (US EPA), and the residents in the Cheverly community. This engagement cultivates a sense of ownership and collective responsibility for environmental well-being, fostering collaborations between communities, local authorities, and scientific organizations.

6. *Long-Term Environmental Improvements*

Through continuous monitoring and tracking of air quality data, the dashboard enables the evaluation of pollution reduction measures and interventions over time. It facilitates evidence-based decision-making for long-term environmental improvements. By assessing the impact of policy changes, technological advancements, and behavioral modifications on air quality and health outcomes, the dashboard guides future strategies for sustainable development and pollution control. The Cheverly community also set a precedent for other small towns and improved environmental health over time.

7. *Support for Research and Scientific Studies*

The air quality dashboard serves as a valuable resource for researchers conducting studies on environmental health. It provides access to real-time and historical air quality data, allowing researchers to analyze trends, identify patterns, and investigate the complex relationships between air pollution and health outcomes for local studies. The availability of comprehensive datasets through the dashboard contributes to the advancement of knowledge in the field.

8. *Data Integration, Communication, and Public Health Education*

Seamless integration of air quality data with other public health information enhances the capabilities of researchers and public health personnel. The air quality dashboard facilitates the dissemination of air quality information, health advisories, and educational materials to the general public. Utilizing interactive features like maps, charts, and guidance, public health professionals convey complex information in a clear and accessible manner. The dashboard supports public health education initiatives, empowering individuals to make informed decisions, adopt healthy behaviors, and actively participate in efforts to improve air quality and protect residents health in the town of Cheverly.

4.4 What Makes the Dashboard Actionable?

This case study explores the advantages of actionable science in environmental health through the lens of a local air quality monitoring dashboard. Specifically, it examines how this dashboard exhibits advantages such as providing information at the right time, reaching the right people, fitting the scenario context, being on point, directly related, accurate, credible, and decisive.

1. *Providing Information at the Right Time*

The local air quality monitoring dashboard delivers information at the right time. By collecting and analyzing real-time air quality data, the dashboard ensures that users have access to up-to-date information regarding the air quality in their specific area. This timely information empowers individuals, communities, and public health authorities to make informed decisions promptly, take necessary precautions, and implement targeted interventions to protect public health.

2. *Reaching the Right People*

The dashboard demonstrates the advantage of reaching the right people effectively. It offers user-friendly interfaces, accessible visualizations, and multi-channel communication methods to disseminate air quality information widely. Whether through interactive maps, mobile applications, social media platforms, or email alerts, the dashboard ensures that information reaches diverse users, including individuals, community organizations, policymakers, and public health personnel. This broad outreach helps engage key stakeholders and facilitates collective action toward improving air quality and safeguarding public health. For example, the Cheverly Community Dashboard targeted users are Cheverly town residents, public health personnel, public policy makers, and researchers. Reaching the right people and providing useful information is one of the key purposes of this dashboard.

3. *Fitting the Scenario Context*

Actionable science in environmental health must be adaptable and context-specific. The local air quality monitoring dashboard excels in fitting the scenario context. It provides localized and customizable information, considering specific geographical areas, pollution sources, and population characteristics. By tailoring the data and recommendations to the local context, the dashboard addresses the unique challenges and needs of different communities, enhancing the relevance and effectiveness of interventions.

4. *Being on Point and Directly Related*

The dashboard's advantage lies in being on point and directly related to the issue. It focuses specifically on air quality and its direct impact on public health. The information presented in the dashboard is concise, targeted, and explicitly addresses air pollution concerns, health risks, and recommended actions. This focused approach enables users to quickly grasp the essential information and take appropriate measures to protect themselves and their communities from the identified air quality risks.

5. *Providing Accurate and Credible Information*

Accuracy and credibility are paramount in actionable science. The local air quality monitoring dashboard ensures the provision of accurate and credible information. It relies on data collected from reputable monitoring stations and reliable sources, undergoing rigorous quality assurance processes. The transparency of data sources, methods, and quality control measures enhances the credibility of the information presented in the dashboard, fostering trust among users and stakeholders.

6. *Making Decisive Actions*

The ultimate goal of actionable science is to drive decisive actions. The local air quality monitoring dashboard facilitates this by offering precise and actionable recommendations based on the air quality data and associated health risks. Users can access specific guidelines for personal protection, receive alerts during high pollution episodes, and obtain information on policy measures and interventions. By providing users with the information required to make informed decisions, the dashboard empowers individuals, communities, and public health authorities to take decisive actions to improve air quality and protect human health.

4.5 What Hinders the Dashboard from Being Actionable?

While air quality monitoring dashboards offer numerous advantages, it is essential to acknowledge and address their limitations and challenges. This section explores the potential drawbacks and hurdles that may arise when utilizing air quality monitoring dashboards. By understanding these limitations, stakeholders can develop strategies to overcome challenges and maximize the effectiveness of these tools in promoting public health and environmental management.

1. *Data Quality and Availability*

One significant limitation is the quality and availability of data. Air quality monitoring dashboards rely on data collected from monitoring stations or sensors, and the quality of these stations may vary. Outdated or insufficient monitoring infrastructure can result in data gaps and inaccuracies. Additionally, accessing real-time data from certain sources or regions may pose challenges, affecting the comprehensiveness and reliability of the information presented on the dashboard.

2. *Spatial and Temporal Resolution*

The spatial and temporal resolution of air quality and public health data can pose limitations in actionable science. Monitoring stations may not encompass every location, resulting in spatial variability and potential disparities across different areas. Similarly, capturing short-term fluctuations in air quality can be challenging, impacting the temporal resolution of data. Moreover, assessing public health data related to air quality in small population groups while preserving privacy may prove challenging. These limitations can influence the precision and accuracy of the information presented through the dashboard, affecting its granularity and reliability.

3. *Interpretation and Communication of Data*

Effectively interpreting and communicating air quality data is crucial for the dashboard's success. However, understanding complex data and translating it into actionable measures may be challenging for some individuals and communities. The dashboard must ensure the information is clear and understandable, accompanied by appropriate guidance and educational resources to facilitate informed decision-making.

4. *Equity and Accessibility*

Ensuring equitable access to the benefits of the dashboard is essential. However, specific populations, such as marginalized communities or those with limited internet access or digital literacy, may need help utilizing the dashboard effectively. Bridging the digital divide and addressing disparities in access and engagement is crucial to ensure that the benefits of the dashboard reach all segments of society.

5. *Technological Infrastructure and Sustainability*

Implementing and maintaining the technological infrastructure to support the dashboard can be resource-intensive. Regular updates, maintenance, and data management require dedicated resources and ongoing funding. Ensuring the long-term sustainability of the dashboard and its infrastructure is crucial to maintain its effectiveness and utility over time.

6. *Dynamic Nature of Air Quality*

Air quality is a dynamic phenomenon influenced by various factors, including weather patterns, human activities, and seasonal variations. The dashboard may face challenges in keeping up with these dynamics and providing real-time information that accurately reflects current air quality conditions. Continual monitoring, data validation, and updating models and algorithms are necessary to address this challenge effectively.

4.6 Looking into the Future: How Can We Enhance the Actionability of the Environmental Health Dashboard?

In this section of the chapter, we have described how GIS and dashboards in environmental science are transformative approaches that significantly enhance the actionability of environmental health data. GIS allows scientists to map environmental and biological hazards, overlay them with population data, and visualize spatial distribution, which aids in identifying populations most vulnerable to certain hazards and developing targeted interventions (Bodenhamer et al. 2015). Coupled with dashboards, this approach offers a user-friendly interface that consolidates and presents data in an easily understandable format, providing scientists and policy-makers the ability to promptly identify, prioritize, and respond to environmental health problems. Environmental health dashboards can be enhanced to provide more actionable insights through several methods, guided by the principles of user-centered design, real-time data and alerts, spatial and temporal resolution, interactive visualizations, contextual information, and integration with actionable resources (Aigner et al. 2007; Jiang et al. 2016b).

Community-engaged and user-centered design is a design philosophy and a process in which the needs, wants, and limitations of communities and end users are given extensive attention at each stage of the design process (van Velsen et al. 2022). For environmental health dashboards, this could mean incorporating features that provide relevant, personalized information in a format that is easy for the user to understand and act upon. During the creation of the Cheverly Community Air Quality Monitoring Dashboard, different stakeholders from the government, university, and other public and private sectors were engaged and provided feedback. Engaging with users to get their feedback and understand their needs can lead to a more intuitive, user-friendly dashboard.

The integration of real-time data feeds and alerts is a crucial feature. This functionality ensures timely communication of environmental conditions like air and water quality, enabling users to make swift, health-preserving decisions. Customizable alerts can be designed to address users' specific needs. Meanwhile, high-resolution data, both spatially and temporally, enable users to make more precise decisions. Providing data that is updated frequently (temporal resolution) and at a localized scale (spatial resolution) – for example, hourly air quality updates for specific neighborhoods – facilitates a nuanced understanding of environmental conditions.

Interactive visualizations and providing contextual information can make the data more engaging and easier to understand. Users could interact with the data, such as by clicking on a map to see detailed data and contextual information for a specific location, or adjusting a slider to see how air quality has changed over time. For example, if the dashboard shows that the AQI is 100, it could also explain what this means (e.g., that this is considered unhealthy for sensitive groups) and provide advice on what actions to take (e.g., limit outdoor activities).

Using air pollution as an example we illustrated how dashboards can offer interactive features that enable users to explore and analyze data within different contexts, further enhancing the interpretability of data and aiding in the identification of causes, trends, and impacts in human populations. The integration of real-time alerts, high-resolution spatial and temporal data, and interactive visualizations, all designed with user-centered principles, allows the environmental data to be more engaging, understandable, and actionable. Additionally, linking the dashboard to practical resources provides an effective conduit to drive decisive action. This dual approach of utilizing GIS and dashboards, backed by principles of timely communication, user-centered design, and actionable linkages, serves to turn environmental science data into effective interventions, making environmental science more practical. This approach could be expanded and implemented for other environmental hazards of human health described previously like environmental pollutants in water, and biological hazards like the emergence of new pandemics including COVID-19.

5 Discussion

Actionable implementation of environmental health is a rapidly evolving strategy that harnesses data to inform interventions and promote public health. By understanding the complex relationship between the environment and human well-being, actionable implementation of environmental health science enables evidence-based strategies to mitigate disease risks. This approach empowers individuals, communities, and policymakers to make informed decisions and take proactive steps toward promoting health and preventing diseases.

While the potential of actionable environmental health is substantial, there are challenges that need to be addressed to fully leverage its benefits. One significant challenge is the limited availability of sufficient data, which hinders the development of effective interventions. Bridging the gap between research findings and practical implementation is another crucial aspect, as even robust data may struggle to influence policy decisions. Furthermore, ensuring data quality and interpretation poses additional challenges due to the multifaceted nature of environmental health issues. Overcoming these challenges requires innovative approaches and collaborations among researchers, policymakers, and communities.

Despite the challenges, there are several reasons for optimism in the functionable implementation in environmental health. Technological advancements like the ones described in this chapter continue to expand data collection methods and improve our understanding of the relationship between environmental hazards and human health. Additionally, advancements in data analytics and modeling techniques enable researchers to analyze large datasets and identify patterns and associations between environmental exposures and health outcomes. Policymakers and decision-makers are increasingly recognizing the importance of evidence based decision-making that creates a favorable environment for actionable strategies in environmental health. The integration of scientific knowledge into policy frameworks is essential for developing effective interventions and regulations. For instance, the adoption of emission standards for pollutants and the implementation of regulations to reduce exposure to hazardous substances are examples of evidence-based policies informed by actionable strategies in environmental health. By incorporating research findings into policy decisions, governments can prioritize public health and promote sustainable practices.

The growing public demand for effective interventions and policies further encourages the implementation of actionable strategies and fosters collaboration among stakeholders. Environmental health advocacy groups, community organizations, and concerned citizens play a crucial role in raising awareness about environmental health risks and advocating for evidence-based solutions. Their efforts create a sense of urgency and drive policymakers to take action in addressing environmental health challenges.

Looking ahead, there are important directions that can further advance the actionable implementation of environmental health. Efforts should focus on developing new technology and emerging cost-effective methods for data collection that provide a comprehensive understanding of environmental hazards and their impact on human health. Standardized protocols and quality control procedures are crucial for ensuring data reliability. Collaborations between researchers and citizen science initiatives can also enhance data collection and expand the reach of environmental health research. Additionally, the interpretation of complex environmental health data requires the development of innovative analytical approaches that can account for multiple factors and interactions. Integrating data from various sources, such as environmental monitoring, biomonitoring, and health records, can provide a more holistic understanding of the relationships between environmental exposures and health outcomes. Advanced statistical methods, machine learning, and modeling techniques can help uncover hidden patterns and associations in large datasets, guiding the development of targeted interventions.

Investment in actionable interventions is also essential for driving progress in environmental health. Governments, private organizations, and foundations should allocate resources to support research initiatives, infrastructure development, and capacity building. By strengthening the application of evidence-based practices, actionable strategies can have a significant impact on improving environmental health outcomes. Funding for research projects, training programs for environmental health professionals, and the establishment of data sharing platforms can enhance the implementation of actionable science and facilitate collaborations among researchers, policymakers, and communities.

6 Conclusions

In conclusion, actionable applications of environmental health is a rapidly evolving strategy that holds immense potential for revolutionizing human health. By utilizing data to answer critical questions about the relationship between the environment and human health, scientists can inform the development of interventions aimed at reducing the risk of disease. This evidence-based approach serves as a foundation for comprehensive public health strategies, empowering individuals, communities, and policymakers to make informed decisions and take proactive steps toward promoting health and preventing disease. As the implementation of actionable uses of environmental health continues to grow, we can anticipate the emergence of new and innovative interventions that have the potential to make a significant difference in the health of populations worldwide.

In this chapter, we have discussed how environmental health draws on diverse scientific fields and technologies to address and mitigate the impacts of environmental hazards, such as air and water pollutants, vector-borne diseases, as well as chemical exposure-related diseases. Cutting-edge technologies like nanotechnology can offer novel solutions to environmental issues, for instance, by developing effective means to remove pollutants from water, or by creating materials that reduce the emission of harmful substances. Likewise, precision environmental health leverages big data and genomics to tailor interventions to individual susceptibilities and exposures, facilitating personalized, effective action to protect health. Other technologies like environmental epigenetics reveal how environmental exposures can leave lasting marks on our genes, enhancing our understanding of the long-term health impacts of environmental hazards, and underscoring the necessity for proactive and preventative interventions. In this chapter, we also explored the integration of GIS and dashboards in environmental health for visualization, analysis, and communication of complex environmental data, illustrating the implementation of this approach using air pollution data, making this information accessible and actionable. GIS empowers the identification and spatial analysis of at-risk populations, while dashboards provide user-friendly, real-time interfaces for data interpretation and decision-making. These advancements in technologies and methodologies are transforming the field of environmental health, making it more actionable by providing precise, timely, and effective tools for mitigating environmental health risks. By addressing the challenges mentioned in this chapter and advancing the field through collaboration, research, and increased investment, the actionable implementation of environmental health will play a leading role in improving and safeguarding the health of populations. Through these collective efforts, we can strive toward a healthier, sustainable future for all.

References

Acheson, D.W.K. 2009. Food and waterborne illnesses. *Encyclopedia of Microbiology*: 365–381. https://doi.org/10.1016/B978-012373944-5.00183-8.

Agusto, Folashade B., S. Bewick, and W.F. Fagan. 2017. Mathematical model of Zika virus with vertical transmission. *Infectious Disease Modelling* 2 (2): 244–267.

Aigner, Wolfgang, Silvia Miksch, Wolfgang Müller, H. Schumann, and Christian Tominski. 2007. Visualizing time-oriented data – A systematic view. *Computers & Graphics* 31 (June): 401–409. https://doi.org/10.1016/j.cag.2007.01.030.

AirNow. n.d. AQI basics | AirNow.Gov. AirNow.gov, U.S. EPA. https://www.airnow.gov/aqi/aqi-basics. Accessed 25 May 2023.

Alexeeff, Stacey E., Kamala Deosaransingh, Noelle S. Liao, Stephen K. Van Den Eeden, Joel Schwartz, and Stephen Sidney. 2021. Particulate matter and cardiovascular risk in adults with chronic obstructive pulmonary disease. *American Journal of Respiratory and Critical Care Medicine* 204 (2): 159–167. https://doi.org/10.1164/rccm.202007-2901OC.

Alderete, Tanya L., Zhanghua Chen, Claudia M. Toledo-Corral, Zuelma A. Contreras, Jeniffer S. Kim, Rima Habre, Leda Chatzi, Theresa Bastain, Carrie V. Breton, and Frank D. Gilliland. 2018. Ambient and traffic-related air pollution exposures as novel risk factors for metabolic dysfunction and type 2 diabetes. *Current Epidemiology Reports* 5 (2): 79–91. https://doi.org/10.1007/s40471-018-0140-5.

American Lung Association. n.d. *Particle pollution.* https://www.lung.org/clean-air/outdoors/what-makes-air-unhealthy/particle-pollution. Accessed 24 May 2023.
APHA. 2022. *Environmental health.* https://www.apha.org/topics-and-issues/environmental-health.
Barzilay, Ezra J., Nicolas Schaad, Roc Magloire, Kam S. Mung, Jacques Boncy, Georges A. Dahourou, Eric D. Mintz, Maria W. Steenland, John F. Vertefeuille, and Jordan W. Tappero. 2013. Cholera surveillance during the Haiti epidemic – The first 2 years. *The New England Journal of Medicine* 368 (7): 599–609. https://doi.org/10.1056/NEJMoa1204927.
Batty, Michael. 2015. A perspective on city dashboards. *Regional Studies, Regional Science* 2 (1): 29–32.
Bodenhamer, David J., John Corrigan, and Trevor M. Harris. 2015. *Deep maps and spatial narratives.* Indiana University Press.
Brook, Robert D., C. Sanjay Rajagopalan, I.I.I. Arden Pope, Jeffrey R. Brook, Aruni Bhatnagar, Ana V. Diez-Roux, Fernando Holguin, et al. 2010. Particulate matter air pollution and cardiovascular disease: An update to the scientific statement from the American Heart Association. *Circulation* 121 (21): 2331–2378.
Burbank, Allison J., and David B. Peden. 2018. Assessing the impact of air pollution on childhood asthma morbidity: How, when and what to do. *Current Opinion in Allergy and Clinical Immunology* 18 (2): 124–131. https://doi.org/10.1097/ACI.0000000000000422.
Butti, Sai Kishore, G. Velvizhi, Mira L.K. Sulonen, Johanna M. Haavisto, Emre Oguz Koroglu, Afsin Yusuf Cetinkaya, Surya Singh, et al. 2016. Microbial electrochemical technologies with the perspective of harnessing bioenergy: Maneuvering towards upscaling. *Renewable and Sustainable Energy Reviews* 53: 462–476.
CDC. 2022a. Lyme disease home | CDC. Centers for Disease Control and Prevention. https://www.cdc.gov/lyme/index.html.
CDC. 2022b. Lead poisoning prevention | Lead | CDC. https://www.cdc.gov/nceh/lead/prevention/default.htm.
Cleveland Clinic. 2022. *Asthma: Types, causes, symptoms, diagnosis & treatment.* Cleveland Clinic. https://my.clevelandclinic.org/health/diseases/6424-asthma.
Das, Arya, Mamata Mohapatra, and Suddhasatwa Basu. 2020. Recent development in cathodic catalyst towards performance of bioelectrochemical systems. In *Novel catalyst materials for bioelectrochemical systems: Fundamentals and applications*, ACS symposium series 1342, vol. 1342, 1–25. American Chemical Society. https://doi.org/10.1021/bk-2020-1342.ch001.
Esposito, Susanna, and Maria Roberta Longo. 2017. Guillain–Barré syndrome. *Autoimmunity Reviews* 16 (1): 96–101.
Feinberg, Andrew P. 2018. The key role of epigenetics in human disease prevention and mitigation. *New England Journal of Medicine* 378 (14): 1323–1334.
Few, Stephen. 2006. *Information dashboard design: The effective visual communication of data.* O'Reilly Media, Inc.
Fraisl, Dilek, Gerid Hager, Baptiste Bedessem, Margaret Gold, Pen-Yuan Hsing, Finn Danielsen, Colleen B. Hitchcock, et al. 2022. Citizen science in environmental and ecological sciences. *Nature Reviews Methods Primers* 2 (1): 1–20. https://doi.org/10.1038/s43586-022-00144-4.
Gehrke, Ilka, Andreas Geiser, and Annette Somborn-Schulz. 2015. Innovations in nanotechnology for water treatment. *Nanotechnology, Science and Applications* 8 (January): 1–17. https://doi.org/10.2147/NSA.S43773.
Goyer, Robert A. 1997. Toxic and essential metal interactions. *Annual Review of Nutrition* 17 (1): 37–50.
He, Zhen, and Largus T. Angenent. 2006. Application of bacterial biocathodes in microbial fuel cells. *Electroanalysis: An International Journal Devoted to Fundamental and Practical Aspects of Electroanalysis* 18 (19–20): 2009–2015.
Hooper, Laura G., and Joel D. Kaufman. 2018. Ambient air pollution and clinical implications for susceptible populations. *Annals of the American Thoracic Society* 15 (Suppl 2): S64–S68. https://doi.org/10.1513/AnnalsATS.201707-574MG.

Iravani, Siavash. 2021. Nanomaterials and nanotechnology for water treatment: Recent advances. *Inorganic and Nano-Metal Chemistry* 51 (12): 1615–1645.

Jiang, Xu-Qin, Xiao-Dong Mei, and Di Feng. 2016a. Air pollution and chronic airway diseases: What should people know and do? *Journal of Thoracic Disease* 8 (1): E31–E40. https://doi.org/10.3978/j.issn.2072-1439.2015.11.50.

Jiang, Qijun, Frank Kresin, Arnold K. Bregt, Lammert Kooistra, Emma Pareschi, Edith van Putten, Hester Volten, and Joost Wesseling. 2016b. Citizen sensing for improved urban environmental monitoring. *Journal of Sensors* 2016 (May): e5656245. https://doi.org/10.1155/2016/5656245.

Joubert, Bonnie R., Siri E. Håberg, Roy M. Nilsen, Xuting Wang, Stein E. Vollset, Susan K. Murphy, Zhiqing Huang, et al. 2012. 450K epigenome-wide scan identifies differential DNA methylation in newborns related to maternal smoking during pregnancy. *Environmental Health Perspectives* 120 (10): 1425–1431.

Kato, Souichiro. 2015. Biotechnological aspects of microbial extracellular electron transfer. *Microbes and Environments* 30 (2): 133–139.

Kelly, Frank J. 2003. Oxidative stress: Its role in air pollution and adverse health effects. *Occupational and Environmental Medicine* 60 (8): 612–616.

Kim, Sun-Young, Esther Kim, and Woo Jin Kim. 2020. Health effects of ozone on respiratory diseases. *Tuberculosis Respiratory Disease* 83 (Supple 1): S6–S11. https://doi.org/10.4046/trd.2020.0154.

Kotloff, Karen L., James P. Nataro, William C. Blackwelder, Dilruba Nasrin, Tamer H. Farag, Sandra Panchalingam, Wu Yukun, et al. 2013. Burden and aetiology of diarrhoeal disease in infants and young children in developing countries (the Global Enteric Multicenter Study, GEMS): A prospective, case-control study. *Lancet (London, England)* 382 (9888): 209–222. https://doi.org/10.1016/S0140-6736(13)60844-2.

Kuerban, Mireadili, Yizaitiguli Waili, Fan Fan, Ye Liu, Wei Qin, Anthony J. Dore, Jingjing Peng, Xu Wen, and Fusuo Zhang. 2020. Spatio-temporal patterns of air pollution in China from 2015 to 2018 and implications for health risks. *Environmental Pollution* 258: 113659.

Kweka, Eliningaya J., Vito Baraka, Leah Mathias, Beda Mwang'onde, Germana Baraka, Lucile Lyaruu, and Aneth M. Mahande. 2018. Ecology of Aedes mosquitoes, the major vectors of arboviruses in human population. In *Dengue fever – A resilient threat face innovation*. IntechOpen.

Landrigan, Philip J., Richard Fuller, Nereus J.R. Acosta, Olusoji Adeyi, Robert Arnold, Niladri Nil Basu, Abdoulaye Bibi Baldé, et al. 2018. The lancet commission on pollution and health. *Lancet (London, England)* 391 (10119): 462–512. https://doi.org/10.1016/S0140-6736(17)32345-0.

Laumbach, Robert J., and Howard M. Kipen. 2012. Respiratory health effects of air pollution: Update on biomass smoke and traffic pollution. *Journal of Allergy and Clinical Immunology* 129 (1): 3–11.

Li, Ning, Constantinos Sioutas, Arthur Cho, Debra Schmitz, Chandan Misra, Joan Sempf, Meiying Wang, Terry Oberley, John Froines, and Andre Nel. 2003. Ultrafine particulate pollutants induce oxidative stress and mitochondrial damage. *Environmental Health Perspectives* 111 (4): 455–460.

Li, Lele, Du Tiantian, and Chi Zhang. 2020. The impact of air pollution on healthcare expenditure for respiratory diseases: Evidence from the People's Republic of China. *Risk Management and Healthcare Policy* 13 (September): 1723–1738. https://doi.org/10.2147/RMHP.S270587.

Lim, Yu Jie, Kunli Goh, Masaru Kurihara, and Rong Wang. 2021. Seawater desalination by reverse osmosis: Current development and future challenges in membrane fabrication – A review. *Journal of Membrane Science* 629: 119292.

Lovley, Derek R. 2006. Microbial fuel cells: Novel microbial physiologies and engineering approaches. *Current Opinion in Biotechnology* 17 (3): 327–332.

Mac Kenzie, W.R., N.J. Hoxie, M.E. Proctor, M.S. Gradus, K.A. Blair, D.E. Peterson, J.J. Kazmierczak, D.G. Addiss, K.R. Fox, and J.B. Rose. 1994. A massive outbreak in Milwaukee of cryptosporidium infection transmitted through the public water supply. *The New England Journal of Medicine* 331 (3): 161–167. https://doi.org/10.1056/NEJM199407213310304.

Madrigano, Jaime, Andrea Baccarelli, Murray A. Mittleman, David Sparrow, I.I.I. Avron Spiro, Pantel S. Vokonas, Laura Cantone, Laura Kubzansky, and Joel Schwartz. 2012. Air pollution and DNA methylation: Interaction by psychological factors in the VA normative aging study. *American Journal of Epidemiology* 176 (3): 224–232.

Messina, Jane P., Oliver J. Brady, Thomas W. Scott, Chenting Zou, David M. Pigott, Kirsten A. Duda, Samir Bhatt, et al. 2014. Global spread of dengue virus types: Mapping the 70 year history. *Trends in Microbiology* 22 (3): 138–146. https://doi.org/10.1016/j.tim.2013.12.011.

National Cancer Institute. 2017. *Asbestos exposure and cancer risk fact sheet – NCI*. CgvArticle. Nciglobalncienterprise. https://www.cancer.gov/about-cancer/causes-prevention/risk/substances/asbestos/asbestos-fact-sheet.

Nieuwenhuijsen, Mark J., David Donaire-Gonzalez, Maria Foraster, David Martinez, and Andres Cisneros. 2014. Using personal sensors to assess the exposome and acute health effects. *International Journal of Environmental Research and Public Health* 11 (8): 7805–7819. https://doi.org/10.3390/ijerph110807805.

Perera, Frederica. 2018. Pollution from fossil-fuel combustion is the leading environmental threat to global pediatric health and equity: Solutions exist. *International Journal of Environmental Research and Public Health* 15 (1): 16. https://doi.org/10.3390/ijerph15010016.

Pohjola, Mikko V., and Jouni T. Tuomisto. 2011. Openness in participation, assessment, and policy making upon issues of environment and environmental health: A review of literature and recent project results. *Environmental Health* 10 (1): 58. https://doi.org/10.1186/1476-069X-10-58.

Reynoso, Glennys V., David N. Gordon, Anurag Kalia, Cynthia C. Aguilar, Courtney S. Malo, Maya Aleshnick, Kimberly A. Dowd, et al. 2023. Zika virus spreads through infection of lymph node-resident macrophages. *Cell Reports* 42 (2): 112126.

Rojas, Elizabeth, Viviana Bastidas, and Christian Cabrera. 2020. Cities-board: A framework to automate the development of smart cities dashboards. *IEEE Internet of Things Journal* 7 (10): 10128–10136.

Russell, Milton, and Michael Gruber. 1987. Risk assessment in environmental policy-making. *Science* 236 (4799): 286–290. https://doi.org/10.1126/science.3563508.

Ryan, Sadie J., Colin J. Carlson, Erin A. Mordecai, and Leah R. Johnson. 2019. Global expansion and redistribution of Aedes-borne virus transmission risk with climate change. *PLoS Neglected Tropical Diseases* 13 (3): e0007213. https://doi.org/10.1371/journal.pntd.0007213.

Steinle, Susanne, Stefan Reis, Clive E. Sabel, Sean Semple, Marsailidh M. Twigg, Christine F. Braban, Sarah R. Leeson, et al. 2015. Personal exposure monitoring of PM2. 5 in indoor and outdoor microenvironments. *Science of the Total Environment* 508: 383–394.

Tim, U.S. 1995. The application of GIS in environmental health sciences: Opportunities and limitations. *Environmental Research* 71 (2): 75–88. https://doi.org/10.1006/enrs.1995.1069.

Troeger, Christopher, Mohammad Forouzanfar, Puja C. Rao, Ibrahim Khalil, Alexandria Brown, Robert C. Reiner, Nancy Fullman, et al. 2017. Estimates of global, regional, and national morbidity, mortality, and aetiologies of diarrhoeal diseases: A systematic analysis for the global burden of disease study 2015. *The Lancet Infectious Diseases* 17 (9): 909–948. https://doi.org/10.1016/S1473-3099(17)30276-1.

Tyler, Christina R., and Andrea M. Allan. 2014. The effects of arsenic exposure on neurological and cognitive dysfunction in human and rodent studies: A review. *Current Environmental Health Reports* 1: 132–147.

US Census Bureau. 2022. *U.S. Census Bureau QuickFacts: Cheverly Town, Maryland*. https://www.census.gov/quickfacts/fact/table/cheverlytownmaryland/PST045222.

US EPA. 2013. *Laws & regulations*. Collections and Lists, January 31. https://www.epa.gov/laws-regulations.

———. 2014. *Particle pollution and respiratory effects*. Collections and Lists, September 15. https://www.epa.gov/pmcourse/particle-pollution-and-respiratory-effects.

———. 2016. *Basic information about NO2*. Overviews and Factsheets, July 6. https://www.epa.gov/no2-pollution/basic-information-about-no2.

van Velsen, Lex, Geke Ludden, and Christiane Grünloh. 2022. The limitations of user-and human-centered design in an EHealth context and how to move beyond them. *Journal of Medical Internet Research* 24 (10): e37341. https://doi.org/10.2196/37341.

Vandenberg, Laura N., Marlene Ågerstrand, Anna Beronius, Claire Beausoleil, Åke Bergman, Lisa A. Bero, Carl-Gustaf Bornehag, et al. 2016. A proposed framework for the systematic review and integrated assessment (SYRINA) of endocrine disrupting chemicals. *Environmental Health* 15: 1–19.

Vayena, Effy, Alessandro Blasimme, and I. Glenn Cohen. 2018. Machine learning in medicine: Addressing ethical challenges. *PLoS Medicine* 15 (11): e1002689. https://doi.org/10.1371/journal.pmed.1002689.

Wani, Ab Latif, Anjum Ara, and Jawed Ahmad Usmani. 2015. Lead toxicity: A review. *Interdisciplinary Toxicology* 8 (2): 55–64. https://doi.org/10.1515/intox-2015-0009.

WHO. 2011. *Burden of disease from environmental noise: Quantification of healthy life years lost in Europe*, 2011. World Health Organization. Regional Office for Europe.

———. 2017. *Guidelines for drinking-water quality, 4th edition, incorporating the 1st addendum.* https://www.who.int/publications-detail-redirect/9789241549950.

———. 2019. *Diarrhoeal disease*. https://www.who.int/news-room/fact-sheets/detail/diarrhoeal-disease

———. 2020. *Fact Sheet about Malaria*. https://www.who.int/news-room/fact-sheets/detail/malaria.

———. n.d.-a. *Types of pollutants*. https://www.who.int/teams/environment-climate-change-and-health/air-quality-and-health/health-impacts/types-of-pollutants. Accessed 24 May 2023.

———. n.d.-b. *Household air pollution*. https://www.who.int/news-room/fact-sheets/detail/household-air-pollution-and-health. Accessed 24 May 2023.

Wild, Christopher Paul. 2012. The exposome: From concept to utility. *International Journal of Epidemiology* 41 (1): 24–32.

Xing, Yu-Fei, Yue-Hua Xu, Min-Hua Shi, and Yi-Xin Lian. 2016. The Impact of PM2.5 on the Human Respiratory System. *Journal of Thoracic Disease* 8 (1): E69–74. https://doi.org/10.3978/j.issn.2072-1439.2016.01.19.

Yoon, Yeo-Chang, Kyoung-Seok Park, and Sam-Dong Kim. 2015. Effects of low preheating temperature for ZnO seed layer deposited by sol–gel spin coating on the structural properties of hydrothermal ZnO nanorods. *Thin Solid Films* 597: 125–130.

Zhang, Yan, Wenjing Tang, Jing Bai, Jinhua Li, Jiachen Wang, Tingsheng Zhou, Xiaohong Guan, and Baoxue Zhou. 2022. Highly efficient removal of total nitrogen and dissolved organic compound in waste reverse osmosis concentrate mediated by chlorine radical on 3D Co3O4 nanowires anode. *Journal of Hazardous Materials* 424: 127662.

Chapter 12
Actionable AI for Climate and Environment

Ziheng Sun

Contents

1 Introduction

AI has gained immense popularity and showcased its power across various domains and societies. Its successful applications have revolutionized industries, transformed the way we live, and enabled groundbreaking advancements (Marr 2019). Within the realm of healthcare, AI finds utility in early disease detection, crafting personalized treatment regimens, and facilitating drug discovery (Johnson et al. 2021). In the

Z. Sun (✉)
Center for Spatial Information Science and Systems, Department of Geography and Geoinformation Science, George Mason University, Fairfax, VA, USA
e-mail: zsun@gmu.edu

Z. Sun (ed.), *Actionable Science of Global Environment Change*,
https://doi.org/10.1007/978-3-031-41758-0_12

financial sector, AI algorithms are instrumental for tasks such as fraud detection, risk evaluation, and algorithmic trading (Nuti et al. 2011). In the domain of transportation, AI fuels the capabilities of autonomous vehicles, optimizes traffic management systems, and enhances predictive maintenance (Iyer 2021). AI has also made significant contributions to entertainment, with recommendation systems, virtual assistants, and immersive experiences (Ali et al. 2022). Moreover, AI has also made an impact in addressing societal challenges. For example, AI-based platforms facilitate personalized learning experiences and intelligent tutoring (Chaipidech et al. 2022). In agriculture, AI facilitates precision farming, monitors crop health (Sun et al. 2020), and optimizes crop yields (Sharma et al. 2022). Additionally, AI algorithms are deployed in the realm of cybersecurity to swiftly identify and mitigate real-time threats (Ali et al. 2022). Besides, maybe the most popular application, ChatGPT (Biswas 2023), is based on the progress of AI in natural language processing, machine translation, and speech recognition, improving communication and accessibility (Hirschberg and Manning 2015). AI has transformed our daily lives through virtual assistants like ChatGPT, Siri and Alexa, smart home automation systems, and personalized digital experiences. It has made significant strides in computer vision, enabling facial recognition, object detection, and augmented reality applications. The popularity and power of AI are reflected in its integration into our everyday devices and services.

AI is considered powerful and highly desired in climate and environment science. Climate and environmental research generate vast amounts of complex data from various sources such as satellite imagery, weather stations, and sensor networks (Sun et al. 2019). AI techniques, such as machine learning (ML) and deep learning, excel at processing and analyzing large datasets, identifying patterns, and extracting valuable insights (Janiesch et al. 2021). AI can help researchers uncover hidden relationships and correlations, enabling a deeper understanding of climate dynamics, ecosystem behavior, and environmental impacts. Another major reason is that AI algorithms can be trained on historical climate and environmental data to develop sophisticated models for predicting future scenarios (Sun et al. 2022, 2023). These models can simulate climate change impacts, forecast extreme weather events, predict species distribution shifts, and assess the effectiveness of mitigation strategies. AI-based prediction models provide decision-makers with valuable information to plan and implement adaptive measures to minimize risks and protect vulnerable ecosystems (Barzegar et al. 2018). Also, AI algorithms can optimize resource allocation and management in climate and environmental domains. They can assist in designing efficient energy systems, optimizing water resource allocation (Sun and Scanlon 2019), and managing waste and pollution. AI techniques enable real-time monitoring, data-driven decision-making, and automated control systems, leading to more sustainable and environmentally friendly practices (Cunningham 2021). AI-powered image and pattern recognition can aid in biodiversity conservation efforts. Meanwhile, AI can analyze images from remote sensing devices, cameras, or drones to identify and monitor endangered species, detect illegal logging or poaching activities, and assess the health of ecosystems (Dauvergne 2020). This helps researchers and conservationists make informed decisions and implement targeted conservation strategies. AI can handle complex tasks with speed and efficiency, allowing for scalable and automated processes. They can process large datasets and perform repetitive tasks more quickly than human experts, saving time and resources. Besides, AI also can facilitate real-time monitoring and decision-making, facilitating rapid response to climate events and environmental emergencies (Chowdhury et al. 2012).

One recent successful example is the use of AI in deforestation monitoring (Shivaprakash et al. 2022). Deforestation is a major environmental issue that contributes to climate change, loss of biodiversity, and other ecological imbalances. AI can detect deforestation by analyzing satellite imagery and other data sources to identify areas at risk and monitor changes in forest cover. Global Forest Watch, a partnership led by the World Resources Institute, utilizes AI algorithms to analyze satellite data and identify forest cover changes in near real-time (Perbet et al. 2019). This information helps governments, organizations, and local communities to take proactive measures to prevent further deforestation by sending enforcement teams to the identified locations, imposing penalties on illegal activities, and engaging local communities in sustainable land management practices. By harnessing the power of AI and ML, we can improve the efficiency and effectiveness of deforestation monitoring and prevention efforts, leading to better conservation outcomes and the preservation of valuable ecosystems.

While AI research has made significant advancements in various fields, there are several reasons why many AI research outputs are not always practical or actionable in real-world decision-making due to a number of restrictions and bottlenecks. First, AI researchers usually lack deep understanding and expertise in specific domains like climate and environment science. This can lead to a disconnect between the AI models developed and the practical needs of decision-makers. For example, an AI model trained to predict climate patterns may produce accurate results, but if it fails to consider the specific needs and constraints of stakeholders, it may not provide actionable insights. In addition, the current AI models heavily rely on data for training and inference. In the context of climate and environment, data may be limited, incomplete, or biased, leading to inaccurate or unreliable predictions. If an AI model is trained on historical climate data that does not reflect recent changes or emerging patterns, its predictions may not be applicable to the current climate scenario. Also, ethical and societal aspects must be comprehensively considered and dealt with before using AI to make any decisions. In climate and environment, decisions often involve trade-offs and value judgments. For instance, an AI model that recommends land-use changes for carbon sequestration may not account for the socioeconomic impacts on local communities or indigenous rights. This lack of ethical considerations can hinder the practicality and acceptability of AI research outputs. Another major challenge is that many AI models, such as deep neural networks, are often considered black boxes, making it challenging to understand and interpret their decision-making processes. This lack of interpretability raises concerns about accountability and trust. Decision-makers may be reluctant to adopt AI solutions if they cannot understand how the models arrive at their recommendations. Addressing these challenges requires close collaboration between AI researchers and domain experts in climate and environment science, policy, and decision-making. By incorporating domain-specific knowledge, ensuring diverse and representative datasets, addressing ethical considerations, and developing interpretable AI models, researchers can bridge the gap between AI research and practical decision-making in climate and environment.

This section basically sets the stage and we will start to explore the role of AI in climate and environmental applications. It highlights the increasing popularity and power of AI technologies in various sectors of society, including climate and environment. It acknowledges the successful applications of AI in areas such as weather forecasting, renewable energy optimization, and biodiversity conservation. However, it also acknowledges that many AI research outputs in this field are not always practical or actionable in real-world decision-making. In the following sections, we will delve into the reasons behind the gap between AI research and practical decision-making in climate and environment. It will discuss the challenges posed by the lack of domain-specific knowledge, data limitations and biases, ethical considerations, and interpretability of AI models. The focus will be on providing insights and strategies to make AI research more actionable and applicable in real-world climate and environmental decision-making. The cutting-edge AI technologies will be introduced, showcasing their potential in addressing climate and environmental challenges. Use cases will be discussed, illustrating how AI has been utilized in areas such as climate modeling, natural disaster prediction and management, environmental monitoring, and sustainable resource management. The chapter will emphasize the need for actionable AI strategies that incorporate domain expertise, ethical considerations, interpretable models, and stakeholder engagement. We will also provide detailed analyses, case studies, and practical recommendations to bridge the gap between AI research and real-world decision-making. It aims to guide researchers, practitioners, and policymakers in harnessing the power of AI to tackle climate and environmental issues effectively and implement actionable solutions.

2 Latest AI Technologies in Daily Practice

2.1 Convolutional Neural Network (CNN)

Convolutional neural networks (CNNs) are a class of deep learning models commonly used for computer vision tasks such as image classification, object detection, and image segmentation (O'Shea et al. 2015). They consist of multiple layers of interconnected neurons that perform convolution and pooling operations to extract relevant features from images. Specifically, a typical CNN usually consists of input layers, hidden convolutional layers, activation and pooling layers, fully connected layers, and a dense layer as output (Fig. 12.1). The input to a CNN is usually an image represented as a grid of pixels with red, green, and blue (RGB) color channels (or more channels if the image is hyperspectral or multi-spectral). The middle

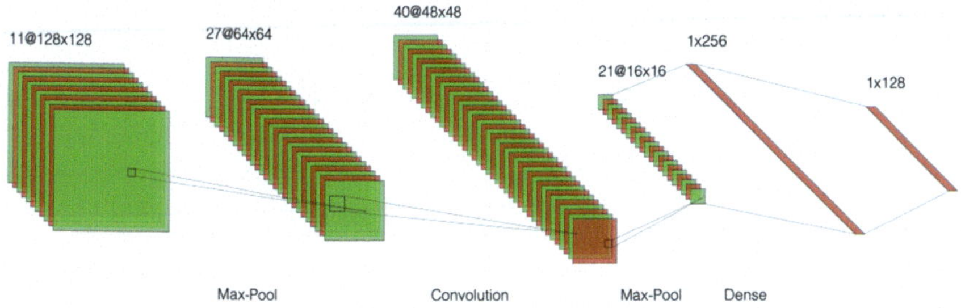

Fig. 12.1 An example CNN

layers of a CNN are convolutional layers that apply a set of filters to the input image. Each filter performs a convolution operation by sliding across the image, extracting local features by computing dot products between the filter weights and the pixel values in the receptive field. The output of this layer is a set of feature maps that capture different aspects of the input image. After each convolutional layer, a non-linear activation function, typically ReLU (Rectified Linear Unit) (Agarap 2018), is applied element-wise to introduce non-linearity and enhance the model's representational power. Pooling layers are then used to downsample the feature maps, reducing their spatial dimensions while retaining important information (Gholamalinezhad et al. 2020). Max pooling is a common pooling operation that selects the maximum value within a pooling window and discards the rest. Once the image features are extracted through convolutional and pooling layers, they are flattened into a 1-dimensional vector. The flattened vector is then connected to one or more fully connected layers, which are traditional artificial neural network layers where each neuron is connected to every neuron in the previous layer. The fully connected layers learn to combine the extracted features to make predictions on the input image, such as classifying it into specific categories. The final layer of the CNN is the output layer, which typically uses a softmax activation function for multi-class classification to produce class probabilities.

During training, the network learns the optimal weights for the filters and fully connected layers by minimizing a loss function, such as categorical cross-entropy, using gradient descent optimization algorithms like backpropagation. Once trained, the CNN can make predictions on new unseen images by forwarding them through the network, and the output with the highest probability corresponds to the predicted class.

CNNs have been successfully used in applications such as image classification, object detection, facial recognition, and medical image analysis. One of the most famous CNN architectures is the VGGNet (Wang et al. 2015), which achieved breakthrough performance in the ImageNet Large-Scale Visual Recognition Challenge (Russakovsky et al. 2015). The VGGNet consists of 16 convolutional layers, 5 max pooling layers, and 3 fully connected layers, with a total of 138 million parameters. It demonstrated the power of CNNs in image classification tasks by achieving state-of-the-art accuracy rates.

2.2 RNN

Recurrent neural networks (RNNs) are a class of neural networks commonly used for sequential data processing tasks (Sun et al. 2019). They have a unique ability to capture dependencies and patterns over time by using recurrent connections within the network. RNNs are particularly effective in tasks involving natural language processing and time series analysis.

Typical RNN include the following components. The input to an RNN is a sequence of data, such as a sentence or a time series. At each time step t, the RNN receives an input x(t) and a hidden state h(t-1) from the previous time step, which captures information from previous steps. At the initial step, h(0) is usually set to zero or initialized randomly. The hidden state h(t-1) is combined with the input x(t) and passed through a non-linear activation function, such as the hyperbolic tangent or the rectified linear unit (ReLU). The output of this activation function becomes the hidden state h(t) at the current time step. It represents a summary of the input sequence up to that point. The hidden state h(t) is then used as the input for the next time step, creating a recurrent connection that allows the RNN to process the sequence iteratively. At each time step, the RNN can produce an output based on the hidden state h(t). For example, in a language model, the output could be a probability distribution over the next word in the sequence. In sequence-to-sequence tasks, such as machine translation, the RNN can produce an output sequence by feeding the output at each time step as the input for the next step. While training, the RNN learns the optimal weights that maximize its predictive performance. This is done by comparing the predicted output with the ground truth labels and adjusting the weights using gradient descent optimization. Backpropagation through time (BPTT) is normally used to calculate the gradients of the loss function with respect to the weights over multiple time steps. It extends the standard backpropagation algorithm to account for the recurrence in the network. The gradients are then used to update the weights using an optimization algorithm such as stochastic gradient descent (SGD).

RNNs have been very successfully used in many applications, including language modeling, sentiment analysis, speech recognition, and machine translation. One popular variant of RNNs is the long short-term memory (LSTM) network (Hochreiter et al. 1997), which addresses the issue of vanishing gradients and allows the network to capture long-term dependencies. LSTM has achieved impressive results in various tasks, such as language translation and speech recognition, and is one of the industry-proven techniques. LSTM networks have been successfully used for language modeling tasks, where the goal is to predict the next word in a sequence of words. A prominent example is Google's Smart Reply feature, which suggests short responses to incoming emails. LSTM models are employed to understand the context and generate relevant replies. LSTM-based language models have also been applied in machine translation systems, improving the accuracy and fluency of generated translations. An example is the listen, attend, and spell (LAS) model (Chan et al. 2016), which uses LSTMs to convert acoustic features of speech into text. LAS has shown remarkable results in automatic speech recognition systems, enhancing transcription accuracy. Also in finance, LSTM models have been used for stock price prediction, enabling traders to make informed decisions. As for art and music, LSTM networks have been used to generate music sequences and compose new melodies. By training on large music datasets, LSTM models can learn musical patterns and create original compositions in various genres. This has led to the development of AI-generated music platforms and tools, such as Jukedeck and Amper Music.

2.3 *Transformers*

Transformers are a type of neural network architecture that has revolutionized natural language processing tasks (Tunstall et al. 2022). They use attention mechanisms to process sequences of data, such as sentences or paragraphs, by attending to different parts of the input. The self-attention mechanism captures dependencies between different words in a sentence or sequence. Each word in the input sequence is represented as a vector, and attention weights are calculated between all pairs of words. These attention weights determine the importance of each word in relation to the others, allowing the model to focus on relevant information. Transformers also consist of an encoder and a decoder. The encoder processes the input sequence, while the decoder generates the output sequence. The encoder's self-attention mechanism captures contextual information from the input sequence, creating rich representations for each word. The decoder's self-attention mechanism helps it attend to previously generated words, ensuring coherence in the generated output. Transformers incorporate positional encoding to account for the sequential order of words in the input sequence. Positional encodings are added to the word embeddings, providing the model with information about the relative positions of words. Transformers employ multi-head attention, where multiple attention heads are used to capture different aspects of the input sequence. Each attention head attends to different parts of the input sequence, allowing the model to capture diverse relationships. Transformers include feed-forward networks to transform the representations obtained from the self-attention mechanism. These networks consist of multiple layers of fully connected neural networks, introducing non-linearity and enabling complex transformations.

ChatGPT is probably the most well-known product of Transformers. It has been trained on a large corpus of text data and can generate coherent and contextually relevant responses in conversational settings. ChatGPT has been used in chatbots, virtual assistants, and other applications that require generating human-like text responses. Transformers also have significantly improved machine translation performance, surpassing traditional approaches. Google's Neural Machine Translation (GNMT) system (Wu et al. 2016) utilizes the Transformer model for high-quality translation between different languages. Transformers have shown effectiveness in capturing long-range dependencies and context, leading to more accurate translations. Other applications include extractive and abstractive text summarization tasks which means selecting important sentences from a document, while abstractive summarization generates concise and coherent summaries. Transformers have shown improvements in generating informative and coherent summaries by leveraging the attention mechanism. Sentiment analysis tasks are another type of work that Transformers can fulfill. They can capture contextual information and dependencies between words, improving sentiment analysis accuracy. Promising results for social media analysis, customer reviews, and other text classification tasks have been successfully obtained via transformers.

2.4 *Reinforcement Learning*

Reinforcement learning (RL) is a branch of AI that focuses on an agent learning to make decisions in an environment to maximize a reward signal (Arulkumaran et al. 2017). It uses trial and error learning through interactions with the environment. The RL process starts with defining an environment that the agent interacts with. The environment can be a simulated environment, a physical system, or a game. It provides feedback to the agent in the form of states, actions, and rewards. The agent is the learner or decision-maker that interacts with the environment. The environment presents the agent with a state, which represents the current situation or observation. The state can be a raw sensory input, a numerical representation, or a combination of various features. Based on the current state, the agent selects an action from a set of available actions. The action determines the agent's behavior or response to the environment. The agent follows a policy, which is a strategy that maps states to actions. After taking an action, the agent receives feedback from the environment in the form of a reward. The reward indicates the desirability or quality of the agent's action. The agent's objective is to learn a policy that maximizes the cumulative reward over time. The agent learns from experience by iteratively interacting with the environment. It updates its policy based on the received rewards to improve its decision-making. The RL algorithms, such as Q-learning and Deep Q-Networks (DQN) (Hester et al. 2018), use different approaches to update the policy and estimate the value of actions.

Reinforcement learning has found success in various applications, including game playing (e.g., AlphaGo), robotics, autonomous vehicles, and recommendation systems. ChatGPT actually used RL in its training extensively. AlphaGo, developed by DeepMind, demonstrated the power of RL in the game of Go (Silver et al. 2016). It defeated the world champion Go player, Lee Sedol, in a five-game match in 2016. AlphaGo used RL techniques, including Monte Carlo Tree Search and deep neural networks, to learn from self-play and make strategic decisions in the game. Another big application field of RL is robotics and RL has been employed to train robots for complex tasks, such as grasping objects and locomotion. For example, OpenAI's robot hand, Dactyl (Akkaya et al. 2019), learned to manipulate objects using RL algorithms. By interacting with the environment and receiving rewards or penalties based on task completion, the robot learned to perform dexterous manipulation tasks. Autonomous vehicles are another major user of RL and it helped the vehicle computers to make decisions in complex driving scenarios. Researchers have applied RL algorithms to optimize driving policies, including lane keeping, decision-making at intersections, and adaptive cruise control. The RL agents learn from simulated or real-world driving experiences and improve their driving performance over time. Another typical use case is training an agent to play the game of Atari Breakout. The agent observes the game screen as the state, selects actions (move paddle left or right), and receives rewards based on its performance (e.g., points for breaking bricks). By interacting with the game environment and receiving rewards, the agent learns to improve its policy and eventually becomes skilled at playing the game.

2.5 *Generative Adversarial Networks (GANs)*

The idea of GANs is highly ingenious (Goodfellow et al. 2020). It consists of two neural networks, a generator and a discriminator, that compete against each other. The generator aims to create realistic data samples, while the discriminator tries to distinguish between real and generated data. The generator takes random noise as input and generates synthetic data samples. The discriminator takes either real or generated data samples as input and predicts their authenticity. The generator generates a batch of synthetic samples by passing random noise through its network. The discriminator is trained on both real and generated samples, learning to classify them correctly. The generator aims to generate samples that are classified as real by the discriminator, fooling it. The discriminator aims to correctly classify real samples as real and generated samples as fake. The training process involves updating the weights of the generator and discriminator using gradient descent. The generator and discriminator are trained iteratively, with the generator trying to improve its generated samples based on the feedback from the discriminator. The goal is to reach a point where the generator can generate highly realistic samples that can fool the discriminator. Deep convolutional GANs (DCGANs) (Radford et al. 2015) are a popular variant of GANs used for image generation. The generator network consists of transposed convolutions that upsample the noise into a realistic image. The discriminator network is a convolutional neural network that classifies between real and generated images. The GAN is trained on a dataset of real images, and the generator learns to generate images that resemble the real ones. Table 12.1 shows the code of a simple version of GAN. If you continue to increase the layer numbers of hidden layers, it will eventually turn into a new DCGAN.

GANs have been successfully used in generating realistic images, synthesizing voice and music, creating deepfakes, and data augmentation for training other models. GANs have been used to generate realistic images that resemble real-world examples. Examples include generating high-resolution images from low-resolution inputs (e.g., super-resolution GANs) and generating new images based on existing

Table 12.1 A simple example of GAN

```
# Import required libraries
import tensorflow as tf
from tensorflow.keras import layers

# Define the generator model
generator = tf.keras.Sequential([
    layers.Dense(7*7*256, input_shape=(100,), use_bias=False),
    layers.BatchNormalization(),
    layers.LeakyReLU(),

    layers.Reshape((7, 7, 256)),
    layers.Conv2DTranspose(128, (5, 5), strides=(1, 1), padding='same', use_bias=False),
    layers.BatchNormalization(),
    layers.LeakyReLU(),

    layers.Conv2DTranspose(64, (5, 5), strides=(2, 2), padding='same', use_bias=False),
    layers.BatchNormalization(),
    layers.LeakyReLU(),

    layers.Conv2DTranspose(1, (5, 5), strides=(2, 2), padding='same', use_bias=False,
activation='tanh')
])

# Define the discriminator model
discriminator = tf.keras.Sequential([
    layers.Conv2D(64, (5, 5), strides=(2, 2), padding='same', input_shape=[28, 28, 1]),
    layers.LeakyReLU(),
    layers.Dropout(0.3),

    layers.Conv2D(128, (5, 5), strides=(2, 2), padding='same'),
    layers.LeakyReLU(),
    layers.Dropout(0.3),

    layers.Flatten(),
    layers.Dense(1)
])

# Compile the discriminator
discriminator.compile(optimizer='adam',
loss=tf.keras.losses.BinaryCrossentropy(from_logits=True))

# Compile the GAN
gan = tf.keras.Sequential([generator, discriminator])
gan.compile(optimizer='adam',
loss=tf.keras.losses.BinaryCrossentropy(from_logits=True))
```

ones (e.g., Pix2Pix). GANs have been employed for style transfer, allowing users to apply the style of one image to another, for example, CycleGAN (Chu et al. 2017) can transfer the style of one domain (e.g., horses) to another (e.g., zebras), and DeepArt enables users to apply artistic styles to their images. GANs have been used to generate images from textual descriptions. GANs is also being experimented to generate synthetic medical images, aiding in data augmentation and addressing privacy concerns, such as generating realistic brain MRI scans, retinal images, and skin lesion images, etc., for image enhancement.

3 AI Research in Climate and Environmental Sciences

3.1 AI for Climate Modeling and Prediction and Impact Assessment

This section overviewed some latest research about using AI in climate modeling and prediction, as well as the climate impact assessment. For example, Kaack et al. (2022) provide a comprehensive framework for understanding the impacts of ML on greenhouse gas (GHG) emissions in the context of climate change mitigation. It emphasizes the need for further research, policy interventions, and organizational actions to ensure that ML is aligned with climate strategies and contributes positively to addressing climate challenges. They introduce a systematic framework for understanding the effects of ML on GHG emissions in the context of climate change mitigation. The framework encompasses three categories: computing-related impacts, immediate impacts of ML applications, and system-level impacts. It addresses the need to holistically account for ML in long-term climate projections and policy design. The article highlights that measuring macro-scale effects of ML is challenging and emphasizes the importance of estimating impacts, understanding dynamics, and prioritizing actions to align ML with climate strategies. The framework provides a comprehensive overview of the different mechanisms through which ML may impact emissions, offering a starting point for research, policymaking, and organizational action. Allawi et al. (2018) explained the importance of accurate simulation models for the effective operation of dam and reservoir systems in water resource management. It emphasizes the role of AI techniques in developing robust models to handle the stochastic nature of hydrological parameters and optimize reservoir operations. The review explores the application of AI in reservoir inflow forecasting, evaporation prediction, and the integration of AI with optimization methods. It also discusses future research directions and proposes a new mathematical procedure for evaluating the performance of optimization models in terms of reliability, resilience, and vulnerability indices. Haupt et al. (2021) discussed the application of artificial intelligence (AI) in post-processing weather and climate model output. It provides a historical overview and highlights the potential of AI in improving numerical weather prediction (NWP) forecasts and climate projections. The article emphasizes the need for trustworthy and interpretable algorithms, adherence to FAIR data practices, and the development of techniques that leverage our physical knowledge of the atmosphere. It also proposes the creation of a repository for datasets and methods to facilitate testing and intercomparison of AI approaches.

Huntingford et al. (2019) discuss the challenges in climate modeling, including the discrepancies between ESMs and the parameterization of sub-grid processes. ML and AI methods are proposed as potential solutions to reduce inter-ESM uncertainty and improve climate projections. The authors also highlight the importance of advanced algorithms in analyzing the increasing amount of climate-related data collected through satellite monitoring. They emphasized the untapped potential of ML and AI in addressing climate change challenges, advocating for their integration into climate research and adaptation planning processes. Crane-Droesch (2018) presents a ML-based approach to modeling crop yields, specifically focusing on corn yield in the US Midwest. The approach combines a semiparametric variant of a deep neural network, which can capture complex nonlinear relationships, with known parametric structures and unobserved cross-sectional heterogeneity. The results demonstrate that this approach outperforms classical statistical methods and fully nonparametric neural networks in predicting yields of withheld years. The study also reveals that the projected impacts of climate change on corn yield are large but less severe than those projected using traditional statistical methods, with a more optimistic outlook for the warmest regions and scenarios. Schultz et al. (2021) investigated the potential of deep learning (DL) methods in the field of meteorology, specifically for weather forecasting. While there is interest in applying DL techniques to improve weather prediction, the authors argue that fundamental breakthroughs are needed before completely replacing current numerical weather models. They highlight challenges such as the lack of explainability of deep neural networks and the need to incorporate physical constraints into DL approaches. Vo et al. (2023) developed a hybrid model, LSTM-CM, for drought prediction by combining long short-term memory (LSTM) and a climate model (CM). The performance of LSTM-CM is compared to standalone LSTM and the climate prediction model GloSea5 (GS5). LSTM-CM demonstrates improved drought predictions by combining the low bias of LSTM-SA and the physical process simulation ability of GS5, resulting in accurate detection of drought events with reduced uncertainty compared to LSTM-SA and GS5. Regarding the general use of AI in the more broad Earth science, Sun et al. (2022) provide an overview of the current status, technology, use cases, challenges, and opportunities of artificial intelligence (AI) in Earth sciences. Led by NASA Earth Science Data Systems Working Groups and Earth science information partners (ESIP) ML cluster, the study aims to improve accuracy, enhance model intelligence, scale up operations, and reduce costs in various subdomains. The paper covers major spheres in the Earth system, investigates representative AI research in each domain, and discusses the challenges and opportunities for Earth AI practitioners.

These AI studies demonstrate the increasing integration of ML and artificial intelligence techniques in various domains of climate and Earth sciences. These studies emphasize the potential benefits and challenges of applying AI in addressing climate change, improving weather forecasting, optimizing reservoir operations, enhancing drought prediction, and advancing crop yield modeling. The research highlights the need for further investigations, policy interventions, and organizational actions to ensure that AI is aligned with climate strategies and contributes positively to tackling climate challenges. The studies also underline the importance of interpretability, data practices, physical knowledge integration, and the development of reliable and scalable AI algorithms in Earth science applications.

3.2 AI for Environmental Monitoring and Conservation

Lamba et al. (2019) looked into the transformative impact of deep learning in the field of artificial intelligence and its potential applications in environmental conservation. It highlights the ability of deep learning to automate the classification of visual, spatial, and acoustic information, thereby enabling large-scale and real-time environmental monitoring. The article also addresses the challenges of resource requirements and data annotation that can hinder the widespread adoption of deep learning in conservation programs. Yang et al. (2021) developed an autonomous indoor environment management approach for smart homes, aiming to ensure a healthy indoor environment with minimized energy costs. The approach formulates the problem as a Markov decision process and proposes a deep reinforcement learning control strategy to make adaptive control decisions based on current observations, without requiring forecast information. Comparative results demonstrate that the proposed approach achieves improved control performance, reducing average daily energy costs while maintaining optimal indoor air quality and temperature. Borowiec et al. (2022) synthesize 818 deep learning studies and highlight the widespread adoption of deep learning in these disciplines since 2019. They discuss the applications, limitations, and future potential of deep learning in ecology and evolution, emphasizing its role in automated species identification, environmental monitoring, genetic analysis, and more. The review also suggests that deep learning will continue to be integrated into biodiversity monitoring, genetic inference, and training programs in the near future. Tuia et al.'s (2022) review work found that advancements in sensor technologies are revolutionizing data acquisition in animal ecology, offering opportunities for large-scale ecological understanding. However, the current processing approaches struggle to efficiently extract relevant information from the vast amount of data collected. Integrating ML with domain knowledge has the potential to enhance ecological models and create hybrid modeling tools, but interdisciplinary collaboration and training are essential for successful implementation in ecology and conservation research. These technological advancements in data collection can address the limitations of conventional methods, providing insights into wildlife diversity, population dynamics, and conservation needs at various spatial and temporal scales.

Deep learning has the potential to revolutionize environmental conservation by automating the classification of visual, spatial, and acoustic data, enabling large-scale and real-time monitoring. However, challenges such as resource requirements and data annotation need to be addressed for widespread adoption in conservation programs. In the context of smart homes, a deep reinforcement learning approach has been proposed for autonomous indoor environment management, reducing energy costs while maintaining optimal air quality and temperature. The adoption of deep learning in ecology and evolution has rapidly increased since 2019, with applications in species identification, environmental monitoring, genetic analysis, and more, and it is expected to continue integrating into biodiversity monitoring and genetic inference.

3.3 AI for Air Quality Prediction and Monitoring

It has been very popular in the past few years among scientists to use AI to monitor and predict air quality. Because of the lightweight and flexibility of AI models (compared to the heavy numerical models), and the availability of large long-time-series datasets, scientists are getting more used to utilizing ML algorithms to analyze various data sources and predict air quality levels in real-time. The training data is collected from air quality monitoring stations, satellite imagery, weather data, and additional environmental parameters and serves as input for training and validating AI models. For instance, the OpenAQ project (https://openaq.org/) collects global air quality data from various sources and makes it accessible for research and analysis. The collected data is then processed and cleaned to remove outliers, fill in missing values, and standardize the format to ensure data consistency and quality for further analysis. Relevant features are extracted from the collected data to represent different aspects of air quality, such as pollutant concentrations, meteorological conditions, geographical factors, and temporal patterns, which will be used as inputs for the AI models. Common target features include particulate matter (PM) concentrations, temperature, wind speed, and humidity from air quality sensor data. The next step is to train AI models, such as regression models, decision trees, support vector machines, or deep learning models, using the preprocessed data. The models will learn the relationships between the input features and the corresponding air quality levels. The trained models are then evaluated using validation data to assess their performance in predicting air quality. Metrics like mean absolute error, root mean square error, or correlation coefficients are commonly used to measure prediction accuracy. The collected data is usually split into training and validation sets, and evaluating the model's performance on the validation set. Once the models are trained and validated, they can be deployed to make real-time air quality predictions based on the latest input data to enable continuous monitoring and timely alerts for potential air quality issues.

There are many recent air quality studies focusing on AI. For example, Alnuaim et al. (2023) have developed a website using AI technology to improve the real time CMAQ ozone products and provide the public with more accurate and reliable ozone forecasting (Fig. 12.2). Indoor air quality monitoring has gained attention due to the COVID-19 pandemic, as indoor spaces can trap pollutants and potentially contribute to virus transmission. Existing monitoring systems lack predictive capabilities, prompting the development of an IoT-based solution that measures multiple pollutants and predicts air quality using ML algorithms. Mumtaz et al. (2021)

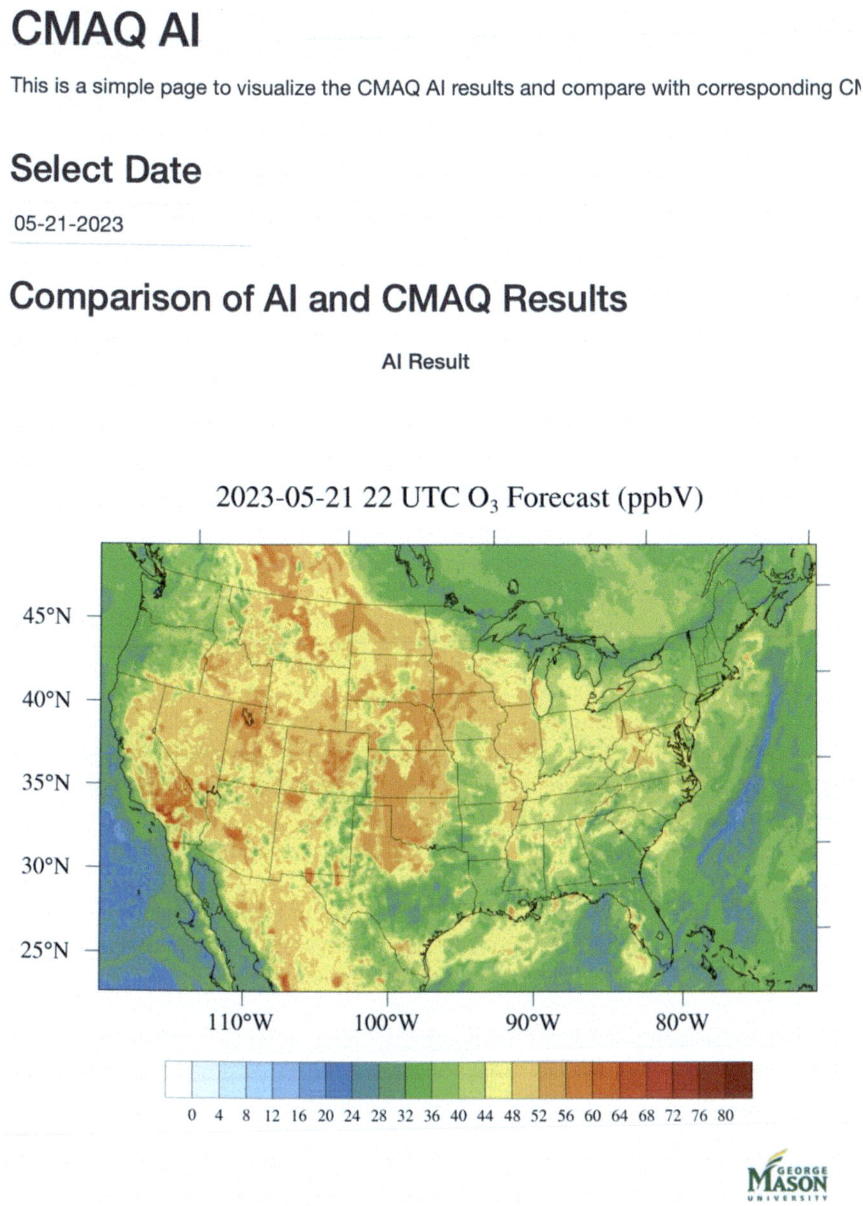

Fig. 12.2 The Community Multiscale Air Quality modeling system (CMAQ) AI website deployed (Alnuaim et al. 2023)

developed a system which can achieve high accuracy in classifying air quality using a neural network model and accurately predicted pollutant concentrations and overall air quality using an LSTM model, offering advantages such as remote monitoring, scalability, and real-time status updates. Kang et al. (2018) explored the use of big data analytics and ML techniques for air quality forecasting. Masih, A. (2019) did a similar study by using support vector regression (SVR) to forecast pollutant and particulate levels and to predict the air quality index (AQI). Vu et al. (2019) used random forest to assess the plan's effectiveness by separating the impact of meteorology on air quality and their results showed that meteorological conditions played a significant role in year-to-year variations in air quality, but the action plan still led to substantial reductions in air pollutants, primarily from coal combustion. Ameer et al. (2019) discussed the challenge of air pollution in city environments and the importance of real-time monitoring using IoT-based sensors. It compares four advanced regression techniques for predicting air quality and evaluates their performance based on Mean Absolute Error (MAE), Root Mean Square Error (RMSE), and processing time. Lee et al. (2020) present a ML approach for predicting PM2.5 concentration in Taiwan and utilized a large-scale database from the Environmental Protection Administration and Central Weather Bureau, incorporating data from 77 air monitoring stations and 580 weather stations. The method shows promising results for 24-hour PM2.5 prediction at most air stations, and proved forecasting accuracy is improved by the method. Lim et al. (2019) used mobile sampling with low-cost air quality sensors to develop land use regression (LUR) models for street-level PM2.5 concentration in Seoul, South Korea. The study collects 169 hours of data from a 3-week campaign using smartphone-based particle counters and incorporates geospatial data from OpenStreetMap. Three statistical approaches are compared, with the stacked ensemble model achieving the highest cross-validation R2 value of 0.80, indicating the potential of mobile sampling and ML for characterizing urban street-level air quality with high spatial resolution, particularly in areas with limited air quality data.

However, besides funding and opportunities, there are many other issues causing these studies to be not immediately actionable. Most of the studies focus on small-scale experiments or specific locations, which may not be easily scalable or replicable in larger areas or diverse contexts. For example, the study by Mumtaz et al. develops an IoT-based system for air quality monitoring and prediction, but its applicability in different indoor environments or regions with varying pollutant sources and characteristics may be limited. They rely on historical data or data collected during specific campaigns, which may not reflect real-time air quality conditions or enable timely interventions. For instance, the research by Ameer et al. (2019) compares regression techniques for air quality prediction but does not address the challenge of real-time monitoring and decision-making. While ML models show promise in predicting air quality, they often neglect external factors that influence pollution levels. For example, Vu et al. assess the effectiveness of an action plan using random forest, but the model's reliance on meteorological conditions alone may overlook other significant contributors to air pollution, such as industrial emissions or traffic patterns. Although some studies explore the effectiveness of air quality improvement measures, the translation of findings into actionable policies or interventions is often overlooked. For instance, while Lee et al. present a ML approach for predicting PM2.5 concentrations, they do not provide concrete recommendations for mitigating pollution or integrating the predictive models into air quality management strategies. Several studies rely on limited data sources or specific geographical locations, which may not represent the complexities and variations of air pollution in different regions or countries. For example, Lim et al.'s study on land use regression models for PM2.5 concentration focuses solely on Seoul, South Korea, making it challenging to generalize the findings to other urban areas with different characteristics and pollutant sources. To ensure the actionable nature of air quality research, it is important to acknowledge early and address these limitations and consider broader factors such as scalability, real-time data availability, policy integration, and generalizability to diverse contexts. Additionally, collaboration between researchers, policymakers, and stakeholders is essential to translate research findings into effective strategies for mitigating air pollution.

3.4 AI for Oceanographic Research

AI has a wide range of applications in oceanography, enabling advancements in various areas such as marine ecosystem monitoring, climate modeling, underwater exploration, and ocean data analysis. AI-powered image recognition algorithms can automate the identification of marine species based on images or video footage. This helps in assessing biodiversity and tracking species distributions. For example, the Fish4Knowledge project (https://homepages.inf.ed.ac.uk/rbf/fish4knowledge/) developed AI algorithms for automated fish species recognition using underwater videos. AI can also predict ocean currents, sea surface temperature anomalies, and harmful algal blooms. For autonomous underwater vehicles (AUVs) and robotics, AI enables them to autonomously navigate, collect data, and perform tasks such as seafloor mapping, oceanographic surveys, and ecosystem monitoring. The REMUS SharkCam (Hawkes et al. 2020), developed by Woods Hole Oceanographic Institution, utilizes AI to track and film sharks in their natural habitats. Ocean acoustic data analysis is another major area for AI to conduct marine mammal detection, underwater noise analysis, and mapping seafloor habitats. On large-scale climate modeling, AI-based models can enhance the accuracy and efficiency of ocean climate and weather predictions by assimilating data from multiple sources. These models can improve storm track predictions, sea surface temperature forecasts, and El Niño/Southern Oscillation (ENSO) predictions. In addition, ocean pollution detection and monitoring is another important application of AI, for example, the DeepSeaVision project developed an AI-based system to detect and track plastic debris in the ocean using satellite images.

In literature, there are waves of AI-related papers published in oceanography journals and conferences. Chen et al. (2019) focused on the remote estimation of surface seawater partial pressure of CO2 (pCO2) in the Gulf of Mexico (GOM) and found that the random forest-based regression ensemble (RFRE) model was the best approach among various modeling techniques. The RFRE model utilized extensive pCO2 datasets collected over 16 years, along with satellite-derived environmental variables such as sea surface temperature, salinity, chlorophyll concentration, and diffuse attenuation of downwelling irradiance. The model has high accuracy, with a root mean square difference (RMSD) of 9.1 μatm, coefficient of determination (R2) of 0.95, and satisfactory performance in both open GOM waters and coastal/river-dominated waters. Niu et al. (2017) used ML algorithms to study source localization in ocean acoustics, including feed-forward neural networks (FNN), support vector machines (SVM), and random forests (RF), to estimate source ranges based on observed acoustic data. Bianco et al. (2019). reviewed that deep learning (DL) has shown promising advancements in acoustics, particularly in tasks such as sound event detection and source localization, can outperform conventional methods, and offer a general framework for acoustics tasks, eliminating the need for specialized algorithms in different subfields. However, a major challenge is the availability of sufficient training data, although synthetic data or data augmentation can help address this limitation. Recent studies have demonstrated the effectiveness of DL architectures, such as convolutional recurrent neural networks (CNNs) and deep residual neural networks (ResNet), in achieving competitive results in sound event detection, direction of arrival (DOA) estimation, and ocean source localization tasks. Gregor et al. (2019) concluded that although advanced statistical inference and ML methods have been used to fill gaps in sparse surface ocean CO2 measurements and constrain the variability in sea-air CO2 fluxes, these methods are reaching their limitations, referred to as "the wall," where pCO2 estimates are constrained by data gaps and scale-sensitive observations. To enhance surface ocean pCO2 estimates, further improvements might be possible by incorporating additional variables, increasing sampling resolution, and integrating pCO2 estimates from alternate platforms. James et al. (2018) trained ML models on iterations of a physics-based wave model to predict ocean conditions. The models, tested on Monterey Bay, replicated wave heights with a root-mean-squared error of 9 cm and correctly identified over 90% of the characteristic periods, achieving efficient computation compared to the physics-based model. Fan et al. (2021) developed OC-SMART, a versatile platform for analyzing data obtained by satellite ocean color sensors, supporting multiple sensors and providing products such as reflectances, chlorophyll concentration, and optical properties. By utilizing extensive radiative transfer simulations and ML techniques, OC-SMART improves the quality of retrieved water products and resolves issues with negative water-leaving radiance. It is claimed to be faster than NASA's SeaDAS platform, includes advanced cloud screening, and can recover valuable data in coastal areas, making it a valuable tool for ocean color analysis. Sinha and Abernathey (2021) explored the use of ML algorithms to infer global surface currents from satellite observable quantities. The ML models are trained using simulated ocean data and show that a neural network (NN) outperforms linear regression models, accurately predicting surface currents over most of the global ocean. By incorporating geographic information and using convolutional filters, their research showed that NN can effectively learn spatial gradients and improve the accuracy of surface flow predictions. Gloege et al. (2022) produced The Lamont Doherty Earth Observatory-Hybrid Physics Data (LDEO-HPD) pCO2 product by using ML to merge observations with global ocean biogeochemical models

(GOBMs) to estimate surface ocean pCO2 and air-sea CO2 exchange. By training an eXtreme Gradient Boosting (XGB) algorithm to correct the model-data mismatch, LDEO-HPD provides a more accurate reconstruction of pCO2 compared to other observation-based products. The results show good agreement with independent pCO2 observations and are consistent with estimates from other products and the Global Carbon Budget.

Similar to other AI applications in climate sciences, while the mentioned research papers present valuable contributions to the field of oceanography and demonstrate the potential, there are certain limitations and challenges that restrict their immediate adoption. For example, the practical implementation of Chen et al.'s model requires extensive pCO2 datasets and satellite data, which may not be readily available or accessible in real-world scenarios. Additionally, the model's performance in different oceanic regions or under different environmental conditions needs to be further evaluated. Niu et al's study will be restricted by the applicability of these algorithms in real-world situations and may be limited by the availability of sufficient training data, which can be a challenge in ocean acoustics. The effectiveness of these algorithms needs to be validated in different acoustic environments and with diverse source characteristics. Although as Bianco et al.'s review revealed and emphasized on the promise of deep learning (DL) in acoustics for tasks such as sound event detection and source localization, the availability of large and diverse training datasets remains a challenge. The reliance on synthetic data or data augmentation techniques may introduce biases or limitations in the generalizability of the trained models. Further research is needed to address these challenges and improve the robustness of DL approaches in acoustics. Gregor et al.'s research acknowledged the "wall" phenomenon which indicates that pCO2 estimates are constrained by data gaps and scale-sensitive observations. The study explicitly suggests potential improvements such as incorporating additional variables, increasing sampling resolution, and integrating data from alternate platforms, practical implementation, and addressing the limitations of sparse data availability and observational constraints remain significant challenges. James et al.'s work focuses on a specific test site, Monterey Bay, and limits the generalizability of the models to different oceanic regions and conditions. Further validation and testing across diverse geographical locations are necessary to assess the models' robustness and applicability in practical oceanographic applications. For Fan et al.'s OC-SMART, the practical adoption may require addressing challenges related to data availability, integration with existing data processing systems, and validation across different sensor platforms and environmental conditions. Sinha et al.'s ML models' performance and generalizability need to be further evaluated across diverse oceanic regions and under different oceanographic conditions. Additionally, incorporating geographic information and using convolutional filters may introduce challenges in terms of data processing requirements and computational complexity. The practical implementation and utilization of the Gloege et al.'s product depend on the availability and accessibility of relevant observational data and the integration of the product into existing carbon cycle research and monitoring frameworks.

4 Analyzing Low Actionability of AI Projects

After reviewing the use cases in the previous section, most AI projects in climate and environmental science face challenges that limit their actionability and practical applicability in real-world decision-making. These challenges arise from the inherent complexities of climate and environmental systems, the presence of uncertainties and incomplete knowledge, limited data availability and quality, as well as the need for model validation and reliability. Additionally, real-world decision-making processes pose their own set of challenges, including policy and governance considerations, communication and stakeholder engagement, and ethical and equity considerations. Climate and environmental systems are characterized by intricate interdependencies and nonlinear dynamics, making them difficult to model accurately. Despite advancements in AI techniques, climate models still struggle to capture all the complexities of the Earth's climate system. Uncertainty and incomplete knowledge are prevalent in climate and environmental science. Predicting the long-term impacts of climate change on specific regions or ecosystems is a complex task due to uncertainties in data, model formulations, and future projections. This limits the reliability of AI-based predictions and decisions. A study by Knutti et al. (2017) emphasizes the importance of quantifying and communicating uncertainties in climate projections to improve decision-making processes.

Another critical challenge is the limited availability and quality of climate and environmental data. Data scarcity in remote or inaccessible regions hampers the development and training of robust AI models. Furthermore, the lack of long-term observations or sparse data for rare events or extreme conditions adds further complexity. Climate Data Records provided by research institutions help bridge the data gaps, but challenges persist in data coverage and quality (Eyring et al. 2016). Validating AI models and ensuring their reliability is crucial for real-world decision-making. The performance of AI models relies on the quality of validation datasets and the ability to reproduce past events accurately. However, validating AI models for future projections, where real-world observations are limited, poses challenges. Models used for long-term climate predictions or ecological forecasts require rigorous validation procedures. Real-world decision-making involves policy and governance considerations. AI projects in climate and environmental science must align with policy frameworks and governance structures to inform decision-making. However, integrating AI-derived information into policy processes and ensuring transparency and interpretability of AI models can be challenging. The United Nations Sustainable Development Goals provide a framework for integrating AI in climate change mitigation strategies and sustainable development initiatives (Lee et al. 2016). Effective communication and stakeholder engagement are essential for AI projects. Communicating complex AI-driven results and uncertainties to policymakers, scientists, and the general public can be challenging. Public perception and understanding of AI predictions related to extreme weather events or biodiversity loss play a crucial role in decision-making processes. Maibach et al. (2015) stress the importance of clear and accessible communication to bridge the gap between scientific findings and public understanding. Ethical considerations and equity concerns are also significant. AI projects should address potential biases, ensure fair distribution of benefits, and avoid excluding marginalized groups in decision-making processes. The impact of AI-driven climate models on vulnerable communities and the potential for AI technologies to reinforce existing inequalities need careful attention. The Climate Justice Research Centre addresses issues of equity and justice in climate and environmental decision-making (Climate Justice Research Centre, https://www.climatejusticecenter.org).

5 How to Make AI Practical?

We believe that every AI project has the potential to be practical and make its positive impacts on climate and environment. This section will list several key strategies that can be employed to make your AI research more action-oriented. However, there are many parties involved when any AI model is going online and making real impacts. The playbooks for each group will be different on how to develop, treat, adapt, utilize, and thrive on the AI application. This section will break down the strategies and give suggestions tailored for them.

5.1 Suggestion for AI Practitioners

AI practitioners should consider the following strategies. First, prioritize understanding the specific needs and context of end-users and stakeholders. This involves engaging with decision-makers, policymakers, and domain experts to identify the key challenges, uncertainties, and decision-making processes. By understanding the user's perspective, AI practitioners can tailor their models and outputs to provide actionable insights. For example, in flood risk management, AI models can be developed to provide real-time flood forecasting and early warning systems that are directly relevant to emergency response agencies and local communities. By aligning the AI models with user needs, the outputs become more actionable and relevant to decision-makers. Second, practitioners should build AI models with the integration of domain knowledge and constraints. This involves collaborating with domain experts to incorporate their insights and understanding of the underlying processes and factors influencing climate and environmental systems. Through combining domain knowledge with AI techniques, practitioners can develop models that are more accurate and reliable. For example, in carbon sequestration projects, AI models can be used to optimize land-use planning by considering ecological constraints, biodiversity conservation, and socioeconomic factors. By accounting for domain-specific knowledge and constraints, AI models become more realistic and feasible for real-world decision-making. Third, AI practitioners should focus on providing outputs that are not only accurate but also actionable and interpretable by end-users.

This can be achieved by translating complex AI outputs into intuitive and understandable formats, such as visualizations or decision-support tools. Additionally, AI models should provide explanations or justifications for their predictions or recommendations to enhance trust and understanding. For example, in climate change impact assessment, AI models can be used to analyze the vulnerability of different regions and provide visualizations that clearly highlight areas at high risk. Also interpretable outputs can help decision-makers better understand the implications of the AI models and make informed decisions. AI practitioners should consider the ethical implications and potential biases associated with their models. Bias in AI can lead to unfair outcomes and exclusion of certain groups or communities. AI practitioners should invest in diverse and representative training data, consider algorithmic fairness techniques, and conduct thorough evaluations for bias. For instance, in environmental justice, AI models can be used to analyze the distribution of environmental burdens and ensure equitable access to environmental resources. By addressing ethical and fairness considerations, AI models become more trustworthy and accountable for real-world decision-making.

AI practitioners should actively reach out to create partnerships for collaboration and interdisciplinary research between AI experts, climate scientists, environmental researchers, and policy stakeholders. This collaboration can help bridge the gap between technical advancements and real-world applications. By working together, different stakeholders can contribute their expertise, validate AI models, provide contextual knowledge, and ensure the relevance and practicality of the AI products. For example, in renewable energy planning, collaborative research between AI experts and energy policymakers can lead to the development of AI-driven tools that optimize renewable energy deployment based on spatial, economic, and environmental considerations. By fostering collaboration, AI practitioners can develop models that address the complex challenges of climate and environmental decision-making.

5.2 *Suggestions for Decision-Makers and Stakeholders*

Although most responsibilities of making AI practical are on AI researchers' shoulders, there is still a lot more that can be done from the user side to help them develop better AI products. AI users should first identify their specific challenges and goals related to climate and environmental issues, like understanding the areas where AI can provide valuable insights or solutions, such as improving resource management, optimizing energy efficiency, or enhancing environmental monitoring. Clearly defining their needs can make users better assess the relevance and potential impact of AI research products. Instead of being pitched by scientists, users with demands should actively reach out to academia. Collaborating with AI experts and researchers can be highly beneficial for users in climate and environment. AI experts can provide guidance on selecting appropriate AI models and methods, tailoring them to specific user requirements, and validating their effectiveness in addressing real-world challenges. This collaboration can involve partnerships with academic institutions, research organizations, or AI consulting firms. Meanwhile, before full-scale adoption, users should consider conducting pilot studies or demonstrations to evaluate the feasibility and benefits of AI research products, such as implementing AI models on a smaller scale or in controlled environments to assess their performance, reliability, and practicality. Pilot studies can help identify any potential limitations or adjustments needed for successful implementation.

As decision-makers and stakeholders often assess the cost-effectiveness and return on investment before adopting AI research products, they should perform a thorough cost-benefit analysis that takes into account the implementation costs, potential savings, improved decision-making capabilities, and long-term benefits. This analysis can help justify the adoption of AI and secure necessary resources. Also, there are a lot of things to consider from the user side besides science integrity. AI users should prioritize addressing privacy, security, and ethical considerations to gain trust and ensure compliance with regulations. This includes safeguarding sensitive data, implementing appropriate security measures, and adhering to ethical guidelines and standards. Users should be transparent about the data sources, model training methods, and potential biases to build confidence among stakeholders.

6 Vision for Earth AI in Future Environment Practice

Our shared vision for AI is very important for bringing scientists and the society together to address pressing issues. Earth AI represents the next generation of systems that can provide innovative solutions for tackling climate and environmental issues in an intelligent and unprecedented manner. This vision is unique in its approach to addressing climate challenges by harnessing advanced technologies and data-driven methodologies to offer actionable insights and innovative solutions for environmental monitoring, conservation, resource management, and climate change mitigation with unparalleled efficiency and capability. Unlike traditional technologies, Earth AI combines the power of artificial intelligence, ML, and big data analytics to unlock the full potential of available environmental data, enabling us to make more informed decisions and take proactive measures in addressing climate issues. Earth AI offers a transformative approach that can enhance our understanding of complex Earth systems, optimize resource allocation, support evidence-based decision-making, and foster collaboration among stakeholders. It is this integration of cutting-edge technologies with environmental stewardship that sets Earth AI apart, making it a crucial component of our future strategy to tackle climate challenges effectively and promote sustainable practices.

Earth AI is expected to provide actionable insights and innovative solutions for environmental monitoring, conservation, resource management, and climate change mitigation, for example, the use of AI technologies to revolutionize environmental monitoring and conservation efforts (Sun et al. 2022). For example, remote sensing data combined with ML algorithms can enable the automated detection and tracking of deforestation, illegal fishing activities, or wildlife poaching in real-time. This allows for more efficient and targeted interventions, such as timely enforcement actions or habitat protection measures. Organizations like Global Forest Watch and Wildlife Insights are already leveraging AI to monitor and protect forests and wildlife. AI can optimize the management of Earth's limited resources, and analyze large datasets from weather sensors, satellite imagery, and agricultural records to optimize water usage, crop yields, and irrigation practices. This helps in reducing water waste, improving food production, and ensuring sustainable resource allocation. It will support climate change mitigation and adaptation strategies by predicting extreme weather events, assessing the impact of climate change on ecosystems, and designing resilient infrastructure. AI-powered models will be able to optimize renewable energy deployment, developing carbon capture and storage technologies, and improving climate risk assessments for vulnerable communities. Projects like ClimateAI and Climate Corporation are actively working on AI-based climate solutions. Earth AI envisions the development of decision-support systems that facilitate collaborative and evidence-based decision-making processes. By integrating AI models, expert knowledge, and stakeholder inputs, these systems can provide policymakers with insights to design effective environmental policies, conservation strategies, and sustainable development plans. They can also simulate the potential outcomes of different policy scenarios to guide decision-makers in making informed choices. Initiatives like AI for Earth by Microsoft and the Earth System Prediction Capability are aiming to provide decision-support tools for environmental management.

7 Conclusion

This chapter listed the need for AI practitioners to bridge the gap between AI research and actionable science in the field of climate and environment, and revealed the challenges faced by AI models and products in terms of their limited actionability and adoption in real-world decision-making processes. It provides valuable insights and future outlooks for AI practitioners to make their products better positioned for actionable science like improving interpretability and transparency of AI models, integrating domain expertise in model development, leveraging interdisciplinary collaborations, focusing on scalability and transferability of models, and addressing data limitations and biases. Through implementing these strategies, AI practitioners can expect to enhance the practicality and relevance of their AI products, ensuring their effective use in addressing climate and environmental challenges and enabling informed decision-making processes.

In future, the guidance for actionable AI can better position our scientists for adoption from the public, by providing better AI products with interpretability and transparency of AI models to gain stakeholders' trust and facilitate understanding of model predictions. We expect this chapter will help Earth AI scientists thrive on integrating domain expertise and involving stakeholders in the model development process to ensure the relevance and applicability of AI solutions, actively reaching out to promoting interdisciplinary collaborations to leverage diverse perspectives and expertise, facilitating the development of holistic and actionable AI products, communicating the benefits and value of AI products effectively to decision-makers, policymakers, and the public, fostering trust, and promoting adoption. By incorporating these future outlooks into their practices, scientists in climate and environment domains can drive the transformation of AI research into actionable science, contributing to effective climate and environmental management and decision-making processes.

References

Agarap, Abien Fred. 2018. Deep learning using rectified linear units (relu). arXiv preprint arXiv:1803.08375.

Akkaya, Ilge, Marcin Andrychowicz, Maciek Chociej, Mateusz Litwin, Bob McGrew, Arthur Petron, Alex Paino et al. 2019. Solving Rubik's cube with a robot hand. arXiv preprint arXiv:1910.07113.

Ali, Sadia, Yaser Hafeez, Mamoona Humayun, Nor Shahida Mohd Jamail, Muhammad Aqib, and Asif Nawaz. 2022. Enabling recommendation system architecture in virtualized environment for e-learning. *Egyptian Informatics Journal* 23 (1): 33–45.

Allawi, Mohammed Falah, Othman Jaafar, Firdaus Mohamad Hamzah, Sharifah Mastura Syed Abdullah, and Ahmed El-Shafie. 2018. Review on applications of artificial intelligence methods for dam and reservoir-hydro-environment models. *Environmental Science and Pollution Research* 25: 13446–13469.

Alnuaim, Ahmed, Ziheng Sun, and Didarul Islam. 2023. AI for improving ozone forecasting. In *Artificial intelligence in Earth science*, 247–269. Elsevier.

Ameer, Saba, Munam Ali Shah, Abid Khan, Houbing Song, Carsten Maple, Saif Ul Islam, and Muhammad Nabeel Asghar. 2019. Comparative analysis of machine learning techniques for predicting air quality in smart cities. *IEEE Access* 7: 128325–128338.

Arulkumaran, Kai, Marc Peter Deisenroth, Miles Brundage, and Anil Anthony Bharath. 2017. Deep reinforcement learning: A brief survey. *IEEE Signal Processing Magazine* 34 (6): 26–38.

Barzegar, Rahim, Asghar Asghari Moghaddam, Ravinesh Deo, Elham Fijani, and Evangelos Tziritis. 2018. Mapping groundwater contamination risk of multiple aquifers using multi-model ensemble of machine learning algorithms. *Science of the Total Environment* 621: 697–712.

Bianco, Michael J., Peter Gerstoft, James Traer, Emma Ozanich, Marie A. Roch, Sharon Gannot, and Charles-Alban Deledalle. 2019. Machine learning in acoustics: Theory and applications. *The Journal of the Acoustical Society of America* 146 (5): 3590–3628.

Biswas, Som S. 2023. Potential use of chat gpt in global warming. *Annals of Biomedical Engineering* 51 (6): 1126–1127.

Borowiec, Marek L., Rebecca B. Dikow, Paul B. Frandsen, Alexander McKeeken, Gabriele Valentini, and Alexander E. White. 2022. Deep learning as a tool for ecology and evolution. *Methods in Ecology and Evolution* 13 (8): 1640–1660.

Chaipidech, Pawat, Niwat Srisawasdi, Tanachai Kajornmanee, and Kornchawal Chaipah. 2022. A personalized learning system-supported professional training model for teachers' TPACK development. *Computers and Education: Artificial Intelligence* 3: 100064.

Chan, William, Navdeep Jaitly, Quoc Le, and Oriol Vinyals. 2016. Listen, attend and spell: A neural network for large vocabulary conversational speech recognition. In *2016 IEEE international conference on acoustics, speech and signal processing (ICASSP)*, 4960–4964. IEEE.

Chen, Shuangling, Hu Chuanmin, Brian B. Barnes, Rik Wanninkhof, Wei-Jun Cai, Leticia Barbero, and Denis Pierrot. 2019. A machine learning approach to estimate surface ocean pCO2 from satellite measurements. *Remote Sensing of Environment* 228: 203–226.

Chowdhury, Mashrur, and Adel W. Sadek. 2012. Advantages and limitations of artificial intelligence. *Artificial Intelligence Applications to Critical Transportation Issues* 6 (3): 360–375.

Chu, Casey, Andrey Zhmoginov, and Mark Sandler. 2017. Cyclegan, a master of steganography. arXiv preprint arXiv:1712.02950.

Crane-Droesch, Andrew. 2018. Machine learning methods for crop yield prediction and climate change impact assessment in agriculture. *Environmental Research Letters* 13 (11): 114003.

Cunningham, Evelyn. 2021. Artificial intelligence-based decision-making algorithms, sustainable organizational performance, and automated production systems in big data-driven smart urban economy. *Journal of Self-Governance and Management Economics* 9 (1): 31–41.

Dauvergne, Peter. 2020. *AI in the wild: Sustainability in the age of artificial intelligence*. MIT Press.

Eyring, Veronika, Sandrine Bony, Gerald A. Meehl, Catherine A. Senior, Bjorn Stevens, Ronald J. Stouffer, and Karl E. Taylor. 2016. Overview of the Coupled Model Intercomparison Project Phase 6 (CMIP6) experimental design and organization. *Geoscientific Model Development* 9 (5): 1937–1958.

Fan, Yongzhen, Wei Li, Nan Chen, Jae-Hyun Ahn, Young-Je Park, Susanne Kratzer, Thomas Schroeder, Joji Ishizaka, Ryan Chang, and Knut Stamnes. 2021. OC-SMART: A machine learning based data analysis platform for satellite ocean color sensors. *Remote Sensing of Environment* 253: 112236.

Gholamalinezhad, Hossein, and Hossein Khosravi. 2020. Pooling methods in deep neural networks, a review. arXiv preprint arXiv:2009.07485.

Gloege, Lucas, Monica Yan, Tian Zheng, and Galen A. McKinley. 2022. Improved quantification of ocean carbon uptake by using machine learning to merge global models and pCO2 data. *Journal of Advances in Modeling Earth Systems* 14 (2): e2021MS002620.

Goodfellow, Ian J., Jean Pouget-Abadie, Mehdi Mirza, Xu Bing, David Warde-Farley, Sherjil Ozair, Aaron C. Courville, and Yoshua Bengio. 2020. Generative adversarial networks. *Communications of the ACM* 63 (11): 139–144.

Gregor, Luke, Alice D. Lebehot, Schalk Kok, Pedro M. Scheel, and Monteiro. 2019. A comparative assessment of the uncertainties of global surface ocean CO_2 estimates using a machine-learning ensemble (CSIR-ML6 version 2019a)–have we hit the wall? *Geoscientific Model Development* 12 (12): 5113–5136.

Haupt, Sue Ellen, William Chapman, Samantha V. Adams, J. Charlie Kirkwood, Scott Hosking, Niall H. Robinson, Sebastian Lerch, and Aneesh C. Subramanian. 2021. Towards implementing artificial intelligence post-processing in weather and climate: Proposed actions from the Oxford 2019 workshop. *Philosophical Transactions of the Royal Society A* 379 (2194): 20200091.

Hawkes, L.A., O. Exeter, S.M. Henderson, C. Kerry, A. Kukulya, J. Rudd, S. Whelan, N. Yoder, and M.J. Witt. 2020. Autonomous underwater videography and tracking of basking sharks. *Animal Biotelemetry* 8: 1–10.

Hester, Todd, Matej Vecerik, Olivier Pietquin, Marc Lanctot, Tom Schaul, Bilal Piot, Dan Horgan, et al. 2018. Deep q-learning from demonstrations. *Proceedings of the AAAI Conference on Artificial Intelligence* 32 (1).

Hirschberg, Julia, and Christopher D. Manning. 2015. Advances in natural language processing. *Science* 349 (6245): 261–266.

Hochreiter, Sepp, and Jürgen Schmidhuber. 1997. Long short-term memory. *Neural Computation* 9 (8): 1735–1780.

Huntingford, Chris, Elizabeth S. Jeffers, Michael B. Bonsall, Hannah M. Christensen, Thomas Lees, and Hui Yang. 2019. Machine learning and artificial intelligence to aid climate change research and preparedness. *Environmental Research Letters* 14 (12): 124007.

Iyer, Lakshmi Shankar. 2021. AI enabled applications towards intelligent transportation. *Transportation Engineering* 5: 100083.

James, Scott C., Yushan Zhang, and Fearghal O'Donncha. 2018. A machine learning framework to forecast wave conditions. *Coastal Engineering* 137: 1–10.
Janiesch, Christian, Patrick Zschech, and Kai Heinrich. 2021. Machine learning and deep learning. *Electronic Markets* 31 (3): 685–695.
Johnson, Kevin B., Wei-Qi Wei, Dilhan Weeraratne, Mark E. Frisse, Karl Misulis, Kyu Rhee, Juan Zhao, and Jane L. Snowdon. 2021. Precision medicine, AI, and the future of personalized health care. *Clinical and Translational Science* 14 (1): 86–93.
Kaack, Lynn H., Priya L. Donti, Emma Strubell, George Kamiya, Felix Creutzig, and David Rolnick. 2022. Aligning artificial intelligence with climate change mitigation. *Nature Climate Change* 12 (6): 518–527.
Kang, Gaganjot Kaur, Jerry Zeyu Gao, Sen Chiao, Lu Shengqiang, and Gang Xie. 2018. Air quality prediction: Big data and machine learning approaches. *International Journal of Environmental Science and Development* 9 (1): 8–16.
Knutti, Reto, Jan Sedláček, Benjamin M. Sanderson, Ruth Lorenz, Erich M. Fischer, and Veronika Eyring. 2017. A climate model projection weighting scheme accounting for performance and interdependence. *Geophysical Research Letters* 44 (4): 1909–1918.
Lamba, Aakash, Phillip Cassey, Ramesh Raja Segaran, and Lian Pin Koh. 2019. Deep learning for environmental conservation. *Current Biology* 29 (19): R977–R982.
Lee, Bandy X., Finn Kjaerulf, Shannon Turner, Larry Cohen, Peter D. Donnelly, Robert Muggah, Rachel Davis, et al. 2016. Transforming our world: Implementing the 2030 agenda through sustainable development goal indicators. *Journal of Public Health Policy* 37: 13–31.
Lee, Mike, Larry Lin, Yu Chih-Yuan Chen, Ting-Hsuan Yao Tsao, Min-Han Fei, and Shih-Hau Fang. 2020. Forecasting air quality in Taiwan by using machine learning. *Scientific Reports* 10 (1): 4153.
Lim, Chris C., M.J. Ho Kim, Ruzmyn Vilcassim, George D. Thurston, Terry Gordon, Lung-Chi Chen, Kiyoung Lee, Michael Heimbinder, and Sun-Young Kim. 2019. Mapping urban air quality using mobile sampling with low-cost sensors and machine learning in Seoul, South Korea. *Environment International* 131: 105022.
Maibach, Edward W., Jennifer M. Kreslake, Connie Roser-Renouf, Seth Rosenthal, Geoff Feinberg, and Anthony A. Leiserowitz. 2015. Do Americans understand that global warming is harmful to human health? Evidence from a national survey. *Annals of Global Health* 81 (3): 396–409.
Marr, Bernard. 2019. *Artificial intelligence in practice: How 50 successful companies used AI and machine learning to solve problems*. Wiley.
Masih, A. 2019. Machine learning algorithms in air quality modeling. *Global Journal of Environmental Science and Management* 5 (4): 515–534.
Mumtaz, Rafia, Syed Mohammad Hassan Zaidi, Muhammad Zeeshan Shakir, Uferah Shafi, Muhammad Moeez Malik, Ayesha Haque, Sadaf Mumtaz, and Syed Ali Raza Zaidi. 2021. Internet of things (IoT) based indoor air quality sensing and predictive analytic—A COVID-19 perspective. *Electronics* 10 (2): 184.
Niu, Haiqiang, Emma Reeves, and Peter Gerstoft. 2017. Source localization in an ocean waveguide using supervised machine learning. *The Journal of the Acoustical Society of America* 142 (3): 1176–1188.
Nuti, Giuseppe, Mahnoosh Mirghaemi, Philip Treleaven, and Chaiyakorn Yingsaeree. 2011. Algorithmic trading. *Computer* 44 (11): 61–69.
O'Shea, Keiron, and Ryan Nash. 2015. An introduction to convolutional neural networks. arXiv preprint arXiv:1511.08458.
Perbet, Pauline, Michelle Fortin, Anouk Ville, and Martin Béland. 2019. Near real-time deforestation detection in Malaysia and Indonesia using change vector analysis with three sensors. *International Journal of Remote Sensing* 40 (19): 7439–7458.
Radford, Alec, Luke Metz, and Soumith Chintala. 2015. Unsupervised representation learning with deep convolutional generative adversarial networks. arXiv preprint arXiv:1511.06434.

Russakovsky, Olga, Jia Deng, Su Hao, Jonathan Krause, Sanjeev Satheesh, Sean Ma, Zhiheng Huang, et al. 2015. Imagenet large scale visual recognition challenge. *International Journal of Computer Vision* 115: 211–252.

Schultz, Martin G., Clara Betancourt, Bing Gong, Felix Kleinert, Michael Langguth, Lukas Hubert Leufen, Amirpasha Mozaffari, and Scarlet Stadtler. 2021. Can deep learning beat numerical weather prediction? *Philosophical Transactions of the Royal Society A* 379 (2194): 20200097.

Sharma, Ashutosh, Mikhail Georgi, Maxim Tregubenko, Alexey Tselykh, and Alexander Tselykh. 2022. Enabling smart agriculture by implementing artificial intelligence and embedded sensing. *Computers & Industrial Engineering* 165: 107936.

Shivaprakash, Kadukothanahally Nagaraju, Niraj Swami, Sagar Mysorekar, Roshni Arora, Aditya Gangadharan, Karishma Vohra, Madegowda Jadeyegowda, and Joseph M. Kiesecker. 2022. Potential for Artificial Intelligence (AI) and Machine Learning (ML) applications in biodiversity conservation, managing forests, and related services in India. *Sustainability* 14 (12): 7154.

Silver, David, Aja Huang, Chris J. Maddison, Arthur Guez, Laurent Sifre, George Van Den Driessche, Julian Schrittwieser, et al. 2016. Mastering the game of Go with deep neural networks and tree search. *Nature* 529 (7587): 484–489.

Sinha, Anirban, and Ryan Abernathey. 2021. Estimating ocean surface currents with machine learning. *Frontiers in Marine Science* 8: 672477.

Sun, Ziheng, and Nicoleta Cristea. 2023. Introduction of artificial intelligence in Earth sciences. In *Artificial intelligence in Earth science*, 1–15. Elsevier.

Sun, Alexander Y., and Bridget R. Scanlon. 2019. How can Big Data and machine learning benefit environment and water management: A survey of methods, applications, and future directions. *Environmental Research Letters* 14 (7): 073001.

Sun, Ziheng, Liping Di, and Hui Fang. 2019. Using long short-term memory recurrent neural network in land cover classification on Landsat and Cropland data layer time series. *International Journal of Remote Sensing* 40 (2): 593–614.

Sun, Ziheng, Liping Di, Hui Fang, and Annie Burgess. 2020. Deep learning classification for crop types in North Dakota. *IEEE Journal of Selected Topics in Applied Earth Observations and Remote Sensing* 13: 2200–2213.

Sun, Ziheng, Laura Sandoval, S. Robert Crystal-Ornelas, Mostafa Mousavi, Jinbo Wang, Cindy Lin, Nicoleta Cristea, et al. 2022. A review of earth artificial intelligence. *Computers & Geosciences*: 105034.

Tuia, Devis, Benjamin Kellenberger, Sara Beery, Blair R. Costelloe, Silvia Zuffi, Benjamin Risse, Alexander Mathis, et al. 2022. Perspectives in machine learning for wildlife conservation. *Nature Communications* 13 (1): 792.

Tunstall, Lewis, Leandro Von Werra, and Thomas Wolf. 2022. *Natural language processing with transformers*. O'Reilly Media, Inc.

Vo, Tuong Quang, Seon-Ho Kim, Duc Hai Nguyen, and Deg-Hyo Bae. 2023. LSTM-CM: A hybrid approach for natural drought prediction based on deep learning and climate models. *Stochastic Environmental Research and Risk Assessment*: 1–17.

Vu, Tuan V., Zongbo Shi, Jing Cheng, Qiang Zhang, Kebin He, Shuxiao Wang, and Roy M. Harrison. 2019. Assessing the impact of clean air action on air quality trends in Beijing using a machine learning technique. *Atmospheric Chemistry and Physics* 19 (17): 11303–11314.

Wang, Limin, Sheng Guo, Weilin Huang, and Yu Qiao. 2015. Places205-vggnet models for scene recognition. arXiv preprint arXiv:1508.01667.

Wu, Yonghui, Mike Schuster, Zhifeng Chen, Quoc V. Le, Mohammad Norouzi, Wolfgang Macherey, Maxim Krikun et al. 2016. Google's neural machine translation system: Bridging the gap between human and machine translation. arXiv preprint arXiv:1609.08144.

Yang, Ting, Liyuan Zhao, Wei Li, Wu Jianzhong, and Albert Y. Zomaya. 2021. Towards healthy and cost-effective indoor environment management in smart homes: A deep reinforcement learning approach. *Applied Energy* 300: 117335.

Chapter 13
Actionable Environmental Science Through Social Media Platforms

Tao Hu, Xiao Huang, and Siqin Wang

Contents

T. Hu (✉)
Department of Geography, Oklahoma State University, Stillwater, OK, USA
e-mail: tao.hu@okstate.edu

X. Huang
Department of Environmental Sciences, Emory University, Atlanta, GA, USA

S. Wang
Spatial Sciences Institute, University of Southern California, Los Angeles, CA, USA

School of Science, Royal Melbourne Institute of Technology (RMIT), Melbourne, Australia

School of Earth and Environmental Sciences, University of Queensland, Brisbane, Australia

Z. Sun (ed.), *Actionable Science of Global Environment Change*,
https://doi.org/10.1007/978-3-031-41758-0_13

1 Introduction

In the face of pressing environmental issues such as climate change, air pollution, deforestation, and loss of biodiversity, it is more important than ever to make actionable science to address those challenges. Today, the vast digital landscapes of social media platforms offer a new avenue to sense the perceptions and engage the public in environmental discourse. This can foster increased understanding and awareness of environmental issues, their causes, consequences, and possible solutions. In addition, through analysis of posts, comments, shares, and other interactions in social media platforms, researchers can gauge public sentiment and opinion on environmental issues. It provides valuable insight into public knowledge, misconceptions, or concerns, helping scientists, policymakers, and educators target their efforts more effectively. Moreover, social media provides a platform for individuals, groups, and organizations to mobilize action on environmental issues. It can range from organizing events and protests, promoting conservation efforts, to sharing tips for sustainable living. Therefore, effective use and understanding of social media can significantly contribute to actionable environmental science by bridging the gap between scientific knowledge, public understanding, and policy-making.

This chapter will first explore how social media have been leveraged to identify public awareness and opinions toward environmental change and identify environmental change awareness. Secondly, the chapter will investigate a significant challenge within the social media platform – the proliferation of misinformation. The rapid spread of false or misleading information about environmental changes can hinder efforts to raise awareness and inspire action. This section explores the prevalence and impact of misinformation, the mechanisms of its spread, and strategies to counteract it. Thirdly, the chapter will introduce numerous tools and applications that can be harnessed to collect, analyze, and utilize data from social media platforms. By understanding how to effectively use these tools, researchers, policymakers, and environmental advocates can gain valuable insights that can inform strategies and policies. Lastly, we will discuss how all these elements can be integrated toward actionable solutions for environmental change. As we delve into the role of influencers, governments, and individuals in disseminating information, we will explore how social media platforms can be leveraged to drive meaningful action for environmental change.

Social media platforms provide many opportunities in actionable environmental science as well as challenges. Throughout this chapter, we hope to provide a comprehensive understanding of how we can leverage the power of social media to create meaningful discourse around environmental change, raise public awareness, and drive action that will lead us toward a more sustainable environment.

2 Sensing Awareness and Opinions on Environmental Changes in Social Media Platforms

2.1 Social Media: A Source of Environmental Change Awareness

Social media's transformative impact on the landscape of information exchange is undeniable, acting as a robust conduit for the propagation of ideas, discussions, and perspectives on a global scale. It has democratized access to information and catalyzed societal discourse on myriad topics. Among these, the role of social media in bolstering environmental consciousness is worth emphasizing due to its remarkable power and potential. Under their expansive reach and instantaneous communication, social media platforms have transformed into instruments for broadcasting environmental dilemmas, elucidating challenges, and promulgating innovative solutions.

Platforms such as Twitter, Weibo, Instagram, Facebook, and TikTok, to name a few, are teeming with a broad spectrum of environmental content. This includes informative material on climatic transformations and biodiversity depletion, and public appeals for initiatives like recycling campaigns or climate strikes, all meticulously curated to enlighten, communicate, and galvanize the masses toward environmental conservation. In social media, hashtags such as #ClimateChange, #GlobalWarming, #Savetheplant, and #PlasticFree has evolved from fleeting trends to potent symbols – effective rallying points for individuals and organizations to augment awareness, disseminate knowledge, and spur collective action. These digital rallying cries have initiated a paradigm shift in how environmental issues are perceived and discussed (Vu et al. 2021).

Furthermore, social media platforms serve as a vital arena for environmental activists, including emerging leaders like Greta Thunberg. Their impassioned messages and real-time activism updates frequently capture the public's imagination, leading to viral circulation and catapulting environmental issues into the mainstream discourse. This amplification effect is equally significant for numerous environmental non-governmental organizations (NGOs) and movements. Social media provides these entities an interactive space to engage with their audience, disseminate progress updates, and orchestrate real-time coordinated actions.

The role of social media is also critical in science communication, particularly for spreading awareness and understanding of environmental science (Osterrieder 2013). Environmental scientists, climatologists, and biologists, among other specialists, are increasingly harnessing the power of social media to convey their research findings in a more digestible and engaging format (Perrin 2015; Auxier and Anderson 2021; Van Eperen and Marincola 2011). This involves transforming complex datasets into easy-to-understand visualizations, memes, or bite-sized videos. Through such innovative methods, these experts are disseminating essential knowledge and combating the rampant spread of myths and misinformation, which are often ubiquitous on digital platforms.

2.2 *Gauging Public Discussions on Environmental Changes in Social Media*

Social media is an abundant repository of information for scholars interested in assessing public consciousness, dispositions, and viewpoints on climate change, pollution, conservation initiatives, and other environmental subjects (Mavrodieva et al. 2019). Scholars typically employ quantitative and qualitative strategies to extract and interpret this data in their research methods. Traditional methodologies consisted of manual content analysis where scholars would painstakingly sift through posts to discern trends and themes in public conversations. While comprehensive, this approach is labor-intensive and frequently falls short of handling the immense data volume on social media platforms (Li et al. 2021).

With technological advancements in recent years, more sophisticated methods have been developed to efficiently retrieve and analyze public sentiments from these platforms. Technologies like Machine Learning and Natural Language Processing (NLP) algorithms are deployed to classify and interpret texts, thus enabling the automatic detection of the sentiment behind social media posts and their categorization according to the specific environmental subject they address. For example, sentiment analysis, a sub-discipline of NLP, is routinely used to decipher if the overall sentiment of a post is positive, negative, or neutral (Loureiro and Alló 2020; Dahal et al. 2019; Cody et al. 2015). This approach has proven highly valuable in assessing public opinion regarding contentious environmental matters. Additionally, topic modeling techniques such as Latent Dirichlet Allocation (LDA) are employed to reveal latent topics in vast volumes of textual data, thereby aiding scholars in identifying the principal themes in public conversations about environmental changes (Dahal et al. 2019; Benites-Lazaro et al. 2018; Al-Rawi et al. 2021). Network analysis is another often-employed technique to understand the spread of information and opinions on environmental topics across social media networks. By analyzing the interconnections among social media users, scholars can identify influential entities driving discussions about environmental changes and discern the pathways these conversations traverse (Williams et al. 2015; Wang et al. 2020).

Nevertheless, these methods are not without their limitations. They frequently encounter difficulties in understanding the subtleties and contexts of human language, notably sarcasm or cultural references (Huang et al. 2022). They may also exhibit bias toward the platform's prevalent language, often English, thereby potentially not representing global perspectives on environmental changes comprehensively (Wang et al. 2023). Issues such as privacy concerns may limit access to social media data, and the presence of "bots" and "trolls" can distort findings (Starbird 2019; Paavola et al. 2016). Despite these obstacles, social media continues to be a potent instrument for assessing public conversations on environmental changes. With the consistent advancements in technology and analytical methodologies, the potential to understand and engage with these conversations is poised for growth.

2.3 Opinions and Initiatives for Environmental Change Advocacy in Social Media

Navigating the terrain of the contemporary digital landscape, it is becoming increasingly clear that social media is not just a mere communication tool, but an indispensable resource for advocacy and change (Guo and Saxton 2014; Pang 2013). A compelling illustration of this transformative potential lies in environmental conservation. In this arena, social media has emerged as a powerful platform to catalyze action, effect change, and shift societal norms. Elaborating on the various opinions and initiatives currently harnessed to advocate for environmental change on these platforms, it becomes evident that the digital advocacy toolkit is diverse and multifaceted. Among the most prominent strategies are awareness-raising campaigns, which have proliferated widely across different social media platforms. These campaigns, in their essence, leverage the power of hashtags and the potential virality of digital content to disseminate critical information about pressing environmental issues, such as climate change (Titifanue et al. 2017), deforestation (Daume et al. 2014), and pollution (Mei et al. 2014). By crafting engaging and shareable content, these campaigns seek to educate the public about the existential threats facing our planet and, crucially, to galvanize them into taking informed, committed action. In addition to these campaigns, another significant development in online environmental advocacy is the ascendance of "eco-influencers." These individuals, often experts in environmental sciences, fervent activists, or well-known public figures dedicated to the cause, wield considerable influence on social media to advance environmental conservation. They do this by sharing ecologically responsible practices, endorsing products that align with sustainable living, and fostering informed discussions around the latest research in the field. The effectiveness of these eco-influencers is markedly augmented by the trust and admiration that they inspire among their followers, a dynamic that often results in concrete action in the real world. Meanwhile, many advocacy groups have innovatively adapted "crowdsourcing" strategies to the social media landscape. These approaches enable the groups to solicit ideas, strategies, and even financial resources from the public. Such open-sourced solutions engender widespread participation in environmental advocacy and foster inclusivity by ensuring that a diverse array of voices and perspectives is represented in the conversation. Lastly, there is a growing trend of virtual protests and online movements, a testament to the unifying power of social media. Notably, digital climate strikes, inspired by the young activist Greta Thunberg, have shown how social media can mobilize the masses toward environmental change. These initiatives are particularly potent as they overcome geographical barriers, enabling individuals worldwide to participate in a powerful demonstration of support for environmental action. Social media platforms offer a cornucopia of opportunities for advocating environmental change, spanning awareness campaigns, eco-influencers, crowdsourcing efforts, and online movements. When effectively harnessed, these initiatives have the potential to significantly amplify the reach and impact of environmental advocacy, underscoring their pivotal role in our collective fight against environmental degradation and climate change.

3 Misinformation in Social Media Platforms

Misinformation on environmental change in social media is a growing concern that impacts public perception, decision-making, and policy-making processes. A common definition of misinformation is referring to information that is false, inaccurate, or misleading. Misinformation can be spread intentionally or unintentionally. When it is spread with the intent to deceive, it is often referred to as disinformation. As the rapid development of social media platforms, misinformation is increasingly shared across these platforms in various forms, such as news, videos, images, memes, and personal communications, among others.

Several studies have documented the widespread prevalence of misinformation related to environmental issues on social media platforms. For example, Vosoughi, Roy, and Aral's study (2018) is a significant contribution to understanding how information propagates through social networks. The authors used a dataset of rumor cascades on Twitter from 2006 to 2017 to study how truthful, false, and mixed news spreads online. Their findings suggest that false news reached more people than the truth, that false news was 70% more likely to be retweeted than the truth, and that false news was significantly more novel than true news. Another study examined the role of political identity in influencing perceptions of climate change on social media (Bergquist et al. 2020). They found that climate change misinformation often gets amplified within politically homogeneous social networks on Facebook.

In addition, researchers have investigated the relationship between media consumption, trust in scientists, and beliefs about environmental changes, such as global warming. Hmielowski et al. (2014) conducted a longitudinal survey of American adults and found that the relationship between media use and perceptions of global warming was mediated by trust in scientists. Van der Linden et al. (2017) used a survey-based experiment to test the inoculation theory, using a fabricated blog post that denied the scientific consensus on climate change as the misinformation "threat." The results showed that the inoculation messages were effective in neutralizing the effects of misinformation, with the specific inoculation proving to be even more effective than the general one. These findings highlight the critical role of media in shaping public perceptions of scientific facts and the potential for misinformation to erode trust in scientific authorities, leading to misperceptions about pressing issues like environmental change.

What can be done about misinformation about environmental change? The first strategy is early detection of malicious accounts in social media platforms, such as bots, and spammers. Social media bots can potentially threat public opinion and democracy toward environmental change. Researchers have designed many methods to detect bots and these methods can be classified as feature-based, graph-based, crowdsourcing-based, and combined approaches (Orabi et al. 2020). Botometer (formerly BotOrNot) is a popular tool using machine learning to predict the likelihood that a Twitter account is a bot (Subrahmanian et al. 2016). It uses a publicly available API[1] and scores accounts based on a range of features, including user metadata, friend's metadata, tweet content and sentiment, and network patterns. While researchers have made efforts to detect and remove malicious bots, it remains a challenging problem due to the ever-evolving tactics of those who create and control these bots. This makes the ongoing work of researchers studying these issues, as well as policy measures aimed at reducing the malicious use of bots, crucial in the fight against misinformation.

The second strategy is social media platforms and government regulations and policies. Social media platforms themselves have a significant role to play in managing the content disseminated on their platforms. This includes developing and enforcing policies to deal with misinformation, providing warnings for disputed content, and promoting more transparency about how their algorithms work. On March 31, 2023, Twitter made its recommendation algorithm open-source in GitHub.[2] At the policy level, regulations can be implemented that require social media platforms to take greater responsibility for the spread of misinformation on their sites. As demonstrated in the study by Van der Linden et al. (2017), proactive measures hold significant potential in the fight against misinformation, particularly in the realm of environmental change. This underlines the importance of communicating accurate scientific information to the public in a manner that anticipates and counteracts the spread of misinformation.

The third strategy involves promoting media literacy and fostering critical thinking skills. There is a rising recognition of the significance of media literacy for deciphering multifaceted environmental messages in the media. By understanding the structural aspects of misinformation and the tactics often used, users can become more adept at spotting false information. This could involve identifying the emotional manipulation often embedded in misinformation or understanding how partial truths are sometimes used to bolster false claims. Alongside knowing about misinformation, it is equally essential to recognize credible information sources. This means being cognizant of the diversity of sources, their potential predispositions, and their overall dependability. Numerous studies highlight the necessity of digital literacy skills, such as the capability to evaluate the trustworthiness of a source, for successfully navigating the complex information landscape of social media. However, it is important to note that media literacy is not a one-size-fits-all solution. It necessitates continuous efforts and the inclusion of media literacy education in school syllabuses, public education initiatives, and possibly even on social media platforms. Nonetheless, as social media increasingly impacts public perceptions of environmental change, cultivating media literacy becomes an essential step toward fostering an informed and engaged population.

---

[1] https://botometer.osome.iu.edu/api

[2] https://github.com/twitter/the-algorithm

4 Tools and Applications for Actional Environmental Insights

4.1 Tools and Applications for Social Media Data Collection

Public tools, platforms, and products for social media data collection and analysis are essential for enhancing actionable environmental change science. These tools provide users with access to large volumes of social media data, enabling them to analyze public perceptions, behaviors, and opinions related to environmental issues. They also offer research-specific features such as advanced data collection methods, data filtering and preprocessing capabilities, and analytical techniques like sentiment analysis and network analysis. Moreover, these platforms foster collaboration among researchers, policy-makers, and stakeholders, prioritize ethical data practices, employ rigorous analysis methods, and provide real-time insights, all of which contribute to the generation of robust and timely evidence for decision-making and policy development in the realm of environmental change. Most social media platforms provide API to access data without requiring a subscription or payment. Some examples of such tools include:

Twitter API[3]: The Twitter API allows developers to access and retrieve public tweets and related metadata. It provides different endpoints and functionalities for data collection, such as searching for specific keywords, retrieving user timelines, and accessing real-time streaming data.

Facebook Graph API[4]: The Facebook Graph API enables access to public posts and data on Facebook. It allows researchers to retrieve public posts from pages, groups, and public profiles, as well as associated engagement metrics.

YouTube Data API[5]: The YouTube Data API allows access to public videos, comments, and related metadata on the YouTube platform. Researchers can collect data on videos related to environmental topics, user interactions, and engagement metrics.

Reddit API[6]: The Reddit API allows retrieval of public posts and comments from Reddit. Researchers can access discussions, threads, and user-generated content related to environmental discussions or specific subreddits.

Although these tools provide access to public data, they may have limitations in terms of data availability, historical data access, and rate limits imposed by the platforms. In addition to data access API, some freely accessible tools are also available for public users to collect data, including:

[3] https://developer.twitter.com/en/products/twitter-api/academic-research

[4] https://developers.facebook.com/docs/graph-api/

[5] https://developers.google.com/youtube/v3

[6] https://www.reddit.com/dev/api/

NCapture[7] and NVivo:[8] These software tools, developed by QSR International, enable researchers to collect, organize, and analyze social media data alongside other qualitative data sources. They provide features for coding, visualizing, and interpreting data, facilitating in-depth analysis of social media content related to environmental issues.

Netlytic:[9] Netlytic is a web-based tool that allows researchers to collect and analyze social media data from platforms such as Twitter, Facebook, and YouTube. It provides features for data collection, text analysis, network analysis, and visualization.

TAGS:[10] The Twitter Archiving Google Sheet (TAGS) is a free tool that enables users to collect and archive tweets based on specified search queries. TAGS is built using Google Sheets and the Twitter API, and it offers features like automated data retrieval and analysis.

4.2 Tools and Applications for Social Media Data Analytics

As social media platforms become increasingly integral to public discourse and sentiment, the importance of analyzing the vast amount of data they generate is critical, particularly in the era of environmental change. Here, we outline some key tools and applications that facilitate this analysis.

Text and sentiment analysis tools are used to understand the sentiment and topics being discussed on social media platforms. They help researchers analyze social media text data to derive public sentiment and opinion on environmental issues. VADER (Valence Aware Dictionary and sEntiment Reasoner) is a popular tool for sentiment analysis, particularly well-suited for analyzing social media text (Hutto and Gilbert 2014). It is a lexicon and rule-based sentiment analysis tool that is explicitly crafted to analyze sentiments in social media contexts (Hu et al. 2021). Other examples include RapidMiner (Verma et al. 2014) and NLTK (Natural Language Toolkit) (Loper and Bird 2002).

Network analysis tools are essential to understand the structure and dynamics of social networks. They can help identify influential users, detect communities, and understand communication patterns, providing insights into how information about environmental issues spreads and evolves over time. Gephi is a popular open-source network visualization and analysis tool (Bastian et al. 2009). It is very user-friendly and supports various types of networks, including social, biological, and Internet networks. Gephi allows for dynamic and hierarchical graphs, and supports a variety

[7] https://help-nv.qsrinternational.com/20/win/Content/ncapture/ncapture.htm

[8] https://help-nv11.qsrinternational.com/desktop/concepts/approaches_to_analyzing_twitter_data.htm

[9] https://netlytic.org/home/

[10] https://tags.hawksey.info/

of layouts, as well as provides statistics and metrics pertinent to network analysis. NetworkX is a Python library used for creating, manipulating, and studying the structure, dynamics, and functions of complex networks. NetworkX provides data structures for representing different types of networks, or graphs, and has many functions and algorithms to analyze these networks (Hagberg et al. 2008). Other network analysis tools include Cytoscape (Shannon et al. 2003), Pajek (Batagelj and Mrvar 2002), and igraph (Csardi and Nepusz 2006).

Machine learning tools are increasingly used to analyze social media data. They can be employed for sentiment analysis, topic modeling, predictive modeling, classification, clustering, and anomaly detection, among other tasks. This can provide deeper insights into public perceptions and behaviors regarding environmental issues. BERT (Bidirectional Encoder Representations from Transformers) is a state-of-the-art deep learning model for natural language processing (NLP), including question answering, sentiment analysis, and more (Devlin et al. 2018). Some powerful open-source libraries for machine learning and deep learning have been widely used in this domain, such as TensorFlow,[11] Scikit-learn,[12] Keras,[13] and Pytorch.[14]

Given that many environmental issues are inherently geographic in nature, geospatial analysis tools are essential to analyze location-based social media data. These tools can provide insights on spatial patterns and variations in public sentiment, behavior, and interaction related to environmental issues. ArcGIS,[15] Google Earth Engine,[16] and QGIS[17] are prominent examples of such tools. Some open-source tools are also available to conduct geospatial analysis and mapping, such as GeoPandas,[18] geemap,[19] and Folium.[20]

5 Toward Actional Design in Environmental Change

5.1 Limitations and Challenges

While social media platforms provide emerging and exclusive data sources to track and monitor people's opinions, discussions, and perspectives on environmental change, existing studies have common limitations and challenges that should be

[11] https://www.tensorflow.org/

[12] https://scikit-learn.org/stable/index.html

[13] https://keras.io/

[14] https://pytorch.org/

[15] https://www.arcgis.com/index.html

[16] https://earthengine.google.com/

[17] https://www.qgis.org/en/site/

[18] https://geopandas.org/en/stable/

[19] https://geemap.org/

[20] https://python-visualization.github.io/folium/

addressed in future action plans to effectively cope with environmental change in practical terms. First, the limited Internet penetration and social media activity in certain geographical areas and among specific social groups result in a selection bias. This bias is particularly pronounced among low-income individuals, those with lower levels of education, the elderly, children, and non-smartphone users (Vu et al. 2021). In many instances, this bias can influence the representation of the silent majority, as people's opinions are influenced by active and influential figures such as political leaders, celebrities, and environmentalists who are prominent in their respective fields (Anderson 2011). To address this bias, it is important to conduct surveys or questionnaires that encompass a wider range of populations across different age groups. This will help calibrate and validate the findings obtained from multiple datasets, ensuring the accuracy and generalizability of the results.

Second, it is important to acknowledge that people's awareness and perspectives on environmental change may be distorted when considering data obtained from social media platforms (Maria and Maria 2020). This is because the data primarily come from users who are inclined to share specific views and are generally more active in contributing to social media discussions. Conversely, individuals with ambivalent opinions may be less active on social media. At the aggregated population level, opinions regarding environmental change are more likely to attribute responsibility for participating in political activities related to environmental protection to developed countries rather than developing or underdeveloped nations (Mavrodieva et al. 2019). This tendency can be influenced by the level of freedom of speech and social institutions, particularly in countries where individuals are encouraged to engage in the political process and express their views on policies (Osterrieder 2013). Therefore, it is crucial to include more global communities that discuss environmental change in non-English speaking nations and in developing or underdeveloped countries. By doing so, a more comprehensive picture can be obtained, enabling the development of actionable strategies in different geographic contexts.

Third, the discussions surrounding climate change on social media tend to focus more on current events rather than past occurrences or future projections. While conveying the urgency of climate change has been effectively incorporated into current actions (Moser and Dilling 2011), there is a growing need to strategically incorporate efficacy or hope appeals into messages. These appeals can help inspire action and engagement with climate change campaigns. Furthermore, public perceptions and engagement are heavily influenced by socio-economic realities and cultural constructs, such as religiosity, hierarchy, gender, age, education level, and social class (Mavrodieva et al. 2019). However, detailed information regarding the socio-demographic profiles of social media users is often absent or difficult to obtain (Correa 2016). This highlights the importance of establishing correlations between the frequent consumption of online information shared through social media and public engagement across various social groups. It is also crucial to address social inequality and injustice, which are significant factors in shaping environmental change actions.

5.2 Future Directions for Actional Environmental Science

Highly Targeted Communication and Dissemination

Highly targeted communication and dissemination play a critical role in actionable environmental science, ensuring that relevant information reaches the right audience and motivates them to take meaningful action (Quevauviller et al. 2005). By tailoring the message to specific groups, stakeholders, policymakers, local communities, industry professionals, or specific interest groups, the communication can address their unique interests, preferences, concerns, and values via conducting research or surveys in social media space to gain insights into the target audience's knowledge, attitudes, and behaviors related to the environmental issue at hand. This understanding will help in crafting messages that resonate with their interests and motivations. Furthermore, engaging visuals (e.g., captivating images and videos), examples and discussions, as well as sharing valuable resources, can significantly enhance the message's reach and facilitate the active participation in relevant environmental communities in the content delivery. It also needs to diverse the communication channels (e.g., community newsletters, local radio stations, environmental conferences, or targeted email campaigns, in addition to social media platforms) in order to ensure that the chosen channels have high engagement and are accessible to the intended audience. Last but not least, fostering two-way communication is important to encourage feedback, questions, and dialogue from the target audience. This can be done through online forums, live chats, public meetings, or surveys – actively listening and responding to their concerns, as well as providing clarifications in the context-specific environmental actions.

Collaboration and Networking with Individuals and Organizations

Collaborating with social media influencers, environmental organizations, and activists can be a powerful way to amplify messages about environmental action and reach a wider audience. By leveraging the influence and reach of these individuals and groups, the message can resonate with a larger number of people and inspire them to take action. First, the action can start with identifying the right social media influencers who align with the environmental cause and have a substantial following via looking for influencers who have demonstrated genuine interest and engagement with environmental issues in the past. Second, reach out to the identified influencers, organizations, and activists through personalized messages by offering mutual value and developing a content strategy that aligns with the environmental message (e.g., videos, articles, infographics, or interactive campaigns). Third, leveraging multiple platforms can maximize the reach and impact of the message, for example, utilizing the influencers' existing platforms (e.g., Instagram, YouTube, or Twitter) as well as the platforms of the environmental organizations and activists involved to increase cross-promotion. Last, it needs to regularly monitor the collaboration's progress and measure the impact on reach, engagement, and audience response, possibly through data monitoring and tracking to adapt the strategy if needed to ensure maximum effectiveness.

Crowdsourcing and Citizen Science

Crowdsourcing and improving citizen science in actionable environmental science hold immense importance for multiple reasons. They enable broader participation and engagement, democratizing the scientific process and allowing individuals from diverse backgrounds to contribute their observations and knowledge (Somerwill and Wehn 2022). This inclusivity helps bridge gaps in data collection, particularly in remote or underrepresented areas, and fosters a deeper understanding of environmental challenges. Furthermore, citizen science generates vast amounts of data, expanding the scope and scale of research efforts. Involving a large number of participants allows for data collection at levels unattainable through traditional scientific methods alone. This extensive data collection leads to more robust analyses and a comprehensive understanding of complex environmental issues. Citizen science also helps build trust and collaboration between scientists, environmentalists, and the public, bridging the gap between academia and society (Somerwill and Wehn 2022). The abundance of user-generated content facilitates online campaigns or challenges, fostering a sense of community and inspiring collective action. Looking ahead, the future direction for citizen science in the context of actionable environmental science involves embracing technological advancements to further enhance data collection, analysis, and dissemination. Mobile applications, sensor networks, and wearable devices offer opportunities for individuals to contribute to data monitoring and respond rapidly to environmental changes effortlessly. Integration with artificial intelligence and machine learning can aid in data quality control, pattern recognition, and predictive climatic modeling. By advancing citizen science in these directions, we can harness the collective power of individuals and communities to effectively address pressing environmental challenges.

Real-Time Monitoring and Response

To facilitate crowdsourcing and citizen science in actionable environmental science, it is crucial to enhance real-time data monitoring and response to environmental events and policies. This involves improving data collection, analysis, and sharing processes. The involvement of a large number of individuals and communities allows for the collection of vast amounts of data from various locations and perspectives, encouraging active participation in environmental stewardship. Crowdsourcing facilitates the pooling of resources and expertise, enabling collaborative analysis that leads to more robust and reliable results. Platforms like AirNow, Earthquake Track, Global Forest Watch, Ocean Observatories Initiative, Global Reef Tracker, and Water Data for the Nation exemplify successful initiatives in this regard (Dong et al. 2015). The diverse perspectives and knowledge contributed by social media users and broader citizens help uncover patterns, trends, and insights that might have otherwise been overlooked. Additionally, sharing the data and findings openly promotes transparency, accountability, and collective learning in environmental actions. This enables scientists, policymakers, and the public to access and utilize the information, fostering evidence-based decision-making and catalyzing effective environmental action.

Engaging Stakeholders and Mobilizing Support

Engaging stakeholders and mobilizing support in environmental actions are critical for creating a collective impact and driving positive change. The key stakeholders in social media, including representatives from local communities, government agencies, industry, non-profit organizations, and academic institutions, have a vested interest in the environmental issue at hand. Purposeful recruitment and retention of stakeholders should be mutually beneficial, with clear expectations regarding costs (e.g., time and effort) and benefits (e.g., influence and results) to avoid engaging stakeholders without a specific purpose (Wesselink and Hoppe 2011). Conducting stakeholder analysis provides insights into their perspectives, motivations, and concerns regarding the environmental issue. Measurable actions are essential to link stakeholders' contributions with specific environmental objectives or deliverables, clarifying when and how to engage with which stakeholders. For example, in online environmental campaigns, stakeholders should be provided with tools, resources, and training to actively participate in environmental actions. Capacity-building workshops, training programs, and educational materials can empower them to contribute effectively. Continuous evaluation and adaptation are necessary to ensure the effectiveness of stakeholder engagement efforts, allowing for sufficient feedback to improve campaign strategies.

6 Conclusion

In the era of digital communication, social media platforms play a significant role in shaping public perceptions and actions related to environmental change. Through these platforms, we have an unprecedented opportunity to sense public awareness, counteract misinformation, and promote accurate and impactful environmental messages. Harnessing these platforms and the tools available for data collection and analysis can lead to actionable environmental insights, informing strategies and policies for effective environmental change. However, the presence of misinformation is a significant challenge that requires sustained efforts in counteracting strategies and the promotion of media literacy.

Overall, understanding and leveraging the potential of social media platforms is crucial for addressing environmental change. The collective action from individuals, influencers, governments, and researchers can create a powerful momentum toward environmental sustainability. As we continue to navigate the complexities of the digital era, it is our shared responsibility to ensure that social media serves as a force for truth, engagement, and positive action for our planet.

References

Al-Rawi, A., O. Kane, and A.J. Bizimana. 2021. Topic modelling of public Twitter discourses, part bot, part active human user, on climate change and global warming. *Journal of Environmental Media* 2 (1): 31–53.

Anderson, A. 2011. Sources, media, and modes of climate change communication: The role of celebrities. *Wiley Interdisciplinary Reviews: Climate Change* 2 (4): 535–546.

Auxier, B., and M. Anderson. 2021. Social media use in 2021. *Pew Research Center* 1: 1–4.

Bastian, M., S. Heymann, and M. Jacomy. 2009. Gephi: An open source software for exploring and manipulating networks. *Proceedings of the International AAAI Conference on Web and Social Media* 3 (1): 361–362.

Batagelj, V., and A. Mrvar. 2002. Pajek—Analysis and visualization of large networks. In *Graph drawing: 9th international symposium, GD 2001 Vienna, Austria, September 23–26, 2001 revised papers 9*, 477–478. Berlin/Heidelberg: Springer.

Benites-Lazaro, L.L., L. Giatti, and A. Giarolla. 2018. Topic modeling method for analyzing social actor discourses on climate change, energy and food security. *Energy Research & Social Science* 45: 318–330.

Bergquist, P., M. Mildenberger, L.C. Stokes, and P. Howe. 2020. Combining climate, economic, and social policy builds political support for climate action in the US. *Environmental Research Letters*. https://doi.org/10.1088/1748-9326/abb8b4.

Cody, E.M., A.J. Reagan, L. Mitchell, P.S. Dodds, and C.M. Danforth. 2015. Climate change sentiment on Twitter: An unsolicited public opinion poll. *PLoS One* 10 (8): e0136092.

Correa, T. 2016. Digital skills and social media use: How internet skills are related to different types of Facebook use among 'digital natives'. *Information, Communication & Society* 19 (8): 1095–1107.

Csardi, G., and T. Nepusz. 2006. The igraph software package for complex network research. *International Journal of Complex Systems* 1695 (5): 1–9.

Dahal, B., S.A. Kumar, and Z. Li. 2019. Topic modeling and sentiment analysis of global climate change tweets. *Social Network Analysis and Mining* 9: 1–20.

Daume, S., M. Albert, and K. von Gadow. 2014. Forest monitoring and social media–complementary data sources for ecosystem surveillance? *Forest Ecology and Management* 316: 9–20.

Devlin, J., M.W. Chang, K. Lee, and K. Toutanova. 2018. Bert: Pre-training of deep bidirectional transformers for language understanding. *arXiv preprint*: arXiv:1810.04805.

Dong, X., R. Li, H. He, W. Zhou, Z. Xue, and H. Wu. 2015. Secure sensitive data sharing on a big data platform. *Tsinghua Science and Technology* 20 (1): 72–80.

Guo, C., and G.D. Saxton. 2014. Tweeting social change: How social media are changing nonprofit advocacy. *Nonprofit and Voluntary Sector Quarterly* 43 (1): 57–79.

Hagberg, A., P. Swart, and D.S. Chult. 2008. *Exploring network structure, dynamics, and function using NetworkX*, No. LA-UR-08-05495; LA-UR-08-5495. Los Alamos: Los Alamos National Lab. (LANL).

Hmielowski, J.D., L. Feldman, T.A. Myers, A. Leiserowitz, and E. Maibach. 2014. An attack on science? Media use, trust in scientists, and perceptions of global warming. *Public Understanding of Science* 23 (7): 866–883.

Hu, T., S. Wang, W. Luo, M. Zhang, X. Huang, Y. Yan, et al. 2021. Revealing public opinion towards COVID-19 vaccines with Twitter data in the United States: Spatiotemporal perspective. *Journal of Medical Internet Research* 23 (9): e30854.

Huang, X., S. Wang, M. Zhang, T. Hu, A. Hohl, B. She, et al. 2022. Social media mining under the COVID-19 context: Progress, challenges, and opportunities. *International Journal of Applied Earth Observation and Geoinformation* 113: 102967.

Hutto, C., and E. Gilbert. 2014. Vader: A parsimonious rule-based model for sentiment analysis of social media text. *Proceedings of the International AAAI Conference on Web and Social Media* 8 (1): 216–225.

Li, Z., X. Huang, X. Ye, Y. Jiang, Y. Martin, H. Ning, et al. 2021. Measuring global multi-scale place connectivity using geotagged social media data. *Scientific Reports* 11 (1): 1–19.

Loper, E., and S. Bird. 2002. NLTK: The natural language toolkit. *arXiv preprint cs/0205028*.

Loureiro, M.L., and M. Alló. 2020. Sensing climate change and energy issues: Sentiment and emotion analysis with social media in the UK and Spain. *Energy Policy* 143: 111490.

Mavrodieva, A.V., O.K. Rachman, V.B. Harahap, and R. Shaw. 2019. Role of social media as a soft power tool in raising public awareness and engagement in addressing climate change. *Climate* 7 (10): 122.

Mei, S., H. Li, J. Fan, X. Zhu, and C.R. Dyer. 2014, August. Inferring air pollution by sniffing social media. In *2014 IEEE/ACM international conference on advances in social networks analysis and mining (ASONAM 2014)*, 534–539. IEEE.

Moser, S.C., and L. Dilling. 2011. Communicating change science: Closing action climate. *The Oxford Handbook of Climate Change and Society. pp.*161.

Orabi, M., D. Mouheb, Z. Al Aghbari, and I. Kamel. 2020. Detection of bots in social media: A systematic review. *Information Processing & Management* 57 (4): 102250.

Osterrieder, A. 2013. The value and use of social media as communication tool in the plant sciences. *Plant Methods* 9: 1–6.

Paavola, J., T. Helo, H. Jalonen, M. Sartonen, and A.M. Huhtinen. 2016. Understanding the trolling phenomenon: The automated detection of bots and cyborgs in the social media. *Journal of Information Warfare* 15 (4): 100–111.

Pang, A. 2013. Social media hype in times of crises: Nature, characteristics and impact on organizations. *Asia Pacific Media Educator* 23 (2): 309–336.

Perrin, A. 2015. Social media usage. *Pew Research Center* 125: 52–68.

Quevauviller, P., P. Balabanis, C. Fragakis, M. Weydert, M. Oliver, A. Kaschl, et al. 2005. Science-policy integration needs in support of the implementation of the EU Water Framework Directive. *Environmental Science & Policy* 8 (3): 203–211.

Shannon, P., A. Markiel, O. Ozier, N.S. Baliga, J.T. Wang, D. Ramage, et al. 2003. Cytoscape: A software environment for integrated models of biomolecular interaction networks. *Genome Research* 13 (11): 2498–2504.

Somerwill, L., and U. Wehn. 2022. How to measure the impact of citizen science on environmental attitudes, behaviour and knowledge? A review of state-of-the-art approaches. *Environmental Sciences Europe* 34 (1): 1–29.

Starbird, K. 2019. Disinformation's spread: Bots, trolls and all of us. *Nature* 571 (7766): 449–450.

Subrahmanian, V.S., A. Azaria, S. Durst, V. Kagan, A. Galstyan, K. Lerman, et al. 2016. The DARPA Twitter bot challenge. *Computer* 49 (6): 38–46.

Titifanue, J., R. Kant, G. Finau, and J. Tarai. 2017. Climate change advocacy in the Pacific: The role of information and communication technologies. *Pacific Journalism Review* 23 (1): 133–149.

Van Eperen, L., and F.M. Marincola. 2011. How scientists use social media to communicate their research. *Journal of Translational Medicine* 9 (1): 1–3.

Van der Linden, S., Anthony L., Seth R., & Edward, M. 2017. Inoculating the public against misinformation about climate change. *Global Challenges* 1 (2): 1600008.

Verma, T., R. Renu, and D. Gaur. 2014. Tokenization and filtering process in RapidMiner. *International Journal of Applied Information Systems* 7 (2): 16–18.

Vosoughi, S., D. Roy, and S. Aral. 2018. The spread of true and false news online. *Science* 359 (6380): 1146–1151.

Vu, H.T., M. Blomberg, H. Seo, Y. Liu, F. Shayesteh, and H.V. Do. 2021. Social media and environmental activism: Framing climate change on Facebook by global NGOs. *Science Communication* 43 (1): 91–115.

Wang, X., Y. Yu, and L. Lin. 2020. Tweeting the United Nations Climate Change Conference in Paris (COP21): An analysis of a social network and factors determining the network influence. *Online Social Networks and Media* 15: 100059.

Wang, S., X. Huang, T. Hu, B. She, M. Zhang, R. Wang, et al. 2023. A global portrait of expressed mental health signals towards COVID-19 in social media space. *International Journal of Applied Earth Observation and Geoinformation* 116: 103160.

Wesselink, A., and R. Hoppe. 2011. If post-normal science is the solution, what is the problem?: The politics of activist environmental science. *Science, Technology, & Human Values* 36 (3): 389–412.

Williams, H.T., J.R. McMurray, T. Kurz, and F.H. Lambert. 2015. Network analysis reveals open forums and echo chambers in social media discussions of climate change. *Global Environmental Change* 32: 126–138.

Chapter 14
Ethics and Accountability of Science in Action

Ziheng Sun

Contents

1 Introduction

Ethics in climate and environmental science involve principles such as integrity, objectivity, and impartiality (Kriebel et al. 2001). The high stakes involved in addressing climate change and environmental degradation make science ethics and accountability very high priority (Grubb 1995). Scientists have a responsibility to ensure that their work is conducted with the highest ethical standards and that the findings are reliable, transparent, and accountable to the wider scientific community and society at large (Von 2013). Scientists must adhere to rigorous research practices, including data collection, analysis, and reporting, to ensure the accuracy and

Z. Sun (✉)
Center for Spatial Information Science and Systems, Department of Geography and Geoinformation Science, George Mason University, Fairfax, VA, USA
e-mail: zsun@gmu.edu

Z. Sun (ed.), *Actionable Science of Global Environment Change*,
https://doi.org/10.1007/978-3-031-41758-0_14

validity of their findings. This is essential in shaping policy decisions and public discourse surrounding climate change and environmental issues. Ethical considerations must ensure that scientific research is not influenced by vested interests or political agendas, and that the results are communicated in a clear and unbiased manner (Krimsky 2004).

Although most of the time, scientists are not the party who are accountable in specific applications, they should be responsible for the methodologies and data used in their research, as well as the implications and potential limitations of their findings (Brewerton and Millward 2001). This involves robust peer review processes, open access to data and methodologies, and clear documentation of research processes. Accountability ensures that scientific work can be replicated, validated, and built upon by other researchers, fostering a culture of transparency and trust in the scientific community. Moreover, accountability ensures that scientists are held responsible for any potential conflicts of interest or ethical breaches, reinforcing the integrity of the field. Examples of the need for ethics and accountability in climate and environmental science can be found in controversies surrounding climate change denial, manipulation of scientific data, or conflicts of interest in research funding (Sarewitz 2004). These instances highlight the importance of maintaining scientific integrity and ensuring that ethical principles are upheld. By embracing ethics and accountability, scientists and researchers can contribute to the development of evidence-based policies and solutions that address climate change and environmental challenges effectively.

The absence of ethics and accountability in climate science will have disastrous consequences that will undermine the integrity and impact of scientific research as its least impact (Gardiner 2010). Without ethical standards and accountability measures, trust in the scientific community and its findings can be eroded. This will inevitably lead to skepticism and public distrust of scientific information on climate change and environmental issues. Without trust, it becomes challenging to mobilize public support for necessary actions and policy changes. There is a risk of misinformation and manipulation of scientific data. This can occur through deliberate distortion or suppression of research findings to serve vested interests or ideological agendas. Such actions can mislead the public, policymakers, and other stakeholders, hindering efforts to address climate change and environmental challenges effectively. Also ethical lapses and lack of accountability can undermine evidence-based policymaking. When scientific research is not conducted with integrity and transparency, policymakers may make decisions based on flawed or biased information, which can result in the implementation of ineffective or insufficient policies that fail to address the urgency and complexity of climate and environmental issues (Glynn et al. 2017). On cost-wise aspects, without ethical standards and accountability, resources may be misallocated or wasted on research that lacks scientific rigor or is influenced by conflicts of interest. This can hinder progress in developing viable solutions or impede the advancement of more robust scientific investigations. Ultimately, limited resources are not optimally utilized to tackle climate change and environmental problems. Last but not least, it can tarnish the reputation of the scientific community as a whole. Instances of ethical breaches or scientific misconduct can lead to the loss of credibility and public confidence in the scientific enterprise. This can have long-term consequences, impeding collaboration, hindering funding opportunities, and damaging the reputation of climate and environmental scientists.

Beside the ethical concerns on the science side, there are many more on the stakeholder or decision-maker side. Ethical and accountability concerns among stakeholders and decision-makers themselves can create barriers to the use of science advice in climate and environmental decision-making (Rickards et al. 2014). Stakeholders involved in decision-making processes may have personal or financial interests that could influence their judgment or compromise their objectivity. For example, a decision-maker who holds shares in a company that could be affected by a proposed environmental policy may be biased in their decision-making process (Hedberg and Ullabeth 2004). Then there are so many lobbying and influencing activities which means powerful interest groups or industry lobbyists may exert undue influence on decision-makers, potentially leading to the manipulation or distortion of scientific advice. This can undermine the integrity and impartiality of the decision-making process. Examples include instances where in history there are stubborn traditional energy companies that have influenced climate change policies to protect their business interests. If decision-making processes lack transparency, it becomes difficult to hold stakeholders accountable for their actions. The lack of transparency can hinder public scrutiny and the ability to identify conflicts of interest or ethical lapses. Decision-makers may prioritize short-term political gains over long-term environmental sustainability. This can result in the neglect or suppression of scientific advice that contradicts political objectives. Political considerations can undermine the credibility and effectiveness of science advice, compromising its utilization in decision-making. They may be influenced by public sentiment that is misinformed or driven by populist narratives which can result in the rejection or distortion of scientific advice that challenges popular opinions or beliefs (Roberts et al. 2002). In such cases, decision-makers may prioritize their own political survival or public support over evidence-based decision-making. Addressing these ethical and accountability concerns requires the establishment of robust governance mechanisms, transparency in decision-making processes, and the enforcement of ethical codes of conduct. Independent oversight bodies and mechanisms to identify and manage conflicts of interest are essential. Creating a culture of accountability and promoting science literacy among decision-makers can enhance their understanding of the importance of utilizing science advice in an ethical and accountable manner.

This chapter aims to shed light on the requirements and challenges related to ethics and accountability in environment science. It provides a comprehensive examination of the ethical considerations that stakeholders and decision-makers should address when utilizing scientific advice to tackle climate and environmental challenges. By highlighting the potential roadblocks and ethical concerns, this chapter serves as a guide to navigate the complexities and ensure the effective utilization of science-guided advice. It will identify the requirements for ethics and accountability across different stakeholders involved in climate science and decision-making processes. It emphasizes the need for transparency, integrity, and independence in the utilization of science advice. By addressing conflicts of interest, lobbying influences, and promoting transparency, decision-makers can overcome ethical concerns and establish a robust framework for incorporating science advice into policy-making (Boston and Lempp 2011). This chapter will explore cases where conflicts of interest have hindered effective decision-making, such as the influence of the fossil fuel industry on climate policies. By learning from these examples and developing strategies to address ethical concerns, decision-makers can ensure that scientific advice is used in an unbiased and accountable manner. Furthermore, it may delve into the importance of public perception, political considerations, and the role of transparency in building trust and public confidence in the decision-making process.

2 Needs of Ethics and Accountability for Actionable Science

2.1 The Needs from the Scientists' Side

On the scientist side, several key needs and considerations revolve around transparency, data sharing, scientific integrity, peer review, and communication (Tennant 2018). Scientists should strive for transparency in their research by clearly documenting their methods, data sources, and analysis techniques. This includes making research data, models, and code openly available for scrutiny and replication. Transparent practices promote accountability and allow for independent verification and validation of scientific findings. Sharing data among scientists and stakeholders is crucial for collaboration, verification, and replication of research findings. Open data policies like FAIR, data management plans, and standardized data formats can facilitate data sharing and enable reproducibility (Alnaim and Sun 2022). For instance, initiatives like the Intergovernmental Panel on Climate Change (IPCC) and the Global Climate Observing System (GCOS) (Plummer et al. 2017) promote data sharing and collaboration in climate science. The relationship between data FAIRness (Findable, Accessible, Interoperable, and Reusable) and scientific integrity in climate science is closely intertwined. FAIR data principles can greatly help the science community uphold scientific integrity by ensuring that data used in research is reliable, transparent, and accessible for scrutiny and validation. They will enhance transparency, promote collaboration, and enable the replication and validation of research findings. By adhering to these principles, scientists can ensure that their research is conducted with rigor, transparency, and accountability.

(1) *Ethics for Traditional Climate Science*

Maintaining scientific integrity is essential for upholding ethical standards in climate science (Edwards and Roy 2017). This involves conducting research with objectivity, rigor, and honesty. Scientists should adhere to sound scientific practices, avoiding biases, conflicts of interest, or the manipulation of data or results to fit preconceived narratives. Rigorous peer review processes, professional codes of conduct, and institutional oversight help ensure scientific integrity. Peer review is a critical component of the scientific process, providing a mechanism for quality control and ensuring that research meets rigorous scientific standards. Peer review helps identify and address any ethical concerns or flaws in study design, methodology, or analysis. It ensures that scientific findings are robust, reliable, and free from bias. Scientists strive to approach their research without personal biases or predetermined conclusions. They employ robust methodologies, rigorous data analysis techniques, and statistical tools to ensure objective interpretation of results. Objectivity in climate science can avoid confirmation bias or cherry-picking data that supports a particular viewpoint. By maintaining objectivity, scientists can contribute to the accuracy and reliability of research findings. For instance, in climate modeling, scientists aim to develop models that accurately represent the physical processes involved in climate change without favoring any specific outcome (Hallegatte 2009). These models undergo extensive validation and evaluation to ensure they objectively capture the complexity of the Earth's climate system. In paleoclimatology, researchers analyze ice cores, tree rings, and other proxy data to reconstruct past climate conditions. Rigorous sampling techniques, precise laboratory measurements, and adherence to established calibration methods can be used to ensure the accuracy and reliability of the reconstructed climate records (Lowe 2001). Scientific integrity also demands honesty in reporting research findings. Scientists should accurately and transparently communicate their methods, results, and limitations. It is essential to avoid data manipulation, selective reporting, or exaggeration of findings. Meanwhile, open and honest discussions of uncertainties and limitations help prevent misinterpretation and maintain public trust in climate science. The Intergovernmental Panel on Climate Change (IPCC) reports must undergo a rigorous review process involving thousands of expert reviewers. The transparency and honesty in the reporting of uncertainties ensure the credibility and integrity of the assessments.

Effective communication of scientific findings is also important for engaging with stakeholders, policymakers, and the public. Scientists should communicate their research in an accessible and transparent manner, emphasizing the uncertainties and limitations associated with their findings. Open dialogue and collaboration with stakeholders can help address ethical concerns, incorporate diverse perspectives, and build trust in the scientific process. Scientists are always encouraged to strive to communicate their research in a way that is accessible to a wide range of audiences. This involves using clear and jargon-free language, visual aids such as graphs and diagrams, and relatable examples to convey complex scientific concepts (Vai and Sosulski 2015). By making their findings understandable to policymakers, stakeholders, and the public, scientists can facilitate informed decision-making and promote the adoption of evidence-based policies. For example, it would be a great channel for climate scientists to often engage in science communication through public lectures, media interviews, and online platforms, where they can use simple language and visualizations to explain the causes and impacts of climate change, helping the general public grasp the scientific concepts involved.

(2) *Ethics for AI*

Very different from the traditional climate research approaches (such as physics modeling and remote sensing), AI has a whole new set of shining tools, which requires additional caution on its ethics when being adopted in practice. The requirement for AI ethics presents unique considerations compared to traditional science ethics due to the distinctive characteristics of AI technologies (Hwang et al. 2020). AI can be susceptible to biases embedded in the data or the algorithms themselves, which can lead to unfair outcomes. Climate and environment scientists should be aware of the potential biases in their data collection and model development processes, ensuring that their AI models are fair and unbiased. They need to be proactive in identifying and mitigating biases, employing techniques like data augmentation, fairness metrics, and algorithmic auditing to promote fairness and mitigate discrimination. For instance, the use of satellite imagery and remote sensing data for environmental monitoring can inadvertently perpetuate biases if certain areas or populations are systematically excluded. Ethical practitioners work toward improving data collection methods and ensuring equitable representation in training datasets.

Unlike traditional scientific methods where the processes and results are often more transparent, AI models can be complex and opaque, making it challenging to understand their decision-making processes. Climate and environment scientists should prioritize explainability and transparency in their AI models, employing techniques such as interpretable machine learning, model visualization, and documentation to ensure that stakeholders can understand and trust the outputs of AI systems (Ganji and Lin 2023). Besides, AI often relies on large-scale data collection and processing, raising concerns about data privacy and security. Climate and environment scientists should ensure that they handle personal and sensitive data in a responsible and ethical manner (Hodson 2003). Implementing robust data protection measures, obtaining informed consent, and anonymizing or aggregating data whenever possible are essential steps to safeguard privacy in AI applications.

In addition, AI systems can have profound non-controllable impacts, therefore there is a pressing need to establish clear lines of accountability and responsibility (Rivas et al. 2023). Climate and environment scientists should consider the potential consequences and unintended effects of AI applications, conducting thorough impact assessments and incorporating mechanisms for ongoing monitoring, evaluation, and accountability (Rillig et al. 2023). This includes regular audits, adherence to ethical guidelines, and mechanisms for addressing ethical concerns raised by stakeholders. Meanwhile, the development and deployment of AI technologies require collaboration and multidisciplinary approaches. Climate and environment scientists should engage with a wide range of stakeholders, including policymakers, industry leaders, and civil society organizations, to ensure that ethical considerations are integrated into the entire AI lifecycle. Collaborative governance frameworks, codes of conduct, and transparency initiatives can help foster responsible and inclusive AI practices.

(3) *Ethics for Human Subject Research*

Some climate science will involve studies that involve human participants or animal subjects. Ethical considerations in these cases include obtaining informed consent from participants, ensuring participant privacy and confidentiality, and minimizing harm or suffering to animals. Research ethics boards and institutional review processes oversee the ethical treatment of human and animal subjects. When conducting surveys on the impacts of climate change on vulnerable communities, researchers must obtain informed consent from participants and protect their privacy and confidentiality. Obtaining informed consent is a required task in human subject research. Participants should have a clear understanding of the study's purpose, procedures, potential risks, and benefits, and voluntarily provide their consent to participate. For example, in studies involving interviews or surveys with individuals impacted by climate change, researchers should ensure participants understand the purpose of the research, how their responses will be used, and any potential risks associated with participation. On the other hand, respecting privacy and confidentiality need to be mandated for all the participating researchers in human subject research and should ensure that participants' personal information is protected and kept confidential. For instance, in studies that involve collecting sensitive information about individuals' experiences with environmental hazards, researchers must handle the data with strict confidentiality and anonymize the data to prevent the identification of participants.

Researchers must conduct a thorough risk-benefit assessment to ensure that the potential risks to participants are minimized, and the benefits of the research outweigh the potential harms. For example, in studies that involve fieldwork in hazardous environmental conditions, researchers should implement appropriate safety measures to mitigate risks to participants' health and well-being (Howe 2022). Special care must be taken when involving vulnerable populations in research, such as children, indigenous communities, or marginalized groups. Researchers should ensure that these populations are not exploited and that their rights and interests are protected. In studies that involve working with vulnerable populations affected by climate change, researchers should engage in meaningful consultation and collaboration, respecting their cultural values, knowledge systems, and rights.

Other requirements include obtaining ethical approval from an Institutional Review Board or similar ethics committees. The IRB ensures that the research design and protocols adhere to ethical guidelines and regulations. Researchers should follow the specific guidelines set by their institutions and obtain the necessary approvals before starting their research projects. Real-world research projects, such as those investigating the impacts of climate change on vulnerable communities or examining the health effects of environmental pollution, often adhere to rigorous ethical standards, such as the studies conducted by the World Health Organization (WHO) on the health impacts of environmental factors prioritize ethical considerations in their research protocols.

2.2 *The Needs from the Stakeholders'/Users' Side*

Stakeholders, who are standing on the other side of the bridge, basically have similar ethical requirements for the science findings. They need transparency in the scientific process to build trust and confidence in the results. They should have access to information about data sources, methodologies, and potential biases or limitations. For example, when receiving the results from the climate models, stakeholders rely on transparent documentation and disclosure of assumptions to understand the basis of projections and make informed decisions (Süsser et al. 2022). Ethical considerations require involving diverse stakeholders and considering their values, needs, and interests in decision-making processes. This includes engaging marginalized communities, indigenous groups, and vulnerable populations to ensure their voices are heard and their rights are respected. For instance, environmental impact assessments must include affected communities in the decision-making process and address any disproportionate impacts.

At the same time, users also expect fairness and equitable outcomes in the distribution of risks, benefits, and burdens associated with climate and environmental actions. Ethical considerations require addressing social, economic, and environmental justice issues. For example, renewable energy projects should consider the potential impacts on local communities, such as land use conflicts or displacement, and ensure fair compensation and benefits (Knox et al. 2022). Users normally can recognize the importance of long-term sustainability and expect science results to contribute to sustainable development. This involves considering the impacts on ecosystems, future generations, and the global commons. For instance, in fisheries management, ethical considerations require balancing short-term economic interests with the long-term health and productivity of fish stocks. Also, as independent individuals, users should adopt science results responsibly and avoid misinterpretation, manipulation, or cherry-picking of data to fit their own agendas. Ethical considerations involve using science to inform evidence-based decision-making and policy development. For example, in climate policy debates, stakeholders should rely on a comprehensive understanding of scientific consensus rather than selectively citing individual studies.

3 Current Law, Policy, and Practice in Society

Law and policy together shaped the frameworks to shape the practice of science in our daily lives. These frameworks provide the legal and regulatory structures that govern activities related to climate change mitigation, adaptation, and environmental protection. They also influence the practices and behaviors of individuals, organizations, and governments in addressing climate and environmental challenges.

3.1 International Agreements and Treaties

International agreements and treaties form the foundation of global efforts to address climate change and protect the environment, for example, the Paris Agreement, which aims to limit global temperature rise and enhance climate resilience, and the Convention on Biological Diversity, which focuses on the conservation and sustainable use of biodiversity. These agreements set the overarching goals and principles that guide national and regional policies and actions. Another recent example is the Kigali Amendment, adopted in 2016, which extends the scope of the Montreal Protocol to include the phase-down of hydrofluorocarbons (HFCs), which are potent greenhouse gases. The Montreal Protocol, established in 1987 (Jansen et al. 2023), aims to protect the ozone layer by phasing out the production and consumption of ozone-depleting substances. HFCs are commonly used as refrigerants in air conditioning and refrigeration systems. While they do not deplete the ozone layer, they have a high global warming potential. The Kigali Amendment sets out a schedule for the gradual reduction of HFCs, with developed countries taking the lead in phasing down HFC production and consumption, followed by developing countries. By reducing the use of HFCs, the Kigali Amendment is expected to make a significant contribution to global efforts to mitigate climate change. The amendment is an example of international cooperation to address a specific climate issue through a legally binding agreement. It reflects the recognition of the global community that coordinated action is necessary to reduce greenhouse gas emissions and limit the warming of the planet. The amendment has gained widespread support, with over 100 countries ratifying or acceding to it as of 2021.

3.2 National Legislation and Regulations

Governments enact laws and regulations to address climate change and environmental issues at the national level. These laws cover a wide range of aspects, such as emissions reductions, renewable energy targets, land and water management, pollution control, and conservation measures. For instance, the Clean Air Act in the United States sets emission standards for pollutants (Belden 2001), while the Renewable Energy Act in Germany promotes the development of renewable energy sources. National policies and regulations provide the legal framework for actions and investments in climate and environmental initiatives. In terms of ethical and accountability implementation, the Clean Air Act establishes a framework that emphasizes transparency, scientific integrity, and public participation. The law requires the Environmental Protection Agency (EPA) to set National Ambient Air Quality Standards (NAAQS) for pollutants that are harmful to human health, such as ozone, particulate matter, carbon monoxide, and sulfur dioxide (Bachmann 2007). These standards are based on scientific research and undergo rigorous review and public comment processes to ensure their integrity and accuracy. The EPA is accountable for enforcing these standards and monitoring air quality across the country. The Clean Air Act also incorporates mechanisms for accountability and compliance. It requires states to develop State Implementation Plans (SIPs) (Reitze 2004) outlining strategies and measures to achieve and maintain air quality standards. The EPA oversees the implementation of SIPs and can take enforcement actions against states or industries that fail to meet the required standards. Additionally, the CAA includes provisions for citizen suits, allowing individuals and organizations to hold violators accountable through legal actions.

However, the ethical considerations in the enforcement of the Clean Air Act involve the protection of vulnerable populations, environmental justice, and the equitable distribution of the benefits and burdens of pollution control measures. The EPA is required to consider the potential impacts of air pollution on marginalized communities and ensure that regulations are not disproportionately affecting disadvantaged groups. Stakeholder engagement and public participation are integral parts of the regulatory process, allowing affected communities to voice their concerns and provide input on decisions that may affect them.

3.3 Regional and Local Policies

Regional and local governments are the primary party that shape the climate and environmental policies. They may adopt specific regulations and initiatives tailored to local conditions and priorities, such as regional emissions trading schemes, municipal waste management programs, and urban planning strategies that promote sustainable transportation and energy-efficient buildings. These policies complement national efforts and allow for more targeted actions in response to regional challenges. Regional emissions trading schemes (ETS) are market-based mechanisms designed to reduce greenhouse gas emissions by setting a cap on total emissions and allowing for the trading of emission allowances among participants. These schemes operate at a regional or subnational level, such as within a specific country or group of countries, and aim to incentivize emission reductions while promoting economic efficiency. One prominent example of a regional emissions trading scheme is the European Union Emissions Trading System (EU ETS). Established in 2005, the EU ETS is the largest international carbon market and covers various sectors, including power generation, industry, and aviation. It sets a cap on carbon dioxide emissions and allows participating entities to buy and sell emission allowances. The scheme has undergone several phases and revisions to strengthen its effectiveness and address challenges.

The United States has the California Cap-and-Trade Program implemented in 2013 (Cushing et al. 2018), covering major sectors like electricity generation, industry, and transportation. The program sets a declining cap on emissions, and companies must hold allowances equal to their emissions. Participants can trade allowances through auctions and secondary markets, providing flexibility and encouraging cost-effective emission reductions. Regional emissions trading schemes offer several advantages. They provide economic incentives for emission reductions by allowing companies to trade allowances, enabling them to find the most cost-effective ways to meet their obligations. These schemes also facilitate the transfer of clean technologies and expertise across regions, promoting international collaboration in climate action. However, the design and implementation of regional emissions trading schemes raise ethical considerations. Ensuring a fair distribution of emission allowances and addressing potential disproportionate impacts on vulnerable communities are crucial ethical concerns. It is important to establish mechanisms that prevent market manipulation, ensure transparency in allowance allocation, and mitigate the potential for carbon leakage (shifting emissions from regulated to unregulated areas).

3.4 Corporate Practices and Voluntary Initiatives

Private sector entities, including companies and industries, have a growing responsibility to address climate change and environmental issues. Many businesses adopt sustainability practices, such as reducing greenhouse gas emissions, implementing eco-friendly production processes, and integrating environmental considerations into their supply chains. Voluntary initiatives, such as the Carbon Disclosure Project and the Global Reporting Initiative, encourage companies to disclose their environmental impacts and take steps to mitigate them. These practices contribute to overall sustainability efforts and can influence broader societal norms and expectations. The Carbon Disclosure Project (CDP) (Hassan et al. 2013) is a global non-profit organization that works with companies, cities, states, and regions to measure and disclose their environmental impacts, particularly their greenhouse gas emissions. It provides a platform for organizations to report their carbon emissions, climate-related risks, and opportunities, and sets a framework for transparency and accountability. The CDP operates through a voluntary reporting system, where companies and other entities respond to an annual questionnaire that assesses their environmental performance. The questionnaire covers areas such as emissions data, climate change strategies, governance, and risk management. Participating organizations can disclose their data on a range of environmental metrics, including energy consumption, water usage, and deforestation. The information collected by the CDP serves multiple purposes. It allows organizations to track their progress in reducing emissions, identify areas for improvement, and compare their performance to industry benchmarks. The data also provides investors, policymakers, and the public with valuable insights into companies' environmental performance, enabling them to make informed decisions and evaluate the climate-related risks and opportunities associated with different entities.

From a policy and legal perspective, the disclosure of environmental information through the CDP helps policymakers assess the effectiveness of existing climate policies and identify areas that require further attention. It provides a basis for evidence-based decision-making and can inform the development of new regulations or incentives to drive emission reductions and sustainable practices. The information collected by the CDP can influence investor behavior and financial markets. Investors increasingly consider environmental factors in their decision-making processes and may use CDP data to assess the sustainability and climate resilience of companies. This can lead to changes in investment patterns, capital allocation, and the integration of climate-related risks and opportunities into financial disclosures and provide a platform for collaboration and knowledge sharing, driving progress toward a low-carbon and sustainable future.

4 Ethical and Accountable Challenges for Actionable Science

4.1 Data Quality and Transparency

Issues related to data quality, reliability, and transparency can arise, hindering the ethics of science in real-world scenarios. Lack of data sharing and open access to research findings can impede transparency and accountability. Robust data management practices, data sharing policies, and open science principles are necessary to address these challenges. Inaccurate or unreliable data can lead to flawed analysis and flawed decision-making. Ensuring data quality requires rigorous data collection methods, appropriate calibration and validation procedures, and adherence to quality control protocols. In climate science, data from weather stations and satellites undergo thorough quality checks to ensure accuracy and consistency. The reliability of data refers to its consistency and stability over time, while reproducibility refers to the ability to obtain the same results when an experiment or analysis is repeated. These aspects are essential for establishing the credibility of scientific findings. Transparent documentation of data collection methods, metadata standards, and sharing data in open repositories can facilitate data reliability and reproducibility. It allows other researchers to verify findings, conduct independent analyses, and build upon previous work. Open data initiatives, such as the Global Biodiversity Information Facility (GBIF) and the Open Data Initiative by the World Bank, aim to make data widely accessible to the scientific community and the public. The lack of standardized data sharing policies and practices can pose challenges to data accessibility. Data owners may be hesitant to share their data due to concerns about intellectual property rights, privacy, or competitive advantage. Encouraging the adoption of data sharing policies and establishing data repositories that facilitate data sharing can help overcome these challenges. The Climate Change Initiative of the European Space Agency promotes data sharing among climate scientists and provides open access to satellite data.

4.2 *Uncertainty and Risk Communication*

Climate and environmental science often deal with complex systems and inherent uncertainties. Communicating scientific findings and uncertainties to policymakers, stakeholders, and the public is a challenge. Ethical communication requires scientists to be transparent about the limitations, assumptions, and uncertainties of their research. This entails acknowledging the complexity of climate and environmental systems and the inherent uncertainties involved in predicting their behavior. Transparent communication helps to avoid the misinterpretation or misrepresentation of research findings, fostering trust in the scientific process. Policymakers, stakeholders, and the public often seek actionable information to inform decision-making. However, it is crucial to strike a balance between providing actionable information and accurately representing uncertainties. Overstating or downplaying uncertainties can lead to misguided policies or misplaced public expectations. Ethical communication should convey the level of certainty or confidence associated with scientific findings, enabling informed decision-making while acknowledging the boundaries of scientific knowledge. Effective communication of scientific findings requires translating complex scientific concepts into clear and accessible language such as avoiding technical jargon and employing effective visualizations or analogies to convey key messages. Ethical communication ensures that scientific information is understandable to a wide range of audiences, including policymakers, stakeholders, and the public. Ethical communication involves engaging with stakeholders throughout the research process including seeking input from stakeholders, incorporating their perspectives, and addressing their concerns. Engaging stakeholders fosters inclusive decision-making processes and enhances the relevance and applicability of scientific findings to real-world challenges. Meanwhile, climate and environmental science often receive significant media attention and scientists have an ethical responsibility to ensure that their research is accurately represented in media coverage. This involves engaging with journalists, providing accurate and contextualized information, and correcting any misinterpretations or misrepresentations that may arise.

5 Guidance on Addressing Ethical Concerns

To address ethical concerns related to transparency and data integrity, researchers should prioritize open data sharing and follow data management best practices, for example, using workflow management tools such as NASA Geoweaver (Sun et al. 2020, 2021, 2022) to document data sources, methodologies, and analytical processes to enhance reproducibility and facilitate collaboration. Real-world examples of initiatives promoting data transparency include the Open Climate Data Initiative and the Global Biodiversity Information Facility (Yesson et al. 2007), which provide open access to climate and biodiversity data, respectively. Ethical considerations demand engaging stakeholders and including their perspectives throughout the research process. This can usually be achieved through participatory approaches, such as involving local communities, indigenous knowledge holders, and non-governmental organizations like the Intergovernmental Panel on Climate Change (IPCC) process, which involves stakeholders in reviewing and synthesizing scientific knowledge for policy making.

Effectively communicating uncertainties is vital to avoid misleading interpretations of scientific findings. Scientists should clearly communicate the limitations, assumptions, and confidence levels associated with their research. Techniques such as probability-based visualizations, scenario-based approaches, and structured expert elicitation can help convey uncertainties. The Climate Futures Toolbox and the International Society for Bayesian Analysis provide resources and guidance on communicating uncertainties in climate science. Modeling and scenario development play a significant role in climate and environmental science. Ethical concerns arise when models and scenarios are used to inform decision-making, as they can have far-reaching implications for society. Researchers should consider ethical dimensions such as distributive justice, intergenerational equity, and fairness in the development and use of models and scenarios. The Shared Socioeconomic Pathways (SSPs) framework (O'Neill et al. 2014) is a high-profile effort to integrate ethical considerations into scenario development. Ethical concerns extend beyond research practices to the conduct of scientists themselves. Researchers should adhere to professional codes of conduct, avoid conflicts of interest, and disclose any financial or institutional affiliations that may influence their work. Ethical leadership is essential for fostering an environment of integrity, trust, and responsible research. Professional societies and organizations like the American Geophysical Union (AGU) provide ethical guidelines and support ethical conduct in climate and environmental science (Marín-Spiotta et al. 2020).

6 Conclusion

This chapter highlights the importance of maintaining ethical standards in scientific research to ensure responsible decision-making. It discusses the current practices, challenges, suggestions, and future outlook on science ethic concerns. The chapter emphasizes the need for transparency, stakeholder engagement, uncertainty communication, ethical modeling, and professional conduct to address ethical challenges in climate and environmental science. Current practices in climate and environmental science involve efforts to promote transparency, data sharing, and open access to research findings. However, challenges arise in ensuring data integrity, addressing conflicts of interest, and effectively communicating uncertainties. To address these challenges, the chapter suggests adopting robust data management practices, promoting stakeholder engagement throughout the research process, and using innovative techniques to communicate uncertainties. The chapter also highlights the importance of ethical leadership and professional conduct among scientists.

In the future we envision a science landscape where ethical considerations are deeply integrated into research practices. This includes enhancing transparency and reproducibility, promoting interdisciplinary collaborations, and addressing societal and distributive justice concerns in modeling and scenario development. By adopting these practices, the scientific community can build trust, engage stakeholders effectively, and ensure that scientific research is used in a responsible and accountable manner to tackle climate and environmental challenges.

References

Alnaim, Ahmed, and Ziheng Sun. 2022. Using Geoweaver to make snow mapping workflow FAIR. In *2022 IEEE 18th international conference on e-science (e-science)*, 409–410. Piscataway: IEEE.

Bachmann, John. 2007. Will the circle be unbroken: A history of the US national ambient air quality standards. *Journal of the Air & Waste Management Association* 57 (6): 652–697.

Belden, Roy S. 2001. *The clean air act*. Chicago: American Bar Association.

Boston, Jonathan, and Frieder Lempp. 2011. Climate change: Explaining and solving the mismatch between scientific urgency and political inertia. *Accounting, Auditing & Accountability Journal* 24: 1000.

Brewerton, Paul M., and Lynne J. Millward. 2001. *Organizational research methods: A guide for students and researchers*. London: Sage.

Cushing, Lara, Dan Blaustein-Rejto, Madeline Wander, Manuel Pastor, James Sadd, Allen Zhu, and Rachel Morello-Frosch. 2018. Carbon trading, co-pollutants, and environmental equity: Evidence from California's cap-and-trade program (2011–2015). *PLoS Medicine* 15 (7): e1002604.

Edwards, Marc A., and Siddhartha Roy. 2017. Academic research in the 21st century: Maintaining scientific integrity in a climate of perverse incentives and hypercompetition. *Environmental Engineering Science* 34 (1): 51–61.

Ganji, Geetha Satya Mounika, and Wai Hang Chow Lin. 2023. Explainable AI for understanding ML-derived vegetation products. In *Artificial intelligence in earth science*, 317–335. Elsevier.

Gardiner, Stephen M. 2010. Ethics and climate change: An introduction. *Wiley Interdisciplinary Reviews: Climate Change* 1 (1): 54–66.

Glynn, Pierre D., Alexey A. Voinov, Carl D. Shapiro, and Paul A. White. 2017. From data to decisions: Processing information, biases, and beliefs for improved management of natural resources and environments. *Earth's Future* 5 (4): 356–378.

Grubb, Michael. 1995. Seeking fair weather: Ethics and the international debate on climate change. *International Affairs* 71 (3): 463–496.

Hallegatte, Stéphane. 2009. Strategies to adapt to an uncertain climate change. *Global Environmental Change* 19 (2): 240–247.

Hassan, Abeer, Andrew Wright, and John Struthers. 2013. Carbon disclosure project (CDP) scores and the level of disclosure on climate change related activities: An empirical investigation of the FTSE 100 companies. *International Journal of Sustainable Economy* 5 (1): 36–52.

Hedberg, Berith, and Ullabeth Sätterlund Larsson. 2004. Environmental elements affecting the decision-making process in nursing practice. *Journal of Clinical Nursing* 13 (3): 316–324.

Hodson, Derek. 2003. Time for action: Science education for an alternative future. *International Journal of Science Education* 25 (6): 645–670.

Howe, Kimberly. 2022. Trauma to self and other: Reflections on field research and conflict. *Security Dialogue* 53 (4): 363–381.

Hwang, Gwo-Jen, Haoran Xie, Benjamin W. Wah, and Dragan Gašević. 2020. Vision, challenges, roles and research issues of artificial intelligence in education. *Computers and Education: Artificial Intelligence* 1: 100001.

Reitze, J., and W. Arnold. 2004. Air quality protection using state implementation plans-thirty-seven years of increasing complexity. *Vill. Envtl. LJ* 15: 209.

Jansen, M. A. K., P. W. Barnes, J. F. Bornman, K. C. Rose, S. Madronich, C. C. White, R. G. Zepp, and A. L. Andrady. 2023 "The Montreal Protocol and the fate of environmental plastic debris." *Photochemical & Photobiological Sciences 1–9.*

Knox, Stephen, Matthew Hannon, Fraser Stewart, and Rebecca Ford. 2022. The (in) justices of smart local energy systems: A systematic review, integrated framework, and future research agenda. *Energy Research & Social Science* 83: 102333.

Kriebel, David, Joel Tickner, Paul Epstein, John Lemons, Richard Levins, Edward L. Loechler, Margaret Quinn, Ruthann Rudel, Ted Schettler, and Michael Stoto. 2001. The precautionary principle in environmental science. *Environmental Health Perspectives* 109 (9): 871–876.

Krimsky, Sheldon. 2004. *Science in the private interest: Has the lure of profits corrupted biomedical research?* Lanham: Rowman & Littlefield.

Lowe, J. John. 2001. Abrupt climatic changes in Europe during the last glacial–interglacial transition: the potential for testing hypotheses on the synchroneity of climatic events using tephrochronology. *Global and Planetary Change 30 (1–2): 73–84.*

Marín-Spiotta, Erika, Rebecca T. Barnes, Asmeret Asefaw Berhe, Meredith G. Hastings, Allison Mattheis, Blair Schneider, and Billy M. Williams. 2020. Hostile climates are barriers to diversifying the geosciences. *Advances in Geosciences* 53: 117–127.

O'Neill, Brian C., Elmar Kriegler, Keywan Riahi, Kristie L. Ebi, Stephane Hallegatte, Timothy R. Carter, Ritu Mathur, and Detlef P. Van Vuuren. 2014. A new scenario framework for climate change research: The concept of shared socioeconomic pathways. *Climatic Change* 122: 387–400.

Plummer, Stephen, Pascal Lecomte, and Mark Doherty. 2017. The ESA climate change initiative (CCI): A European contribution to the generation of the global climate observing system. *Remote Sensing of Environment* 203: 2–8.

Rickards, Lauren, John Wiseman, and Yoshi Kashima. 2014. Barriers to effective climate change mitigation: The case of senior government and business decision makers. *Wiley Interdisciplinary Reviews: Climate Change* 5 (6): 753–773.

Rillig, Matthias C., Marlene Ågerstrand, Mohan Bi, Kenneth A. Gould, and Uli Sauerland. 2023. Risks and benefits of large language models for the environment. *Environmental Science & Technology* 57 (9): 3464–3466.

Rivas, Pablo, Christopher Thompson, Brenda Tafur, Bikram Khanal, Olawale Ayoade, Tonni Das Jui, Korn Sooksatra, Javier Orduz, and Gissella Bejarano. 2023. AI ethics for earth sciences. In *Artificial intelligence in earth science*, 379–396. Amsterdam: Elsevier.

Roberts, Julian V., Loretta J. Stalans, David Indermaur, and Mike Hough. 2002. *Penal populism and public opinion: Lessons from five countries*. Oxford: Oxford University Press.

Sarewitz, Daniel. 2004. How science makes environmental controversies worse. *Environmental Science & Policy* 7 (5): 385–403.

Sun, Ziheng, Liping Di, Annie Burgess, Jason A. Tullis, and Andrew B. Magill. 2020. Geoweaver: Advanced cyberinfrastructure for managing hybrid geoscientific AI workflows. *ISPRS International Journal of Geo-Information* 9 (2): 119.

Sun, Ziheng, and Nicoleta Cristea. 2021. Geoweaver for Automating ML-based High Resolution Snow Mapping Workflow. In AGU Fall Meeting Abstracts, vol. 2021, pp. IN11C-07.

Sun, Ziheng, Laura Sandoval, Robert Crystal-Ornelas, S. Mostafa Mousavi, Jinbo Wang, Cindy Lin, Nicoleta Cristea et al. 2022. A review of earth artificial intelligence. Computers & Geosciences 159 (2022): 105034.

Süsser, Diana, Hannes Gaschnig, Andrzej Ceglarz, Vassilis Stavrakas, Alexandros Flamos, and Johan Lilliestam. 2022. Better suited or just more complex? On the fit between user needs and modeller-driven improvements of energy system models. *Energy 239: 121909.*

Tennant, Jonathan P. 2018. The state of the art in peer review. *FEMS Microbiology Letters* 365 (19): fny204.

Vai, Marjorie, and Kristen Sosulski. 2015. *Essentials of online course design: A standards-based guide.* New York: Routledge.

Von Schomberg, Rene. 2013. A vision of responsible research and innovation. In *Responsible innovation: Managing the responsible emergence of science and innovation in society*, 51–74. New York: Wiley.

Yesson, Chris, Peter W. Brewer, Tim Sutton, Neil Caithness, Jaspreet S. Pahwa, Mikhaila Burgess, W. Alec Gray et al. 2007. How global is the global biodiversity information facility?. PloS One 2 (11): e1124.

Index

Z. Sun (ed.), *Actionable Science of Global Environment Change*,
https://doi.org/10.1007/978-3-031-41758-0

The manufacturer's authorised representative in the EU is Springer Nature Customer Service Centre GmbH, Europaplatz 3, 69115 Heidelberg, Germany. If you have any concerns regarding our products, please contact ProductSafety@springernature.com

Printed and bound by CPI Group (UK) Ltd, Croydon, CR0 4YY
01/10/2026
02225855-0013